Integral Representations and Residues in Multidimensional Complex Analysis

TRANSLATIONS OF
MATHEMATICAL
MONOGRAPHS

VOLUME 58

L. A. Aĭzenberg
A. P. Yuzhakov

Integral Representations and Residues in Multidimensional Complex Analysis

American Mathematical Society

ИНТЕГРАЛЬНЫЕ ПРЕДСТАВЛЕНИЯ И ВЫЧЕТЫ В МНОГОМЕРНОМ КОМПЛЕКСНОМ АНАЛИЗЕ

Л. А. АЙЗЕНБЕРГ И А. П. ЮЖАКОВ

ИЗДАТЕЛЬСТВО «НАУКА»
СИБИРСКОЕ ОТДЕЛЕНИЕ
Новосибирск 1979

Translated from the Russian by H. H. McFaden
Translation edited by Lev J. Leifman

2000 *Mathematics Subject Classification.* Primary 32-XX.

Abstract. This book deals with integral representations of holomorphic functions of several complex variables, the multidimensional logarithmic residue, and the theory of multidimensional residues. Applications are given to implicit function theory, systems of nonlinear equations, computation of the multiplicity of a zero of a mapping, and computation of combinatorial sums in closed form. Certain applications in multidimensional complex analysis are considered.

The monograph is intended for specialists in theoretical and applied mathematics and theoretical physics, and for post graduate and graduate students interested in multidimensional complex analysis or its applications.

Library of Congress Cataloging in Publication Data

Aĭzenberg, Lev Abramovich, 1937–

Integral representations and residues in multidimensional complex analysis.

(Translations of mathematical monographs; v. 58)

Translation of: Integral'nye predstavleniîa i vychety v mnogomernom kompleksnom analize.

Bibliography: p.

Includes indexes.

1. Holomorphic functions. 2. Integral representations. 3. Congruences and residues. I. Îuzhakov, Aleksandr Petrovich. II. Title. III. Series.

QA331.A46413 1983 515.9'8 83-15549

ISBN 0-8218-4511-X

11 10 9 8 7 6 5 4 3 2 06 05 04 03 02 01

Table of Contents

Preface

Integral representations and the theory of residues occupy an important place in multidimensional complex analysis. At the present time these areas are continuing to develop intensively and are finding ever newer applications in function theory itself and in other areas of mathematics, as well as in theoretical physics. For example, the Khenkin-Ramírez de Arellano integral representation for functions holomorphic in strictly pseudoconvex domains has led to a number of important results in the theory of holomorphic functions of several complex variables.

We mention also a certain "nonsymmetry" in the integral representations of holomorphic functions of several complex variables that existed before 1967: functions were represented by integration of exterior differential forms. This "nonsymmetry" was removed by the Martinelli-Bochner-Koppelman formula (forms are represented by integration of forms), which has found many serious applications in multidimensional complex analysis. In particular, Khenkin used this formula to get an integral representation for a solution of the $\bar{\partial}$-problem in strictly convex domains, and this led to a new proof of the main theorems of Oka without the use of sheaf theory or the theory of differential operators.

There recently appeared a Leray-Koppelman integral formula of very general form that includes both all possible integral representations of holomorphic functions and formulas for the multidimensional logarithmic residue (see §3). Therefore, Chapter I contains both integral representations of general form and formulas for the logarithmic residue. We remark that integral representations and residues are closely connected, since they are based on Stokes' formula.

The theory of multidimensional residues, whose beginning goes back to Poincaré, has developed in various directions and presently is far from complete. In generalizing the theory of residues to the multidimensional case there arise serious difficulties of a topological nature, since for holomorphic functions of several complex variables the role of isolated singular points is played by analytic sets, which can have a complicated topological structure. To overcome these difficulties it is necessary to resort to the tools of algebraic topology.

Multidimensional residues, besides having well-known applications in the investigation of Feynman integrals by homological methods and the investigation of fundamental solutions of hyperbolic partial differential equations (Leray), have recently been applied to implicit functions, systems of nonlinear equations, and combinatorial analysis. These new applications have not yet been reflected in the existing texts of function theory. Moreover, in these texts integral representations and residues of functions of several complex variables are not presented in a sufficiently complete manner. The present book is intended to fill this gap. It arose from special courses taught by the aughors at Krasnoyarsk University in 1965–1974 (see [26] and [218]).

As already mentioned, Chapter I contains integral representations of holomorphic functions of several complex variables and formulas for the multidimensional logarithmic residue, presented here from a unified point of view.

In §2 we define the multiplicity of a zero of a holomorphic mapping and prove the main theorems on zeros of holomorphic mappings, using only the properties of integrals of closed forms and not using results in the local theory of analytic sets. In particular, a compact set of zeros of a holomorphic mapping is proved to be discrete. The theorem of Osgood asserting that a one-to-one holomorphic mapping is biholomorphic is obtained as a corollary in §4.

Moreover, in Chapter I we present the Andreotti-Norguet integral representation for the derivatives of a holomorphic function (§5) and the Martinelli-Bochner-Koppelman integral representation for exterior differential forms (§7).

Chapter II deals with a special form of integral representations derived from Leray's formula. The most important of them are the Bergman-Weil formula (§9) for functions holomorphic in analytic polyhedra and the Khenkin-Ramírez de Arellano formula (§10) for functions holomorphic in strictly pseudoconvex domains. §12 stands somewhat isolated; there we consider an integral representation for holomorphic functions with nonnegative real part (a generalization of the classical Riesz-Herglotz formula).

In this book integral representations are considered only for bounded domains, a reflection of the personal preferences of the authors (see, for example, [52]–[54], [60]–[62], [224] and [236] about integral representations in unbounded domains).

The theory of multidimensional residues is presented in Chapter III. We formulate the problem of this theory (§13) and indicate some ways of solving it that follow from Stokes' formula, the Cauchy-Poincaré integral theorem, the

Alexander-Pontryagin topological duality (§14), and the de Rham duality (§15). Then (§§16 and 17) we give the main results in the theory of residues on a complex analytic manifold (the Leray theory of residues). The chapter concludes with an application of the Alexander-Pontryagin duality and the de Rham duality to the computation of residues of meromorphic and rational forms (§§18 and 19).

This book does not reflect results on extension of the Leray theory of residues to forms having singularities on submanifolds of codimension greater than 1, nor residues of currents, nor the abstract theory of residues which studies dual homomorphisms of the homology and cohomology groups of a closed subset and its complement in a manifold, nor Grothendieck residues, nor Dolbeault residues of semimeromorphic forms.

Chapter IV deals with so-called "external" applications (found during the last few years mainly by Krasnoyarsk mathematicians) of various formulas for the multidimensional logarithmic residue and simply for residues.

In §20 we consider applications of the multiple logarithmic residue for representing implicit functions of several variables determined by systems of equations, in particular, for inverting a holomorphic mapping in the form of a multiple power series or a function series (multidimensional analogues of the Lagrange expansion and the Bürmann-Lagrange expansion), with estimates given of the domain of convergence for these series. We consider some cases in which single-valued branches of multi-valued implicit functions are determined, and obtain effective formulas for the coefficients of a Weierstrass pseudopolynomial.

In §21 the formulas for the multiple logarithmic residue are applied to systems of nonlinear equations, in particular, to the solution of the problem of eliminating unknowns in systems of algebraic equations. We obtain multidimensional analogues of the Waring formula which expresses power sums of the roots of equations in terms of their coefficients.

The multiplicity of a zero of a system of holomorphic functions is determined from the Taylor expansions of these functions in §22. Here the main tool of the investigation is again the theory of multiple residues.

In §23 multidimensional residues are used to find generating functions and to compute various combinatorial sums in closed form.

Chapter V contains some "internal" applications, mainly of integral representations to problems in multidimensional function theory. There are discussions of holomorphic extension from the boundary of a domain to the whole domain and of expansion of holomorphic functions in series of polynomials or "partial fractions" (§24). In §25 we prove the main theorems of Oka without using sheaf theory or the theory of differential operators. The text in this section was written by Sh. A. Dautov on the basis of an unpublished manuscript which G. M. Khenkin kindly made available to the authors. In general, Khenkin's ideas had a large influence on the contents of Chapters I, II, and V. §26 contains information about the description of forms orthogonal to holomorphic functions when integrated over the boundary of a domain, about the general form of formulas for integral representation of holomorphic functions and for the logarithmic residue, and about the solvability of the Lewy equation. §27 deals with bases in spaces of holomorphic functions with fixed algebraic singularities.

The limited space in the book did not permit the authors to present many other applications of integral representations and residues. Here we should mention first of all the work of Leray on differential equations, as well as the applications of residues to the investigation of Feynman integrals. A number of other references are mentioned in the section *Brief historical comments*.

We note that §§1, 3, 5–12, 21, and 24 were written by L. A. Aĭzenberg, §§2, 4, 13–20, 23, and 27 by A. P. Yuzhakov, §§25 and 26 by Sh. A. Dautov, and §22 by A. K. Tsikh. The authors thank G. M. Khenkin and the editor, Sh. A. Dautov, for numerous useful improvements and changes in the text.

Furthermore, the authors use this opportunity to thank V. A. Bolotov, A. M. Kytmanov, E. S. Mkrtchyan, M. A. Mkrtchyan, V. A. Stepanenko, and A. K. Tsikh for help in preparing the manuscript for printing.

The authors

Preliminaries

1°. Integration of differential forms. Stokes' formula. We assume that the reader is acquainted with the concepts of a differentiable manifold and a complex analytic manifold (see, for example, [133], [144] or [189]). A *differential form* ω *of degree* p (a p-*form*) *on a differentiable manifold* ([1]) X can in a neighborhood of each point $a \in X$ be represented in local coordinates $x = (x_1, \dots, x_n)$ in the form

$$\omega(x) = \sum_J a_J(x)\, dx_{i_1} \wedge \cdots \wedge dx_{i_p}, \tag{0.1}$$

where $J = (i_1, \dots, i_p)$, $1 \leqslant i_j \leqslant n$, the $a_J(x)$ are real or complex functions of x, and $\wedge$ is the symbol for exterior multiplication, which satisfies the anticommutation condition

$$dx_i \wedge dx_j = -dx_j \wedge dx_i, \quad i \neq j, \qquad dx_i \wedge dx_i = 0. \tag{0.2}$$

The representation (0.1) of a p-form ω as given is unique if $1 \leqslant i_1 < i_2 < \cdots < i_p \leqslant n$. In passing to other local coordinates $y = (y_1, \dots, y_n)$ the expression (0.1) is transformed to

$$\sum_{1 \leqslant j_1 < \cdots < j_p \leqslant n} \left[\sum_J a_J(x(y)) \frac{\partial(x_{i_1}, \dots, x_{i_p})}{\partial(y_{j_1}, \dots, y_{j_p})} \right] dy_{j_1} \wedge \cdots \wedge dy_{j_p}.$$

Differential forms can be added and multiplied by functions. The operation of *exterior multiplication* is defined for them; in local coordinates it is realized according to the rule for multiplying polynomials, with the condition (0.2) taken into account. This condition implies the following property of the exterior product: $\omega \wedge \varphi = (-1)^{pq} \varphi \wedge \omega$, where p and q are the degrees of ω and φ. We say that ω is a *differential form of class* $C^{(r)}$ if its coefficients $a_J(x)$ in the given local representation (0.1) are r times continuously differentiable.

([1]) The exact definition of a differential form, along with a more detailed presentation of the material in this section, can be found in the books [102], [133], [158], [161] and [218].

The form ω is said to be *regular* if $r = \infty$. The *linear space of p-forms of class $C^{(r)}$ on the manifold X* is denoted by $C_p^{(r)}(X)$. Following Leray's book [113], we often denote the *space $C_p^{(\infty)}(X)$ of regular p-forms* also by $\Omega^p(X)$.

The operator d of *exterior differentiation* assigns to each form $\omega \in C_p^{(r)}(X)$ the form $d\omega \in C_{p+1}^{(r-1)}(X)$ determined in local coordinates by the formula

$$d\omega = \sum_J da_J(x) \wedge dx_J, \tag{0.3}$$

where

$$da_J = \sum_{i=1}^{n} \frac{\partial a_J}{\partial x_i} dx_i, \qquad dx_J = dx_{i_1} \wedge \cdots \wedge dx_{i_p}.$$

The exterior differential has the following properties: 1) $d(\omega_1 + \omega_2) = d\omega_1 + d\omega_2$; 2) $d(\omega \wedge \varphi) = d\omega \wedge \varphi + (-1)^p \omega \wedge d\varphi$, where p is the degree of the form ω; 3) $d\,d\omega = 0$. A differential form ω is said to be *closed* if $d\omega = 0$, and *exact* if $\omega = d\varphi$ for some form φ. The set of closed regular p-forms is denoted by $Z^p(X)$, and the set of exact regular p-forms by $B^p(X)$. The properties 1)–3) of the exterior differential imply the inclusions of groups (vector spaces) $B^p(X) \subset Z^p(X) \subset \Omega^p(X)$.

The factor group $H^p(X) = Z^p(Z)/B^p(X)$ is called the *p-dimensional de Rham cohomology group of the manifold X*. Its elements are called *cohomology classes*. Two forms ω_1 and ω_2 that belong to the same cohomology class are said to be *cohomologous* ($\omega_1 \approx \omega_2$).

An infinitely differentiable mapping of manifolds

$$f\colon X \to Y \tag{0.4}$$

induces group homomorphisms

$$\tilde{f}\colon \Omega^p(Y) \to \Omega^p(X), \qquad f^*\colon H^p(Y) \to H^p(X)$$

defined as follows. Let $x = (x_1, \ldots, x_n)$ and $y = (y_1, \ldots, y_n)$ be local coordinates in neighborhoods U_a and V_b of the points $a \in X$ and $b = f(a) \in Y$, and suppose that the form $\omega \in \Omega^p(Y)$ is represented as $\omega(y) = \sum_J a_J(y)\, dy_{i_1} \wedge \cdots \wedge dy_{i_p}$. Then in U_a

$$\tilde{f}(\omega)(x) = \omega(f(x)) = \sum_J a_J(y(x))\, dy_{i_1}(x) \wedge \cdots \wedge dy_{i_p}(x),$$

where $dy_i(x) = \sum_{j=1}^n (\partial y_i/\partial x_j) dy_j$. Since $d \circ \tilde{f} = \tilde{f} \circ d$, $\tilde{f}$ induces a factor homomorphism f^*.

If Z is a complex analytic manifold with local coordinates $z_j = x_j + iy_j$, $j = 1, \ldots, n$, then instead of dx_j and dy_j we take the linear combinations $dz_j = dx_j + i dy_j$ and $d\bar{z}_j = dx_j - i dy_j$ of them as basis differentials. A form $\omega \in C_{p+q}^{(\infty)}(Z)$ is said to be a *form of type (p, q)* if in an arbitrary local system of coordinates $z = (z_1, \ldots, z_n)$ it looks like $\omega = \Sigma_{J,\mathcal{J}} a_{J,\mathcal{J}}(z)\, dz_J \wedge d\bar{z}_{\mathcal{J}}$, where $dz_J = dz_{i_1} \wedge \cdots \wedge dz_{i_p}$ and $d\bar{z}_{\mathcal{J}} = d\bar{z}_{j_1} \wedge \cdots \wedge d\bar{z}_{j_q}$. The class of these forms is denoted by $C^{(\infty)}_{p,q}(Z)$. The class $C^{(r)}_{p,q}(Z)$ is introduced similarly. If $r = 0$, then we write simply $C_{p,q}(Z)$. A form is said to be *holomorphic* if it is of type $(p, 0)$ and its coefficients are holomorphic functions.

We shall consider integrals of differential forms over piecewise smooth oriented surfaces on a manifold X. In order that the apparatus of algebraic topology can be applied, the surface is assumed to be broken up into parts: singular simplexes. A *p-dimensional smooth singular simplex* (a *p-simplex*) *on a differential manifold* X is defined to be a pair $\sigma_p = (\Delta_p, g)$, where Δ_p is a rectilinear simplex in $\mathbf{R}^p$ (for example, $\Delta_p = \{t = (t_1, \ldots, t_p) \in \mathbf{R}^p : t_j \geqslant 0, t_1 + \cdots + t_p \leqslant 1\}$), and $g : \Delta_p \to X$ is a continuously differentiable mapping. Moreover, the simplex is assigned some *orientation*, determined by the order of the coordinates $t_1, \ldots, t_p$. If other coordinates $\tau_1, \ldots, \tau_p$ are chosen in a neighborhood of Δ_p, then they determine the same orientation if $\partial t/\partial \tau = \partial(t_1, \ldots, t_p)/\partial(\tau_1, \ldots, \tau_p) > 0$, and the opposite orientation if $\partial t/\partial \tau < 0$.

The simplex σ_p taken with the opposite orientation is denoted by $-\sigma_p$. A finite linear combination

$$c_p = \sum_i m_i \sigma_p^{(i)} \tag{0.5}$$

of oriented smooth singular p-simplexes, where the m_i are integers([2]), is called a *smooth singular p-chain*. The set of p-chains on a manifold X forms an abelian group $C_p(X)$. The *carrier of the simplex* $\sigma_p = (\Delta_p, g)$ is defined to be the set $|\sigma_p| = g(\Delta_p)$, and the *carrier of the chain* (0.5) is the set

$$|c_p| = \bigcup_{m_i \neq 0} |\sigma_p^{(i)}|.$$

The $(p-1)$-simplex $\sigma_{p-1}^{(i)} = (\Delta_{p-1}^{(i)}, g|_{\Delta_{p-1}^{(i)}})$, where $\Delta_{p-1}^{(i)}$ is a $(p-1)$-dimensional face of the rectilinear simplex Δ_p, is called a *face of the singular p-simplex* σ_p. We choose a system of coordinates $t_1, \ldots, t_p$ in $\mathbf{R}^p \supset \Delta_p$ determining the orientation of σ_p in such a way that the face $\Delta_{p-1}^{(i)}$ lies in the plane $t_1 = 0$, and $t_1 \leqslant 0$ at the points $t \in \Delta_p$. Then the parameters $t_2, \ldots, t_p$ determine the *orientation of the face* $\sigma_{p-1}^{(i)}$ *coherent with that of* σ_p. The *boundary of an oriented simplex* σ_p is defined to be the $(p-1)$-chain formed by summing its coherently oriented faces: $\partial\sigma_p = \Sigma_{i=0}^p \sigma_{p-1}^{(i)}$. The *boundary of the chain* (0.5) is defined by the formula $\partial c_p - \Sigma_i m_i \partial\sigma_p^{(i)}$. The following property holds:

$$\partial\partial c_p = 0. \tag{0.6}$$

([2]) It is sometimes convenient to consider chains with real or complex coefficients (see §15).

A chain $\gamma \in C_p(X)$ is called a *cycle* if $\partial\gamma = 0$. The property (0.6) means that the boundary of a chain is a cycle. A cycle γ is said to be *homologous to zero* ($\gamma \sim 0$) if there exists a chain c such that $\partial c = \gamma$. Let $Z_p(X) = \{\gamma \in C_p(X) : \partial\gamma = 0\}$ and $B_p(X) = \{\partial c : c \in C_{p+1}(X)\}$. The factor group $H_p(X) = Z_p(X)/B_p(X)$ is called the *p-dimensional smooth singular homology group of the manifold X*. For differentiable manifolds the smooth singular homology groups are isomorphic to homology groups defined in other ways (see [114], Appendix A). The definition of smooth singular homology given here is also suitable for a piecewise smooth manifold. A cycle γ is said to be *weakly homologous to zero* ($\gamma \approx 0$) if $k\gamma \sim 0$ for some integer k. The factor group of $Z_p(X)$ by the subgroup of cycles weakly homologous to zero is called a *weak homology group*. It coincides with $H_p(X)$ in the cases of interest to us. We denote it too by $H_p(X)$. If $H_p(X)$ is a group with finitely many generators, then its dimension is called the *p-dimensional Betti number of the manifold X*. A system $\{\gamma_j\} \subset Z_p(X)$ of cycles is called a *p-dimensional homology basis* if any cycle $\gamma \in Z_p(X)$ can be uniquely represented in the form $\gamma \approx \sum_j m_j\gamma_j$, where the m_j are integers. A mapping (0.4) of manifolds assigns to each simplex $\sigma_p = (\Delta_p, g)$ in X a simplex $f(\sigma_p) = (\Delta_p, f \circ g)$ in Y. This determines a homomorphism $f\colon C_p(X) \to C_p(Y)$ and, since $f \circ \partial = \partial \circ f$, also a homomorphism $f_*\colon H_p(x) \to H_p(Y)$.

The *integral of a p = form ω of class $C^{(0)}$ over a smooth p-simplex* $\sigma_p = (\Delta_p, g)$ is defined by the equality

$$\int_{\sigma_p} \omega = \int_{\Delta_p} \tilde{g}(\omega) = \int_{|\Delta_p|} A(t)\, dt_1 \cdots dt_p, \tag{0.7}$$

where $A(t)\, dt_1 \wedge \cdots \wedge dt_p = \tilde{g}(\omega)$ is a form on $\Delta_p \subset \mathbf{R}^p$. If the form is represented as in (0.1), then

$$A(t) = \sum_J a_J(x(t)) \frac{\partial(x_{i_1}, \ldots, x_{i_p})}{\partial(t_1, \ldots, t_p)}.$$

The last integeral in (0.7) is understood as an ordinary p-fold integral in $\mathbf{R}^p$. Obviously, the definition (0.7) is invariant with respect to the choice of local coordinates on the manifold and the parametrization of the simplex σ_p. The *integral of a form ω over the chain* (0.5) is defined by the equality

$$\int_{c_p} \omega = \sum_i m_i \int_{\sigma_p^{(i)}} \omega. \tag{0.8}$$

We remark that every piecewise smooth oriented compact surface on X can be broken up into smooth simplexes, i.e., can be represented as a chain $c_p \in C_p(X)$.

From the definition of the homomorphism $\tilde{f}$ and $\underset{\sim}{f}$ and formulas (0.7) and (0.8) we get

THEOREM 0.1 (Change-of-variables formula). *If $f\colon X \to Y$ is a mapping of manifolds and $\gamma \in C_p(X)$, then for every form ω of class $C^{(0)}$ on Y*

$$\int_\gamma \tilde{f}(\omega) = \int_{\underset{\sim}{f}(\gamma)} \omega.$$

Thus, the homomorphisms $\tilde{f}$ and $\underset{\sim}{f}$ are dual with respect to integration. The duality of the homomorphisms d and ∂ is established by

THEOREM 0.2 (Stokes' formula). *If ω is a p-form of class $C^{(1)}$ on the manifold X and $c \in C_{p+1}(X)$, then*

$$\int_{\partial c} \omega = \int_c d\omega.$$

COROLLARY 0.3. *The integral of an exact form over a cycle is equal to zero.*

COROLLARY 0.4. *The integral of a closed form over a cycle that is weakly homologous to zero is equal to zero.*

COROLLARY 0.5. *If $\gamma_1 \approx \gamma_2$ and $\omega_1 = \omega_2 + d\varphi$, then $\int_{\gamma_1} \omega_1 = \int_{\gamma_2} \omega_2$.*

Corollary 0.5 enables us to define the integral of a cohomology class $h^* = \{\omega\} \in H^p(X)$ over a homology class $h_* = \{\gamma\} = H_p(X)$ by the equality

$$\int_{h_*} h^* = \int_\gamma \omega.$$

In what follows we shall often have to consider bounded domains D in the complex space $\mathbf{C}^n$ of variables $z = (z_1, \dots, z_n)$, where $z_k = x_k + iy_k$, $k = 1, \dots, n$. We assume that $\mathbf{C}^n$ is oriented in such a way that

$$\int_D dx_1 \wedge \cdots \wedge dx_n \wedge dy_1 \wedge \cdots \wedge dy_n$$
$$= \left(-\frac{i}{2}\right)^n \int_D d\bar{z}_1 \wedge \cdots \wedge d\bar{z}_n \wedge dz_1 \wedge \cdots \wedge dz_n > 0.$$

The boundary ∂D of D is assigned the orientation induced by that of D. If $D = \{z : \rho(z) < 0\}$, where ρ is a real-valued function of class $C^{(r)}$ in a neighborhood of ∂D such that $\operatorname{grad} \rho = (\rho'_{z_1}, \dots, \rho'_{z_n}) \neq 0$ on ∂D, then we write $\partial D \in C^{(r)}$. According to Stokes' formula,

$$(-i)^n \int_{\partial D} \sum_{k=1}^n (-1)^{k-1} \bar{\zeta}_k d\bar{\zeta}_1 \wedge \cdots \wedge d\bar{\zeta}_{k-1} \wedge d\bar{\zeta}_{k+1}$$
$$\wedge \cdots \wedge d\bar{\zeta}_n \wedge d\zeta_1 \wedge \cdots \wedge d\zeta_n > 0.$$

We remark that the orientation of $\mathbf{C}^n$ adopted in [51] and [172] is such that

$$\int_D dx_1 \wedge dy_1 \wedge \cdots \wedge dx_n \wedge dy_n$$
$$= \left(-\frac{i}{2}\right)^n \int_D d\bar{z}_1 \wedge dz_1 \wedge \cdots \wedge d\bar{z}_n \wedge dz_n > 0.$$

It is easy to see that the integrals computed with this orientation differ from those computed with out orientation by the factor $(-1)^{n(n-1)/2}$. A particular case of Corollary 0.4 is

THEOREM 0.6 (Cauchy-Poincaré). *Suppose that the function $f(z)$ is holomorphic in a domain $D \subset \mathbf{C}^n$. Then for any $(n+1)$-chain σ in D*

$$\int_{\partial\sigma} f(z)\, dz = \int_{\partial\sigma} f(z_1, \ldots, z_n)\, dz_1 \wedge \cdots \wedge dz_n = 0.$$

From Stokes' formula and a property of the exterior differential we get the following *formulas for integration by parts.*

PROPOSITION 0.7. *Suppose that $\varphi \in \Omega^p(X)$, $\psi \in \Omega^q(X)$, and $\gamma \in C_{p+q+1}(X)$. Then*

$$\int_\gamma d\varphi \wedge \psi = \int_{\partial\gamma} \varphi \wedge \psi - (-1) \int_\gamma \varphi \wedge d\psi.$$

If $\gamma \in Z_{p+q+1}(X)$, then $\int_\gamma d\varphi \wedge \psi = (-1)^{p-1} \int_\gamma \varphi \wedge d\psi$.

PROPOSITION 0.8. *Suppose that the functions $f(z)$ and $\varphi(z)$ are holomorphic in a domain $D \subset \mathbf{C}^n$ and $\gamma \in Z_n(D)$. Then*

$$\int_\gamma \frac{\partial f}{\partial z_j} \varphi\, dz = -\int_\gamma f \frac{\partial \varphi}{\partial z_j} dz, \qquad \int_\gamma \frac{\partial^{|\alpha|} f}{\partial z^\alpha} \varphi\, dz = (-1)^{|\alpha|} \int_\gamma f \frac{\partial^{|\alpha|} \varphi}{\partial z^\alpha} dz,$$

where $\alpha = (\alpha_1, \ldots, \alpha_n)$, $|\alpha| = \alpha_1 + \cdots + \alpha_n$, and $\partial^{|\alpha|} f / \partial z^\alpha = \partial^{\alpha_1 + \cdots + \alpha_n} f / \partial z_1^{\alpha_1} \cdots \partial z_n^{\alpha_n}$.

Let D be a bounded domain in $\mathbf{C}^n$ with piecewise smooth boundary ∂D, and let the domains D_1 and D_1' be such that $D_1' \Subset D_1 \Subset D$ (see 5° for the notation). On $\bar{D}_1 \setminus \bar{D}_1'$ we are given an exterior differential form α of dimension $2n-1$ that is continuous in $\bar{D} \setminus \bar{D}_1'$ and closed in $D \setminus \bar{D}_1'$. The domain D can be approximated from the inside by polyhedrons D_m $(m = 2, 3, \ldots)$, $D_1 \subset D_m$, such that

$$\lim_{m \to \infty} \int_{\partial D_m} \beta = \int_{\partial D} \beta,$$

for every form β of degree $2n-1$ that is continuous in $\bar{D}\setminus\bar{D}_1'$. On the other hand, $\int_{\partial D_m}\alpha = \int_{\partial D_l}\alpha$ by Corollary 0.5, and so we have

COROLLARY 0.9. *Under the above assumptions,* $\int_{\partial D_1}\alpha = \int_{\partial D}\alpha$.

This corollary is frequently used below for proving the validity of various integral representations.

We shall sometimes encounter double forms $\omega(\zeta, z)$, i.e., forms in z whose coefficients are (ordinary) forms in ζ, or forms in ζ whose coefficients are forms in z. In this connection it is assumed that the differentials of the variables z commute with those of the variables ζ. See [144] for more details on double forms.

2°. Some propositions from analysis and algebra. Let $C^{(m,\lambda)}$ be the class of m times continuously differentiable functions whose derivatives of order m all satisfy a Hölder condition with exponent λ. The condition $\partial D \in C^{(m,\lambda)}$ is defined analogously, where ∂D is the boundary of a bounded domain in $\mathbf{C}^n$.

PROPOSITION 0.10. *Suppose that* $\partial D \in C^{(m+1,\lambda)}$ *and that* $f \in C^{(m,\lambda)}\partial D$, $m \geqslant 0$ *and* $0 < \lambda \leqslant 1$. *Then the functions defined by the integral*

$$\int_{\partial D_\zeta} \frac{f(\zeta)\, d\bar{\zeta}_1 \wedge \cdots \wedge d\bar{\zeta}_{k-1} \wedge d\bar{\zeta}_{k+1} \wedge \cdots \wedge d\bar{\zeta}_n \wedge d\zeta_1 \wedge \cdots \wedge d\zeta_n}{\left(|\zeta_1 - z_1|^2 + \cdots + |\zeta_n - z_n|^2\right)^{n-1}}$$

inside and outside the domain can be extended to the closures of the corresponding open sets as functions of class $C^{(m+1,\lambda')}$, *and the functions defined by the integral*

$$\int_{D_\zeta} \frac{f(\zeta)d\bar{\zeta}_1 \wedge \cdots \wedge d\bar{\zeta}_n \wedge d\zeta_1 \wedge \cdots \wedge d\zeta_n}{\left(|\zeta_1 - z_1|^2 + \cdots + |\zeta_n - z_n|^2\right)^{n-1}}$$

can be extended as functions of class $C^{(m+2,\lambda')}$, *where* $0 < \lambda' < \lambda$ (*see, for example*, [69], *Chapter* II, §19, *Theorem* 4, §20, *Theorem* 2).

Let X and Y be differentiable manifolds and $f\colon X \to Y$ a mapping of class $C^{(1)}$. A point $p \in X$ is called a *critical point of* f if $\min(\dim X, \dim Y) > \operatorname{rank}(df)_p$, where $(df)_p$ is the differential of f at p. The range of f on the critical points is called the *wrinkle* and denoted by $\operatorname{wr}(f)$.

THEOREM 0.11 (Sard). *If* $\dim X = \dim Y$, *then the Lebesgue measure of* $\operatorname{wr}(f)$ *in* Y *is zero. The Lebesgue measure of* $\operatorname{wr}(f)$ *in* Y *is also zero if* $X, Y \in C^\infty$ *and* $f \in C^{(\infty)}$.

Only the operations of addition and multiplication appear in a numerical determinant; therefore, we can introduce a determinant whose elements are members of an arbitrary ring. Such a determinant can be expanded according to the usual rules, with the place of each factor in the terms of the sum fixed by the number of the column to which this factor belongs. It is convenient to introduce the following notation in connection with the way in which a determinant is expanded: if $\theta^1,\ldots,\theta^m$ are n-dimensional vectors, then $D_{\nu_1,\ldots,\nu_m}(\theta^1,\ldots,\theta^m)$ is the nth-order determinant with the vector θ^1 in the first ν_1 columns, the vector θ^2 in the next ν_2 columns,..., and the vector θ^m in the last ν_m columns, where $\nu_1+\cdots+\nu_m=n$.

We give some properties of determinants needed for what follows.

PROPERTY 0.12. *If two rows are interchanged, then the determinant changes sign. If all the elements in some column commute with all the elements in a neighboring column, then the determinant changes sign when these columns are interchanged, but if the corresponding elements anticommute, then the determinant does not change.*

Suppose now that the determinant is composed of elements of a ring of double exterior differential forms. Its rows are said to be *linearly dependent* if some linear combination of them with function coefficients vanishes, and for each point of the domain in which the determinant is defined at least one coefficient does not vanish.

PROPERTY 0.13. *If all the elements of some column (row) are multiplied by one and the same function, then the determinant is multiplied by this function. If the rows of the determinant are linearly dependent, then it is equal to* 0.

If the forms α_i $(i=1,\ldots,m)$ are of class $C^{(1)}$ and degree q_i in z, then

$$\bar\partial_z(\alpha_1\wedge\cdots\wedge\alpha_m)=\sum_{i=1}^{m}(-1)^{q_1+\cdots+q_{i-1}}\alpha_1\wedge\cdots\wedge\alpha_{i-1}\wedge\bar\partial_z\alpha_i\wedge\alpha_{i+1}\wedge\cdots\wedge\alpha_m.$$

A consequence of this is (here and below, $\bar\partial_z\theta=(\bar\partial_z\theta_1,\ldots,\bar\partial_z\theta_n)$)

PROPERTY 0.14. *If the elements of the columns θ^i $(i=1,\ldots,n)$ of a determinant are double forms of class $C^{(1)}$ and degree q_i in z, then*

$$\bar\partial_z D_{1,\ldots,1}(\theta^1,\ldots,\theta^n)=\sum_{i=1}^{n}(-1)^{q_1+\cdots+q_{i-1}}D_{1,\ldots,1}(\theta^1,\ldots,\theta^{i-1},\bar\partial_z\theta^i,\theta^{i+1},\ldots,\theta^n).$$

In what follows we shall use elementary concepts from measure theory without explaining, for example, the words "finite Borel measure" or "sup with respect to a given measure μ" (denoted by ess $\sup_\mu$). We also use the concept of a partition of unity subordinate to a given cover [133].

3°. Some facts from function theory. It is assume that the reader is familiar with the elementary properties of holomorphic functions of several complex variables. They can be found in the first chapters of the textbooks [51], [57], [148], [173], [189], [201] and [26]. For example, it is assumed known that a function holomorphic in a domain can be expanded in a multiple power series in a neighborhood of each point of the domain, and that a holomorphic mapping with nonzero Jacobian is locally invertible. It is assumed also that the reader is acquainted with such concepts as the envelope of holomorphy $\mathcal{H}(D)$ of a domain D, a complex line, a complex hyperplane, a circular domain, and an n-circular domain.

Recall that if $f(z_1,\ldots,z_n)$ is a holomorphic mapping of a domain $D \subset \mathbf{C}^n$ into a space of the same complex dimension, $f = (f_1,\ldots,f_n)$, $f_k = u_k + iv_k$, $z_k = x_k + iy_k$, $k = 1,\ldots,n$, then

$$\left|\frac{\partial(f_1,\ldots,f_n)}{\partial(z_1,\ldots,z_n)}\right|^2 = \frac{\partial(u_1,\ldots,u_n,v_1,\ldots,v_n)}{\partial(x_1,\ldots,x_n,y_1,\ldots,y_n)}.$$

If the boundary ∂D is piecewise smooth, the form $\alpha \in C_{n,n-1}(\partial D)$ is $\bar{\partial}$-exact, i.e., $\alpha = \bar{\partial}\beta$ on ∂D for some $\beta \in C^{(1)}_{n,n-2}(\partial D)$, and f is holomorphic on $\bar{D}$, then by Stokes' formula (see Corollary 0.3)

$$\int_{\partial D} f\alpha = \int_{\partial D} f\bar{\partial}\beta = \int_{\partial D} \bar{\partial}(f\beta) = \int_{\partial D} d(f\beta) = 0.$$

Suppose that f is holomorphic in a neighborhood of a point a. The *order* $d_a(f)$ *of a zero of* f *at* a is defined to be the smallest degree (in the variables jointly) of the terms with nonzero coefficients in the Taylor series of this function at the given point.

4°. Some concepts and facts from algebraic topology(3) will be needed in §§14, 16, and 18 of Chapter III. The *singular homology groups* are defined in topology for arbitrary topological spaces in exactly the same way as the smooth singular homology groups were defined in 2°, with only the difference that in the definition of a singular simplex $\sigma_p = (\Delta_p, g)$ the mapping g is assumed to be continuous, and the orientation of σ_p is given by the order of the vertices of the simplex Δ_p.

A topological space X is said to be a *polyhedron* if it admits a *triangulation*, i.e., it can be broken up into nondegenerate simplexes(4) $\sigma_p^{(i)}$ ($i = 1,2,\ldots,p = 0,1,\ldots$) with disjoint interiors. The family $K(X) = \{\sigma_p^{(i)}\}$ is called a *simplicial*

(3) See, for example [30], [114], [157], [163] or [175].

(4) A simplex $\sigma_p = (\Delta_p, g)$ is nondegenerate if $g: \Delta_p \to X$ is a homeomorphism.

complex of the polyhedron X. For a polyhedron we introduce the concept of a *homology group of a simplicial complex*, defined in the usual way, but only on the simplexes appearing in $K(X)$. The homology groups of a simplicial complex of a polyhedron are isomorphic to the singular homology groups of the polyhedron and, consequently, are invariant with respect to triangulation. A differentiable manifold admits a smooth triangulation, so all three definitions given above for the homology groups are equivalent for it. Instead of a simplicial complex it is sometimes more convenient to consider a *cell complex*: a partition of the polyhedron into cells $\sigma_p = (M_p, g)$, where M_p is a polytope in $\mathbf{R}^p$, and $g: M_p \to X$ is a continuous mapping that is a homeomorphism interior to M_p.

For a simplicial complex and a cell complex of a compact polyhedron X we have the *Euler-Poincaré formula*

$$\sum_{q=1}^{n} (-1)^q \alpha_q = \sum_{q=1}^{n} (-1)^q \beta_q, \tag{0.9}$$

where α_q is the number of q-dimensional simplexes (cells of the complex $K(X)$), β_q is the q-dimensional Betti number of X, and $n = \max\{q : \alpha_q \neq 0\}$ is the dimension of X.

From the definition of homology groups it is easy to get the following propositions.

PROPOSITION 0.15. *If X is an arcwise connected space, then $H_0(X) \simeq \mathbf{Z}$, where $\mathbf{Z}$ is the group of integers.*

PROPOSITION 0.16. *If $X = \bigcup_{\alpha \in A} X_\alpha$, where the X_α ($\alpha \in A$) are the arcwise connected components of X, then $H_p(X) = \bigoplus_{\alpha \in A} H_p(X_\alpha)$.*

PROPOSITION 0.17. *The homology groups for the n-dimensional sphere $S^n = \{x \in \mathbf{R}^{n+1} : |x| = 1\}$ are*

$$H_p(X) = \begin{cases} \mathbf{Z} & \text{if } p = 0, n, \\ 0 & \text{if } p \neq 0, n. \end{cases}$$

It can be shown that a 2-dimensional orientable manifold S is homeomorphic to some 4ρ-gon $\sigma_2 = a_1 b_1 a_1^{-1} b_1^{-1} \cdots a_\rho b_\rho a_\rho^{-1} b_\rho^{-1}$ in which the first and third and the second and fourth sides in each quadruple are identified pairwise, with the opposite orientation with respect to a circuit along the boundary of σ_2 (a sphere with ρ "handles"). The integer ρ is called the *genus of the manifold* (surface).

PROPOSITION 0.18. *The homology groups of a 2-dimensional orientable manifold S of genus ρ are*

$$H_0(S) \simeq H_2(S) \simeq \mathbf{Z}, \qquad H_1(S) \simeq \underbrace{\mathbf{Z} \oplus \cdots \oplus \mathbf{Z}}_{2\rho};$$

the 1-dimensional cells $a_1, b_1, \ldots, a_\rho, b_\rho$ *form a basis for the 1-dimensional homology of S, and are called "canonical cuts" of S.*

Continuous mappings $f, g: X \to Y$ are said to be *homotopic*, $f \approx g$, if there exists a continuous mapping $F(x, t): X \times [0, 1] \to Y$, such that $F(x, 0) = f(x)$ and $F(x, 1) = g(x)$.

A subset A of X is called a *retract* of X if there exists a continuous mapping $f: X \to A$ such that $f|_A = 1_A$ (the identity mapping). The set A is called a *deformation retract* of X if $f \circ i \approx 1_X: X \to X$, where i is the imbedding $A \subset X$.

THEOREM 0.19. *If* $f \approx g: X \to Y$, *then* $f_* = g_*: H_p(X) \to H_p(Y)$.

COROLLARY 0.20. *If A is a deformation retract of X, then* $H_p(A) \simeq H_p(X)$.

Let A be a subspace of X. The factor group $C_p(X, A) = C_p(X)/C_p(A)$ is called the *group of relative chains of the pair* (X, A). The boundary homomorphism $\partial: C_p(X) \to C_{p-1}(X)$ induces a homomorphism $\partial: C_p(X, A) \to C_{p-1}(X, A)$ in the natural way. Consequently, the *relative homology group* $H_p(X, A)$ is defined. We define a homomorphism $\partial_*: H_p(X, A) \to H_{p-1}(A)$ that assigns to each class $\{c_p\} \in H_p(X, A)$ the class $\{\partial c_p\} \in H_{p-1}(A)$. Observe that $c_p \in C_p(X)$ determines a relative cycle if $\partial c_p \in A$.

A sequence of group homomorphisms

$$\cdots \to A_{k-1} \xrightarrow{f_{k-1}} A_k \xrightarrow{f_k} A_{k+1} \to \cdots$$

is said to be *exact* if the kernel of each homomorphism coincides with the image of the preceding homomorphism ($\operatorname{Ker} f_k = \operatorname{Im} f_{k-1}$).

THEOREM 0.21. *The sequence of homomorphisms*

$$\cdots \xrightarrow{\partial_*} H_p(A) \xrightarrow{i_*} H_p(X) \xrightarrow{j_*} H_p(X, A) \xrightarrow{\partial_*} H_{p-1}(A) \xrightarrow{i_*} \cdots,$$

where i_* *and* j_* *are the homomorphisms induced by the imbedding* $i: A \to X$ *and the factor mapping* $j: C_p(X) \to C_p(X, A)$, *is exact.*

Let A be an abelian group. We denote the group of homomorphisms from A to $\mathbf{C}^1$ (the group of complex-valued linear functions on A) by $A^* = \{\lambda: A \to \mathbf{C}^1\}$. A homomorphism $\lambda: A \to B$ of abelian groups induces a dual homomorphism $\lambda^*: B^* \to A^*$, defined as follows: if $b^* \in B^*$, then $\lambda^* b^*(a) = b^*(\lambda(a))$ for any $a \in A$.

THEOREM 0.22. *If a sequence of homomorphisms*

$$\cdots \to A_{k-1} \overset{\varphi_{k-1}}{\to} A_k \overset{\varphi_k}{\to} A_{k+1} \to \cdots$$

of abelian groups is exact, then the sequence of dual homomorphisms

$$\cdots \leftarrow A^*_{k-1} \overset{\varphi^*_{k-1}}{\leftarrow} A^*_k \overset{\varphi^*_k}{\leftarrow} A^*_{k+1} \leftarrow \cdots$$

is also exact.

5°. Main notation. The space of n complex variables is denoted by $\mathbf{C}^n$, and its points are denoted by z, ζ, z^0, ζ^0, etc. If $z = (z_1,\dots,z_n)$, then $\bar{z} = (\bar{z}_1,\dots,\bar{z}_n)$. For $z, \zeta \in \mathbf{C}^n$ let $\langle z, \zeta\rangle = z_1\zeta_1 + \cdots + z_n\zeta_n$, $|z| = \sqrt{\langle z, \bar{z}\rangle}$, and $\|z\| = \max_{1\leqslant i\leqslant n}|z_i|$. We introduce notation for the ball $B_r = B_r(z^0) = B(z^0, r) = \{z : |z - z^0| < r\}$, the polydisk $U = U(z^0, r) = \{z : |z_i - z_i^0| < r, i = 1,\dots,n\}$, and the generalized ball $B_r^k(z^0) = \{z : |z_1 - z_1^0|^{2k_1} + \cdots + |z_n - z_n^0|^{2k_n} < r^2\}$, where $k = (k_1,\dots,k_n)$. In general, the expression $\{z : \cdots\}$ denotes a set in $\mathbf{C}^n$, with other cases stipulated in special ways. The space of n real variables is denoted by $\mathbf{R}^n$, and its points are denoted by x, y, etc.

If J and $\mathcal{J}$ are vectors with nonnegative integer coordinates (multi-indices), then $|J| = i_1 + \cdots + i_n$; the inequality $J \leqslant \mathcal{J}$ means that $i_k \leqslant j_k$, $k = 1,\dots,n$ where $J = (i_1,\dots,i_n)$ and $\mathcal{J} = (j_1,\dots,j_n)$. The expression $k \notin J$ means that $k \neq i_l$, $l = 1,\dots,n$. Further, $J! = i_1!\cdots i_n!$,

$$z^J = z_1^{i_1}\cdots z_n^{i_n}, \qquad D^J = \frac{\partial^{|J|}}{\partial z_1^{i_1}\cdots \partial z_n^{i_n}},$$

$$D^{J,\mathcal{J}}_{w,z} = \frac{\partial^{|J|+|\mathcal{J}|}}{\partial w_1^{i_1}\cdots \partial w_n^{i_n}\partial z_1^{j_1}\cdots \partial z_n^{j_n}},$$

$$dz = dz_1 \wedge \cdots \wedge dz_n, \qquad dz_J = dz_{i_1} \wedge \cdots \wedge dz_{i_p},$$

where $J = (i_1,\dots,i_p)$; and $dz_{|J|} = dz_{j_1} \wedge \cdots \wedge dz_{j_{n-p}}$. The notation [ ] will often be used in the same sense; for example, $a_1,\dots,[j_1,\dots,j_k],\dots,a_n$ means that the elements $a_{j_1},\dots,a_{j_k}$ are omitted, and $dz^{(\alpha)}_{[k]} = dz_1^{\alpha_1} \wedge \cdots [k] \cdots \wedge dz_n^{\alpha_n}$.

A neighborhood of a point $a \in \mathbf{C}^n$ is denoted by $U = U_a$ (sometimes by other letters).

Let M and N be subsets of $\mathbf{C}^n$; then ∂M is the boundary of M,

$$\rho(z, M) = \inf_{\zeta\in M} |\zeta - z|, \qquad \rho(M, N) = \inf_{\substack{\zeta\in M\\ z\in N}} |\zeta - z|,$$

$\overline{M}$ is the closure of M, int M is the interior of M, and $M \Subset N$ means that $\overline{M}$ is compact and $\overline{M} \subset \operatorname{int} N$. If $w = (w_1,\dots,w_n)$, and $\varphi = (\varphi_1,\dots,\varphi_n)$, then

$$\langle w, d\varphi\rangle = w_1 d\varphi_1 + \cdots + w_n d\varphi_n,$$

and

$$\langle w, \varphi^{(\alpha+\mathrm{I})}\rangle = w_1\varphi_1^{\alpha_1+1} + \cdots + w_n\varphi_n^{\alpha_n+1},$$

where $\alpha = (\alpha_1,\ldots,\alpha_n)$ and $\mathrm{I} = (1_1,\ldots,1)$. Further, $0 = (0,\ldots,0)$ is the origin of coordinates, and $d\varphi/\varphi = d\varphi_1/\varphi_1 \wedge \cdots \wedge d\varphi_n/\varphi_n$. The Jacobian of a mapping $f = f(z) = (f_1,\ldots,f_n)$ is denoted by

$$\frac{\partial(f_1,\ldots,f_n)}{\partial(z_1,\ldots,z_n)} = \frac{\partial f}{\partial z},$$

$(\partial f/\partial z)|_a$ is the value of this Jacobian at a point a, and grad $\Phi = (\Phi'_{z_1},\ldots,\Phi'_{z_n})$. We frequently write $\psi(z)$ in place of $\psi(z,\bar z)$ for a function. The symbol $\oplus$ denotes a direct sum, δ_{ik} is the Kronecker symbol, $\|g_{ij}\|$ is the matrix with elements g_{ij}, $\det\|g_{ij}\|$ is the determinant of this matrix, and $\partial z = (dz_1,\ldots,dz_n)$.

Let F be a closed subset of $\mathbf{C}^n$ and Ω an open subset. Then $C_r^{(m)}(F)$ $(m = 0,1,2,\ldots,\infty)$ is the class of r-dimensional forms(5) whose coefficients can be extended to a neighborhood of F as m times continuously differentiable functions. The subclass of $C_{p+q}^{(m)}(F)$ consisting of forms of type (p,q) is denoted by $C_{p,q}^{(m)}(F)$. When $m = 0$, this index will be omitted. The projective limit of the classes $C_{p,q}^{(m)}(F)$, $F \Subset \Omega$, is denoted by $C_{p,q}^{(m)}(\Omega)$. For

$$\alpha = \sum_{J,\mathcal{J}}{}' a_{J\mathcal{J}}\, dz_J \wedge d\bar z_{\mathcal{J}} \in C_{p,q}^{(m)}(\Omega),$$

we set

$$\bar\partial\alpha = \sum_{J,\mathcal{J}}{}' \sum_{k=1}^{n} \frac{\partial a_{J\mathcal{J}}}{\partial \bar z_k}\, d\bar z_k \wedge dz_J \wedge dz_{\mathcal{J}}$$

(for $m = 0$ the differentiation is understood as differentiation in the space of generalized functions). Here the prime on the summation sign indicates summation over $1 \leqslant i_1 < \cdots < i_p \leqslant n$, $1 \leqslant j_1 < \cdots < j_q \leqslant n$, where $J = (i_1,\ldots,i_p)$ and $\mathcal{J} = (j_1,\ldots,j_q)$.

The class of functions holomorphic in a domain D (on a compact set K) is denoted by $A(D)$ (by $A(K)$). Furthermore, $A_C(D) = A(D) \cap C(\bar D)$, and $A^n(D)$ is the class of mappings $f = (f_1,\ldots,f_n)$ that are holomorphic in D. Moreover, $\operatorname{Re} A_C(D)$ is the class of functions that are real parts of functions in $A_C(D)$. Let $Ph(D)$ be the class of pluriharmonic functions in D, and $R(D)$ the class of holomorphic functions in D with nonnegative real part. Finally, $Ph_C(D) = Ph(D) \cap C(\bar D)$, and $R_C(D) = R(D) \cap C(\bar D)$. If a measure μ is given on ∂D, then $[\operatorname{Re} A_C(D)]_\mu^\perp$ is the set of $\varphi \in C(\partial D)$ such that $\int_{\partial D} f\varphi\, d\mu = 0$

(5) The word "form" is used instead of "exterior differential form" here and frequently below.

for all $f \in \operatorname{Re} A_C(D)$. The closure of $A_C(D)$ in the metric of $L^2_{d\mu}$ is denoted by $H^2_{d\mu}$.

An index ζ (or z) on a domain of integration indicates that ζ (respectively, z) is the variable of integration; $|\partial D|$ is the image of ∂D under the mapping $z \to (|z_1|,\ldots,|z_n|)$, and $\Delta_\zeta = \{z : |z_i| = |\zeta_i|,\ i = 1,\ldots,n\}$. Finally, $A \times B$ denotes the topological product of sets A and B. Some other notation was introduced in 1°–4°.

If $\alpha \in C_{p,q}(\Omega)$, then $\operatorname{supp} \alpha = \overline{\{z \in \Omega : \alpha(z) \neq 0\}}$. Suppose that the form α is defined in some set $U \subset X$, and that a function χ is defined on the manifold X and has $\operatorname{supp} \chi \Subset U$; then, as is common, we regard $\psi = \chi\alpha$ as defined on X by the equalities $\psi = \chi\alpha$ for $x \in U$ and $\psi = 0$ for $x \notin U$.

The symbol □ marks the end of a proof.

CHAPTER I

Integral Representations and the Logarithmic Residue

§1. The Martinelli-Bochner integral representation

1°. THEOREM 1.1. *Suppose that $f \in A_C(D)$, where D is a bounded domain in $\mathbf{C}^n$ with piecewise smooth boundary ∂D. Then the Martinelli-Bochner formula holds*:

$$\int_{\partial D} f(\zeta)\omega(\zeta - z, \bar{\zeta} - \bar{z}) = \begin{cases} f(z), & z \in D, \\ 0, & z \notin \bar{D}, \end{cases} \tag{1.1}$$

where

$$\omega(\zeta - z, \bar{\zeta} - \bar{z}) = \frac{(n-1)!}{(2\pi i)^n} \frac{\sum_{k=1}^n (-1)^{k-1}(\bar{\zeta}_k - \bar{z}_k) d\bar{\zeta}_{[k]} \wedge d\zeta}{|\zeta - z|^{2n}}. \tag{1.2}$$

PROOF. 1. It is easy to establish by direct calculation that the Martinellil-Bochner kernel ω is closed with respect to ζ on the set $\mathbf{C}^n \setminus \{z\}$. From this and Corollary 0.4 we get (1.1) for the case $z \notin \bar{D}$. But if $z \in D$, then, by the same Corollary 0.4, the integration over ∂D can be replaced by integration over the sphere $\partial B_r(z)$, where the radius r is sufficiently small.

2. In a sufficiently small neighborhood of a point $z \in D$ the holomorphic function $f(\zeta)$ can be expanded in a multiple power series about z; hence, to prove (1.1) it suffices to establish that

$$\frac{(n-1)!}{(2\pi i r^2)^n} \int_{\partial B_r} (\zeta - z)^\beta \sum_{k=1}^n (-1)^{k-1}(\bar{\zeta}_k - \bar{z}_k) d\bar{\zeta}_{[k]} \wedge d\zeta = \begin{cases} 0 & \text{if } \beta \neq 0, \\ 1 & \text{if } \beta = 0. \end{cases} \tag{1.3}$$

3. Note that under the change of variables $\zeta_k - z_k = e^{it_k}(\zeta_k' - z_k)$ with all the t_k real ($k = 1, \dots, n$) the set $\partial B_r(z)$ is carried into itself, and the integrand form in (1.3) is multiplied by $e^{it_1\beta_1} \cdots e^{it_n\beta_n}$; therefore, (1.3) holds for the case $\beta \neq 0$.

4. If $\beta = 0$, then, by Stokes' formula, the left-hand side of (1.3) is equal to

$$\frac{n!}{(2\pi i r^2)^n}\int_{B_r} d\bar{\zeta} \wedge d\zeta$$

$$= \frac{n!2^n}{r^{2n}}\int_{|\zeta_1 - z_1|^2 + \cdots + |\zeta_n - z_n|^2 < r^2} |\zeta_1 - z_1| \cdots |\zeta_n - z_n| d|\zeta_1 - z_1| \wedge \cdots \wedge d|\zeta_n - z_n|$$

$$= n!\int_{\tau_1 + \cdots + \tau_n < 1, \tau_i \geqslant 0} d\tau_1 \wedge \cdots \wedge d\tau_n = 1. \quad \square$$

2°. For $n > 1$ the Martinelli-Bochner kernel $\omega(\zeta - z, \bar{\zeta} - \bar{z})$ is not holomorphic in z but only harmonic; nevertheless, the following assertion holds.

THEOREM 1.2 (Aronov and Kytmanov). *If D is a bounded domain with piecewise smooth boundary and the Martinelli-Bochner integral representation* (1.1) *is valid for $z \in D$ for a particular $f \in C^{(1)}(\bar{D})$, then $f \in A(D)$.*

To prove this theorem we need some lemmas.

LEMMA 1.3 (The Martinelli-Bochner formula for smooth functions). *If $f \in C^{(1)}(\bar{D})$, then*

$$\int_{\partial D} f(\zeta)\omega(\zeta - z, \bar{\zeta} - \bar{z}) - \int_D \bar{\partial} f(\zeta) \wedge \omega(\zeta - z, \bar{\zeta} - \bar{z}) = \begin{cases} f(z), & z \in D, \\ 0, & z \notin \bar{D}. \end{cases} \quad (1.4)$$

This lemma can be proved in the same way as Theorem 1.1, except that the integral of $d_\zeta(f(\zeta)\omega(\zeta - z, \bar{\zeta} - \bar{z})) = \bar{\partial}_\zeta f(\zeta) \wedge \omega(\zeta - z, \bar{\zeta} - \bar{z})$, over D also appears, and it is not possible to expand $f(\zeta)$ in a neighborhood of z in a Taylor series with respect to powers of $\zeta - z$, but is instead necessary to approximate f in this closed neighborhood by a polynomial in $\zeta - z$ and $\bar{\zeta} - \bar{z}$. The formula (1.4) enables us to easily compute the saltus of an integral of Martinelli-Bochner type in passing across the boundary of a domain D.

LEMMA 1.4. *Let $f \in C^{(1)}(\partial D)$, and define*

$$F^{\pm}(z) = \int_{\partial D} f(\zeta)\omega(\zeta - z, \bar{\zeta} - \bar{z}), \qquad z \in D^{\pm}, \quad (1.5)$$

where $D^+ = D$ *and* $D^- = \mathbf{C}^n \setminus \bar{D}$. *Then the functions* $F^{\pm}(z)$ *are continuous all the way up to the boundary of* $D^{\pm}$, *respectively, and*

$$F^+(z) - F^-(z) = f(z), \qquad z \in \partial D. \tag{1.6}$$

PROOF. Suppose that $F \in C^{(1)}(\bar{D})$ and $F|_{\partial D} = f$. The function

$$\varphi(z) = \int_D \bar{\partial} F(\zeta) \wedge \omega(\zeta - z, \bar{\zeta} - \bar{z})$$

is continuous in $\mathbf{C}^n$ (see Theorem 0.10). By (1.4),

$$F^+(z) = F(z) + \varphi(z), \quad z \in D^+,$$
$$F^-(z) = \varphi(z), \qquad z \in D^-,$$

and this and the continuity of φ and F give us the continuity of $F^{\pm}(z)$ all the way up to the boundary of $D^{\pm}$, along with equality (1.6). □

We shall also make use of the following result of Keldysh and Lavrent'ev, which we present in the even-dimensional case we need.

LEMMA 1.5 (Keldysh and Lavrent'ev). *The linear manifold generated by the fractions of the form*

$$\frac{1}{|\zeta - z|^{2n-2}}, \qquad \zeta \in K, z \in \mathbf{C}^n \setminus K,$$

where K *is a compact set in* $\mathbf{C}^n$ *of total Lebesgue measure zero and* $n > 1$, *is dense in the space* $C(K)$.

PROOF. By the Stone-Weierstrass theorem, it can be assume that $f \in C^{(1)}(K)$. We extend f as a smooth function to some closed ball $\bar{B}$ with $B \supset K$, and we represent it by formula (1.4), where $D = B$. If we now replace the domain of integration B in the second integral in (1.4) by $B \setminus U$, where U is a neighborhood of K of sufficiently small volume, and the integrals themselves by appropriate integral sums, then, using the fact that the coefficients of the form $\omega(\zeta - z, \bar{\zeta} - \bar{z})$ are the derivatives of the fundamental solution

$$g(\zeta, z) = (1 - n)^{-1} |\zeta - z|^{2-2n} \tag{1.7}$$

of the Laplace equation and replacing the derivatives of g by difference quotients, we obtain linear combinations of the desired fractions that uniformly approximate f on K. □

PROOF OF THEOREM 1.2. Consider the integrals of Martinelli-Bochner type (1.5); then $F^+ \equiv f$ in D. It now follows from (1.6) that $F^-|_{\partial D} = 0$. We note that F^- is harmonic in D^- and has a zero of order $2n - 1$ at infinity. By the uniqueness theorem for harmonic functions,

$$F^-(z) \equiv 0, \qquad z \in D^-. \tag{1.8}$$

Since f is representable by formula (1.1), it is harmonic in D. Let us represent it by Green's formula for harmonic functions:

$$\int_{\partial D}\left[f\omega + g(\zeta, z)\nu_f(\zeta)\right] = \begin{cases} f(z), & z \in D, \\ 0, & z \notin \overline{D}, \end{cases} \tag{1.9}$$

where

$$\nu_f(\zeta) = (-1)^{n-1}\frac{(n-1)!}{(2\pi i)^n}\sum_{k=1}^{n}(-1)^{k-1}\frac{\partial f}{\partial \bar{\zeta}_k}d\zeta_{[k]} \wedge d\bar{\zeta}.$$

This "complex" Green's formula can be proved in the same way as its "real" analogue, the ordinary Green's formula (see also the proof of Lemma 1.3). It follows from (1.1) and (1.9) that

$$\int_{\partial D}\frac{\nu_f(\zeta)}{|\zeta - z|^{2n-2}} = 0 \tag{1.10}$$

for $z \in D$. Comparing (1.8) and (1.9), we get that (1.10) is also valid for points $z \in D^-$. Lemma 1.5, applied to the compact set ∂D, now shows that $\nu_f(\zeta)|_{\partial D} = 0$.

The function f is harmonic in D; therefore, $d\nu_f(\zeta) = c\Delta f\, d\zeta \wedge \overline{d\zeta} = 0$, i.e., the form ν_f is closed in D, and so $d(\overline{f(\zeta)} \times \nu_f(\zeta))$, extends continuously to $\overline{D}$. Stokes' formula gives us that

$$(2i)^n\int_D\sum_{k=1}^{n}\left|\frac{\partial f}{\partial \bar{\zeta}_k}\right|^2 dV = \int_D d\left(\overline{f(\zeta)} \cdot \nu_f(\zeta)\right) = \int_{\partial D}\overline{f(\zeta)}\,\nu_f(\zeta) = 0.$$

From this and the continuity of the derivatives $\partial f/\partial\bar{\zeta}_k$ $(k = 1, \ldots, n)$ it follows that these derivatives are equal to zero in D, i.e., f is holomorphic in D. □

The following result is harder to prove, and we present it without proof.

THEOREM 1.6 (Kytmanov and Aĭzenberg). *If a function $f \in C(\overline{D})$ can be represented in D by the Martinelli-Bochner integral* (1.1), *then it is holomorphic in D in each of the following cases*:

a) $\partial D \in C^{(2)}$;

b) $n = 2$, *and the boundary ∂D is connected and belongs to the class $C^{(1)}$*.

§2. Multiplicity of a zero of a holomorphic mapping. The Rouché principle

Suppose that the mapping

$$w = f(z) \tag{2.1}$$

is holomorphic in the domain $D \subset \mathbf{C}^n$, where $w = (w_1, \ldots, w_n)$ and $f = (f_1, \ldots, f_n)$. A point $a \in D$ is called a *zero of the mapping* (2.1) if $f(a) = 0$. Let

E_f denote the set of zeros of f. If the Jacobian of (2.1) satisfies $(\partial f/\partial z)|_a \neq 0$ at an isolated point a of E_f, then a is called a *simple zero of the mapping*.

We have the following statement, which we prove a little later.

PROPOSITION 2.1. *If the closure of a neighborhood U_a of a zero a of the mapping* (2.1) *does not contain other zeros, then there exists an $\varepsilon > 0$ such that for almost all $\zeta \in B_\varepsilon$ the mapping*

$$w = f(z) - \zeta \tag{2.2}$$

has only simple zeros in U_a, and their number depends neither on ζ nor on the choice of the neighborhood U_a.

The number of zeros of (2.2) in U_a indicated in this proposition is called the *multiplicity of the zero a of the mapping* (2.1), and denoted by $\mu_a(f)$.([1])

EXAMPLE. The point (0,0) is a zero of multiplicity 2 for the mapping $w_1 = z_1$, $w_2 = z_2^2 + z_1^2$. Indeed, if $|\zeta|$ is small and $\zeta_1^2 \neq \zeta_2$, then the mapping $w_1 = z_1 - \zeta_1, w_2 = z_2^2 + z_1^2 - \zeta_2$ has the two simple zeros $(\zeta_1, \sqrt{\zeta_2 - \zeta_1^2})$ and $(\zeta_1, -\sqrt{\zeta_2 - \zeta_1^2}$ in a neighborhood of this point.

The next result follows from the local invertibility of a holomorphic mapping at points where $\partial f/\partial z \neq 0$.

PROPOSITION 2.2. *The multiplicity of a simple zero is equal to* 1.

We pass to the case of a zero that is not simple.

PROPOSITION 2.3. *If a is an isolated zero of* (2.1) *and* $(\partial f/\partial z)|_a = 0$, *then its multiplicity $\mu_a(f)$ is greater than* 1.

This statement justifies calling an isolated zero of (2.1) *multiple* if $(\partial f/\partial z)|_a = 0$.

Just as the logarithmic residue expresses the number of zeros of a holomorphic function in the theory of functions of a single complex variable, its multidimensional analogues are connected with the number of zeros of a holomorphic mapping. The following multidimensional analogue of the classical logarithmic residue of Cauchy was constructed with the help of the Martinelli-Bochner integral representation (1.1).

THEOREM 2.4. *Let $f \in A^n(\overline{D})$, where the domain D is bounded, and its boundary ∂D is piecewise smooth and does not contain zeros of f. Then f has only*

([1]) This is the so-called dynamic definition of multiplicity. See §22 about the connection of $\mu_a(f)$ with the coefficients of the Taylor expansion of (2.1) at the point a.

isolated zeros in D, and their number, with each zero counted as many times as its multiplicity,(2) *is expressed by the formula*

$$N = N(f, D) = \int_{\partial D} \omega(f(z), \overline{f(z)}), \tag{2.3}$$

where

$$\omega(f, \bar{f}) = \frac{(n-1)!}{(2\pi i)^n} \frac{1}{|f|^{2n}} \sum_{j=1}^{n} (-1)^{j-1} \bar{f}_j d\bar{f}_{[j]} \wedge df.$$

THEOREM 2.5. (The Rouché principle). *Suppose that the mapping* (2.1) *and the domain D satisfy the conditions of Theorem* 2.4, *and let $g \in A^n(\overline{D})$. If for each point $z \in \partial D$ there is at least one index j ($j = 1, \ldots, n$) such that*

$$|g_j(z)| < |f_j(z)|, \tag{2.4}$$

then (2.1) *and the mapping*

$$w = f(z) + g(z) \tag{2.5}$$

have the same number of zeros in D.

REMARK. Condition (2.4) holds, for example, if $|g| < |f|$ or $\|g\| < \|f\|$ on ∂D. This condition can be relaxed. In fact, we can require only that $f(z) + tg(z) \neq 0$ for $z \in \partial D$ and $0 \leqslant t \leqslant 1$. It suffices that at each point $z \in \partial D$ there is at least one index j such that $\operatorname{Re} g_j(z) < \operatorname{Re} f_j(z)$ or $\operatorname{Im} g_j(z) < \operatorname{Im} f_j(z)$.

We need some lemmas to prove these assertions.

LEMMA 2.6. *If $f \in A^n(D)$, then for any piecewise smooth cycle*

$$\gamma \in Z_{2n-1}(D \setminus E_f)$$

the integral $\int_\gamma \omega(f, \bar{f})$ is an integer.

PROOF. Since (2.1) maps the domain $D \setminus E_f$ into $\mathbf{C}^n \setminus \{0\}$, and the sphere ∂B_r forms a basis for the $(2n-1)$-dimensional homology, the image of the cycle γ under the mapping (2.1) is homologous to $N \cdot \partial B_r$, where N is an integer. Using Theorem 0.1, Corollary 0.5, and the Martinelli-Bochner formula (1.1), we get that

$$\int_\gamma \omega(f, \bar{f}) = \int_{f(\gamma)} \omega(\zeta, \bar{\zeta}) = N \int_{\partial B_r} \omega(\zeta, \bar{\zeta}) = N. \quad \square$$

---

(2) Henceforth we adhere to this convention wherever the number of zeros is concerned.

LEMMA 2.7. *Suppose that the mapping* (2.1) *and the domain D satisfy the conditions of Theorem* 2.4. *If, moreover, all the zeros of* (2.1) *in D are simple, then for any function $\varphi \in A_C(D)$*

$$\int_{\partial D} \varphi\omega(f, \bar{f}) = \sum_{a \in E_f} \varphi(a). \tag{2.6}$$

In particular (*for $\varphi \equiv 1$*), (2.3) *holds.*

PROOF. The form $\varphi\omega(f, \bar{f})$ is closed in $\bar{D} \setminus E_f$; this can be verified directly as in the proof that the Martinelli-Bochner form (1.2) is closed. According to Corollary 0.5,

$$\int_{\partial D} \varphi\omega(f, \bar{f}) = \sum_{a \in E_f} \int_{\partial U_a} \varphi\omega(f, \bar{f}), \tag{2.7}$$

where U_a is a sufficiently small neighborhood of a. We take U_a to be the connected component of the set $\{z : z \in D,\ |f(z)| < \varepsilon\}$ containing a. Since $(\partial f/\partial z)|_a \neq 0$, f is biholomorphic in $\bar{U}_a$ for sufficiently small ε; therefore, by Theorem 0.1 and the Martinelli-Bochner formula (1.1),

$$\int_{\partial U_a} \varphi\omega(f, \bar{f}) = \int_{\partial B_\varepsilon} \varphi(f^{-1}(\zeta))\omega(\zeta, \bar{\zeta}) = \varphi(f^{-1}(0)) = \varphi(a).$$

From this and (2.7) we get (2.6). □

LEMMA 2.8. *If the conditions of Theorem* 2.5 *hold, then*

$$\int_{\partial D} \omega(f, \bar{f}) = \int_{\partial D} \omega(f + g, \overline{f + g}). \tag{2.8}$$

PROOF. According to (2.4), $f(z) + tg(z) \neq 0$ for $z \in \partial D$ and $0 \leq t \leq 1$. Consequently, the integral

$$J(t) = \int_{\partial D} \omega(f + tg, \overline{f + tg})$$

is a continuous function of t on $[0, 1]$, and, by Lemma 2.6, its values are integers. Therefore, $J(t)$ is constant, and so $J(0) = J(1)$, i.e., (2.8) is valid. □

PROOF OF PROPOSITION 2.1. Let $\varepsilon > 0$ be such that $\varepsilon > \min_{z \in \partial U_a} |f(z)|$. Then, by Lemma 2.8,

$$\int_{\partial U_a} \omega(f) = \int_{\partial U_a} \omega(f - \zeta, \overline{f - \zeta}) \tag{2.9}$$

for $\zeta \in B_\varepsilon$. By Sard's Theorem 0.11, meas $f(\{z : z \in D,\ \partial f/\partial z = 0\}) = 0$; consequently, for almost all $\zeta \in B_\varepsilon$ the mapping (2.2) has only simple zeros. According to Lemma 2.7, the integral on the right-hand side of (2.9) is equal to

$N(f-\zeta, U_a)$ for such ζ. The equality (2.9) implies that

$$N(f-\zeta, U_a)=\int_{\partial U_a}\omega(f,\bar{f});$$

hence, $N(f-\zeta, U_a)$ does not depend on the choice of $\zeta\in B_\varepsilon$. If U_a' is another neighborhood of a, then, by Corollary 0.5,

$$\int_{\partial U_a'}\omega(f,\bar{f})=\int_{\partial U_a}\omega(f,\bar{f}).\quad\square$$

In passing we have proved that the multiplicity of an isolated zero of a holomorphic mapping (2.1) can be expressed by the formula

$$\mu_a(f)=\int_{\partial U_a}\omega(f,\bar{f}).\quad\square \tag{2.10}$$

PROOF OF THEOREM 2.4. 1. Assume first that $\partial f/\partial z\not\equiv 0$. We show first of all that (2.1) can have only isolated zeros in D. Take ε such that $0<\varepsilon\leqslant\min_{\partial D}|f|$. From Sard's theorem and Lemma 2.8 it follows that for almost all $\zeta\in B_\varepsilon$ the system (2.2) has only simple zeros in D, and their number $k\geqslant 0$ does not depend on $\zeta\in B_\varepsilon$.

If $E_f\neq\varnothing$, then for any point $a^{(1)}\in E_f$ we can choose a sequence of points $\zeta^{(m)}\to 0$, $\zeta^{(m)}\in B_\varepsilon$, such that for $\zeta=\zeta^{(m)}$ the mapping (2.2) has the simple zeros $\{z^{m,1},\ldots,z^{m,k}\}$, $k\geqslant 1$, and $z^{m,1}\to a^{(1)}$ as $m\to\infty$. Indeed, since $\partial f/\partial z\not\equiv 0$, there is a point $z^{m,1}$ at which $\partial f/\partial z\not\equiv 0$ in any neighborhood of $a^{(1)}$. Take $\zeta^{(m)}=f(z^{m,1})$; then $z^{m,1}$ is a simple zero of (2.2) for $\zeta=\zeta^{(m)}$. According to Sard's theorem, by varying $\zeta^{(m)}$ slightly we can make the remaining zeros of this mapping also simple. It is clear that $\zeta^{(m)}\to 0$ as $z^{m,1}\to a^{(1)}$.

On the other hand, we show that for any choice of $\zeta^{(m)}\to 0$ the simple zeros $\{z^{m,1},\ldots,z^{m,k}\}$ have a single finite set([3]) $\{a^{(1)},\ldots,a^{(k)}\}\subset D$ of limit points; consequently, $E_f=\{a^{(1)},\ldots,a^{(k)}\}$. Indeed, replacing the sequences $\{z^{(m,j)}\}_{m=1}^{\infty}$, $j=1,\ldots,k$, by subsequences, we can make them converge to some points $a^{(j)}\in D, j=1,\ldots,k$ (because there are no zeros on ∂D). The set E_f consists of these points alone. Suppose not: suppose that there exists a point $b^{(1)}\in E_f$ with $b^{(1)}\neq a^{(j)}$, $j=1,\ldots,k$. Then just as before we get a sequence of points $\zeta^{(m)}\to 0$ such that for $\zeta=\xi^{(m)}$ the system (2.2) has only simple zeros $\{w^{m,1},\ldots,w^{m,k}\}$, and $w^{m,j}\to b^{(j)}$, $j=1,\ldots,k$, as $m\to\infty$. We choose a $\varphi\in H_C(D)$ such that $\varphi(a^{(1)})=\cdots=\varphi(a^{(k)})=\varphi(b^{(2)})=\cdots=\varphi(b^{(k)})=0$, and

([3]) Here some of the points $a^{(1)},\ldots,a^{(k)}$ may coincide.

$\varphi(b^{(1)}) = 1$. By Lemma 2.7,

$$\int_{\partial D} \varphi\omega\left(f - \zeta^{(m)}, \overline{f - \zeta^{(m)}}\right) = \sum_{j=1}^{k} \varphi(z^{m,j});$$

$$\int_{\partial D} \varphi\omega\left(f - \xi^{(m)}, \overline{f - \xi^{(m)}}\right) = \sum_{j=1}^{k} \varphi(w^{m,j}).$$

Passing to the limit as $m \to \infty$, we get a contradiction:

$$\int_{\partial D} \varphi\omega(f, \bar{f}) = \sum_{j=1}^{k} \varphi(a^{(j)}) = 0;$$

$$\int_{\partial D} \varphi\omega(f, \bar{f}) = \sum_{j=1}^{k} \varphi(b^{(j)}) = 1.$$

Thus, E_f is finite.

Let us prove (2.3). We enclose each point $a \in E_f$ in a neighborhood U_a in such a way that $\overline{U}_a \subset D$ and $\overline{U}_a \cap \overline{U}_b = \varnothing$ for $a \neq b$, $a, b \in E_f$. Since the form $\omega(f, \bar{f})$ is closed in $D \setminus E_f$, Corollary 0.5 and (2.10) give us that

$$\int_{\partial D} \omega(f, \bar{f}) = \sum_{a \in E_f} \int_{\partial U_a} \omega(f, \bar{f}) = \sum_{a \in E_f} \mu_a(f) = N(f, D).$$

Moreover, if $E_f \neq \varnothing$, then $N(f, D) = k > 0$.

2. Suppose now that $\partial f/\partial z \equiv 0$. In this case the integral on the right-hand side of (2.3) is equal to zero, since $df = (\partial f/\partial z)\, dz$ appears in the form $\omega(f, \bar{f})$. We show that $N(f, D)$ is also equal to zero, i.e., f has no zeros in D. Suppose the opposite: $f(a) = 0$ for some point $a \in D$. Since $f|_{\partial D} \neq 0$, there is a holomorphic mapping $g\colon \overline{D} \to \mathbf{C}^n$ such that $g(a) = 0$, $|g| < |f|$ on ∂D, and $\partial(f + g)/\partial z \not\equiv 0$ (for example, we can take the linear mapping $g(z) = \lambda(z - a)$, where $\lambda = \|\lambda_{jk}\|$ is a matrix with sufficiently small elements such that $\operatorname{rank}\|\partial f_j(a)/\partial z_k + \lambda_{jk}\| = n$).

$$\int_{\partial D} \omega(f, \bar{f}) = \int_{\partial D} \omega(f + g, \overline{f + g}) > 1.$$

This is a contradiction. □

The proof of Theorem 2.5 follows from Theorem 2.4 and Lemma 2.8. □

Proof of Proposition 2.3. By the definition of the multiplicity of an isolated zero, $\mu_a(f) \geq 0$. Observe that $\partial f/\partial z \not\equiv 0$, for otherwise there would not be any isolated zeros at all (see part 2 of the proof of Theorem 2.4). There is a point $z_0 \in U_a$ such that $(\partial f/\partial z)|_{z_0} \neq 0$ and $|f(z_0)| < \min_{\partial U_a} |f|$. It now follows from Proposition 2.2 and Theorem 2.5 that $\mu_a(f) \geq 1$. If $\mu_a(f) = 1$, then for any $\zeta \in B_\varepsilon$, where ε is sufficiently small, (2.2) has exactly one zero in

U_a, i.e., (2.1) is one-to-one in a neighborhood of a. Then Osgood's theorem (see Theorem 4.15 below) gives us that this mapping is biholomorphic in a neighborhood of a, and $(\partial f/\partial a)|_a \neq 0$. □

We explain the geometric meaning of formula (2.3). Consider the $(n-1)$-dimensional closed piecewise smooth surface (cycle) $S = \{x : x = x(t),\ t \in I^{n-1}\}$ in $\mathbf{R}^n \setminus \{0\}$, where

$$I^{n-1} = \{t : t = (t_1,\ldots,t_{n-1}),\ 0 \leqslant t_j \leqslant 1, j = 1,\ldots,n-1\},$$
$$x(t_1,\ldots,0,\ldots,t_{n-1}) = x(t_1,\ldots,1,\ldots,t_{n-1}),$$
$$j = 1,\ldots,n-1.$$

The *index of the surface S with respect to the origin* is defined to be the number

$$J_0(S) = \frac{1}{\Sigma_n}\int_S \frac{1}{|x|^n}\sum_{k=1}^n (-1)^{j-1} x_j\, dx_{[j]}$$
$$= \frac{1}{\Sigma_n}\int_{I^{n-1}} \frac{1}{|x(t)|^n}\begin{vmatrix} x_1(t) & \cdots & x_n(t) \\ x'_{11}(t) & \cdots & x'_{n1}(t) \\ \cdots & \cdots & \cdots \\ x'_{1n-1} & \cdots & x'_{nn-1}(t)\end{vmatrix} dt, \qquad (2.12)$$

where $\Sigma_n = n\pi^{n/2}/\Gamma(n/2+1)$ is the measure (volume) of the unit sphere in $\mathbf{R}^n$. The integrand form in (2.12) expresses the measure of the projection of a surface element onto the unit sphere about zero (the solid angle under which this element is "seen" from the origin). The integral (2.3) can be represented in the form

$$\int_{\partial D}\omega(f,\bar f) = \int_\gamma \omega(\zeta,\bar\zeta) = \frac{(n-1)!}{(2\pi i)^n}\int_\gamma \frac{1}{|\zeta|^{2n}}\sum_{j=1}^n (-1)^{j-1}\bar\zeta_j\, d\bar\zeta_{[j]} \wedge d\zeta$$
$$= \frac{1}{\Sigma_{2n}}\int_\gamma \frac{1}{|\zeta|^{2n}}\Bigg[\sum_{k=1}^n (-1)^{j-1} u_j\, du_{[j]} \wedge dv + \sum_{j=1}^n (-1)^{j+n-1} v_j\, du \wedge dv_{[j]}$$
$$+ i\left(\sum_{j=1}^n (-1)^{j-1} v_j\, du_{[j]} \wedge dv + \sum_{j=1}^n (-1)^{j+n-1} u_j\, du \wedge dv_{[j]}\right)\Bigg] = J_0(S),$$

where $\gamma = f(\partial D)$ and $\zeta_j = u_j + iv_j, j = 1,\ldots,n$; here

$$\int_\gamma \frac{1}{|\zeta|^{2n}}\left(\sum_{j=1}^n (-1)^{j-1} v_j\, du_{[j]} \wedge dv + \sum_{j=1}^n (-1)^{j+n-1} u_j\, du \wedge dv_{[j]}\right)$$
$$= \int_\gamma \frac{1}{|\zeta|^{2n}}\, d\left(\frac{\sum_{j=1}^n du_{[j]} \wedge dv_{[j]}}{(2n-2)|\zeta|^{2n-2}}\right) = 0,$$

by Corollary 0.3. We can thus formulate the following statement.

THEOREM 2.9 (The argument principle). *If the conditions of Theorem* 2.4 *hold, then the number of zeros of the mapping f in* (2.1) *in the domain D is equal to the index with respect to the origin of the image $f(\partial D)$ of the boundary of D under f (the number of "circuits" of the surface $f(\partial D)$ around* 0).

This fact can also be obtained from topological considerations. Since the mapping (2.1) has no zeros on ∂D, the image $f(\partial D)$ of the cycle ∂D under this mapping is in $Z_{2n-1}(\mathbf{C}^n \setminus \{0\})$. Consequently, $f(\partial D) \sim N \cdot \partial B_\varepsilon$, where N is an integer indicating how many times the cycle $f(\partial D)$ "goes around" zero. From (1.1), Theorem 0.1, and Corollary 0.5 it now follows that $N = N(f, D)$. □

§3. The fundamental integral formula of Leray and Koppelman

1°. THEOREM 3.1 (Yuzhakov and Roos). *Suppose that D is a bounded domain with piecewise smooth boundary, and that f is in $A^n(\overline{D})$ and does not have zeros on ∂D. Then for any $\varphi \in A_C(D)$*

$$\int_{\partial D} \varphi\omega(f, \bar{f}) = \sum_{a \in E_f} \mu_a(f) \cdot \varphi(a). \tag{3.1}$$

PROOF. Choose $\varepsilon > 0$ such that $\varepsilon < \min_{\partial D}|f|$. For almost all $\zeta \in B_\varepsilon$ the mapping (2.2) has only simple zeros in D, and, by Lemma 2.7,

$$\int_{\partial D} \varphi\omega(f - \zeta, \overline{f - \zeta}) = \sum_{a_\zeta \in E_{f-\zeta}} \varphi(a_\zeta). \tag{3.2}$$

If we now let ζ tend to zero, then (see part 1 of the proof of Theorem 2.4) the zeros of (2.2) tend to the zeros of (2.1), and, moreover, the number of simple zeros of $f - \zeta$ tending to any particular zero of f is equal to its multiplicity. If we pass to the limit in (3.2) as $\zeta \to 0$, we get (3.1). □

2°. Let us consider the following exterior differential form (important for what follows), which depends on a holomorphic mapping f, a continuous vector-valued function $w^{(0)}$, and continuously differentiable vector-valued functions $w^{(1)}, \ldots, w^{(n-1)}$:

$$\Omega(w^{(0)}, w^{(1)}, \ldots, w^{(n-1)}, f) = \frac{(-1)^{n(n-1)/2}}{(2\pi i)^n} \frac{\langle w^{(0)}, df \rangle}{\langle w^{(0)}, f \rangle} \wedge d\frac{\langle w^{(1)}, df \rangle}{\langle w^{(1)}, f \rangle} \wedge \cdots \wedge d\frac{\langle w^{(n-1)}, df \rangle}{\langle w^{(n-1)}, f \rangle}. \tag{3.3}$$

THEOREM 3.2 (Fundamental integral formula). *Let D be a bounded domain with piecewise smooth boundary, and suppose that the mapping $f \in A^n(\overline{D})$ does not have zeros on ∂D and that the vector-valued functions $w^{(0)} \in C(\partial D)$ and*

$w^{(j)} \in C^{(1)}(\partial D), j = 1,\ldots,n-1$, *satisfy the condition*

$$\langle w^{(j)}(z), f(z)\rangle \neq 0, \qquad z \in \partial D, j = 0, 1, \ldots, n-1. \tag{3.4}$$

Then any function $\varphi \in A_C(D)$ *satisfies the formula*

$$\int_{\partial D} \varphi \Omega(w^{(0)}, w^{(1)}, \ldots, w^{(n-1)}, f) \sum_{a \in E_f} \mu_a(f) \cdot \varphi(a). \tag{3.5}$$

We indicate other ways of writing the kernel Ω. Let

$$u^{(j)} = \frac{w^{(j)}}{\langle w^{(j)}, f\rangle}, \qquad j = 0, 1, \ldots, n-1;$$

then (3.3) gives us

$$\Omega = \frac{(-1)^{n(n-1)/2}}{(2\pi i)^n} \langle u^{(0)}, df\rangle \wedge d\langle u^{(1)}, df\rangle \wedge \cdots \wedge d\langle u^{(n-1)}, df\rangle, \tag{3.6}$$

and (3.4) becomes

$$\langle u^{(j)}(z, \bar{z}), f(z)\rangle = 1, \qquad z \in \partial D, j = 0, 1, \ldots, n-1. \tag{3.7}$$

Further, (3.6) leads to

$$\Omega = \frac{1}{(2\pi i)^n} \sum_{(i_1, \ldots, i_{n-1})} \begin{vmatrix} u_1^{(0)} u_{1i_1}^{(1)} \cdots u_{1i_{n-1}}^{(n-1)} \\ \cdots\cdots\cdots \\ \cdots\cdots\cdots \\ u_n^{(0)} u_{ni_1}^{(1)} \cdots u_{ni_{n-1}}^{(n-1)} \end{vmatrix} d\bar{z}_{i_1} \wedge \cdots \wedge d\bar{z}_{i_{n-1}} \wedge df, \tag{3.8}$$

where $u_{ji_l}^{(k)} = \partial u_j^{(k)}/\partial \bar{z}_{i_l}$, and the summation is over all possible $(n-1)$-tuples $(i_1, \ldots, i_{n-1})$ of numbers taken from the set $\{1, 2, \ldots, n\}$.

We mention also that $\Omega(\bar{f}, \ldots, \bar{f}, f) = \omega(f, \bar{f})$, i.e., (3.1) is a particular case of (3.5), and condition (3.4) means that $|f|^2 \neq 0$ on ∂D.

To prove Theorem 3.2 we need some simple properties of the kernel Ω.

LEMMA 3.3. $\Omega(w^0, w^1, \ldots, w^{n-1}, f)$ *does not depend on the vector-valued function* w_0.

PROOF. Let Ω_1 and Ω_2 be two forms (3.6) depending on $u^{(0)}, u^{(1)}, \ldots, u^{(n-1)}$ and $v^{(0)}, u^{(1)}, \ldots, u^{(n-1)}$, respectively. We must show that $\Omega_1 = \Omega_2$. Let $s^{(0)} = u^{(0)} - v^{(0)}$. Then

$$\langle s_0(z, \bar{z}), f(z)\rangle = 0, \qquad z \in \partial D. \tag{3.9}$$

Using the representation of Ω_1 and Ω_2 in the form (3.8), we get that the form $\Omega_1 - \Omega_2$ has the determinants

$$\begin{vmatrix} s_1^{(0)} u_{1i_1}^{(1)} \cdots u_{1i_{n-1}}^{(n-1)} \\ \cdots\cdots\cdots\cdots \\ \cdots\cdots\cdots\cdots \\ \cdots\cdots\cdots\cdots \\ s_n^{(0)} u_{ni_1}^{(1)} \cdots u_{ni_{n-1}}^{(n-1)} \end{vmatrix}, \tag{3.10}$$

as coefficients. Further, (3.7) gives us that

$$\left\langle \frac{\partial u^{(j)}(z,\bar z)}{\partial \bar z_{i_k}}, f(z) \right\rangle = 0, \qquad z \in \partial D, j = 1, \dots, n-1,$$

and this and (3.9) imply that the determinants (3.10) vanish. □

LEMMA 3.4. *If the vector-valued functions* $w^{(1)}, \dots, w^{(n-1)}$ *and* $p^{(1)}, \dots, p^{(n-1)}$ *are in the class* $C^{(2)}(\partial D)$, *then the difference*

$$\Omega(w^{(0)}, w^{(1)}, \dots, w^{(n-1)}, f) - \Omega(p^{(0)}, p^{(1)}, \dots, p^{(n-1)}, f) \tag{3.11}$$

is a $\bar\partial$*-exact form.*

PROOF. The difference (3.11) can be represented as a sum of analogous differences of forms that differ only in one argument, so it suffices to prove that the difference

$$\Omega(w^{(0)}, w^{(1)}, \dots, w^{(n-1)}, f) - \Omega(w^{(0)}, p^{(1)}, w^{(2)}, \dots, w^{(n-1)}, f)$$

is $\bar\partial$-exact. Indeed,

$$\bar\partial\left\{ \frac{(-1)^{n(n-1)/2}}{(2\pi i)^n} \frac{\langle w^{(1)}, df\rangle}{\langle w^{(1)}, f\rangle} \wedge \frac{\langle p^{(1)}, df\rangle}{\langle p^{(1)}, f\rangle} \wedge d\frac{\langle w^{(2)}, df\rangle}{\langle w^{(2)}, f\rangle} \wedge \cdots \wedge d\frac{\langle w^{(n-1)}, df\rangle}{\langle w^{(n-1)}, f\rangle} \right\}$$

$$= \frac{(-1)^{n(n-1)/2}}{(2\pi i)^n} \left\{ \left(d\frac{\langle w^{(1)}, df\rangle}{\langle w^{(1)}, f\rangle} \right) \wedge \frac{\langle p^{(1)}, df\rangle}{\langle p^{(1)}, f\rangle} \wedge \cdots - \frac{\langle w^{(1)}, df\rangle}{\langle w^{(1)}, f\rangle} \wedge d\frac{\langle p^{(1)}, df\rangle}{\langle p^{(1)}, f\rangle} \wedge \cdots \right\}$$

$$= \Omega(p^{(1)}, w^{(1)}, w^{(2)}, \dots, w^{(n-1)}, f) - \Omega(w^{(1)}, p^{(1)}, w^{(2)}, \dots, w^{(n-1)}, f)$$

$$= \Omega(w^{(0)}, w^{(1)}, w^{(2)}, \dots, w^{(n-1)}, f) - \Omega(w^{(0)}, p^{(1)}, w^{(2)}, \dots, w^{(n-1)}, f),$$

where Lemma 3.3 was used in passing to the last equality. □

PROOF OF THEOREM 3.2. If $w^{(1)}, \ldots, w^{(n-1)}$ belong to the class $C^{(2)}(\partial D)$, then the difference of the kernels in (3.1) and (3.5) is a $\bar{\partial}$-exact form, by Lemma 3.4, i.e., this difference is orthogonal to holomorphic functions when integration is considered over ∂D (see Corollary 0.3). Therefore, (3.5) follows from (3.1). But if $w^{(j)} \notin C^{(2)}(\partial D)$, then we can approximate them on ∂D by smoother vector-valued functions with preservation of the condition (3.4), write the formula (3.5) for the approximating functions, and pass to the limit. □

We mention some immediate corollaries of (3.5).

COROLLARY 3.5 (Leray and Koppelman). *If $w^{(0)}, \ldots, w^{(n-1)}$ satisfy the condition*

$$\langle w^{(j)}(z, a), z - a \rangle \neq 0, \qquad z \in \partial D, a \in D, j = 0, 1, \ldots, n-1,$$

then every $\varphi \in A_C(D)$ satisfies the generalized Cauchy-Fantappiè formula

$$\int_{\partial D} \varphi \Omega(w^{(0)}, w^{(1)}, \ldots, w^{(n-1)}, z - a) = \varphi(a). \tag{3.12}$$

COROLLARY 3.6 (Leray). *If $w \in C^{(1)}(\partial D)$ and $\langle w(z, a), z - a \rangle \neq 0$ for $z \in \partial D$ and $a \in D$, then every $\varphi \in A_C(D)$ can be represented by the Cauchy-Fantappiè formula*

$$\int_{\partial D} \varphi \omega(z - a, w) = \varphi(a), \tag{3.13}$$

where

$$\omega(z - a, w) = \frac{(n-1)!}{(2\pi i)^n} \frac{\sum_{k=1}^n (-1)^{k-1} w_k dw_{[k]} \wedge dz}{\langle w, z - a \rangle^n}.$$

PROOF. From (3.6) we get that

$$\Omega(w, \ldots, w, z - a) = \frac{(n-1)!}{(2\pi i)^n} \sum_{k=1}^n (-1)^{k-1} u_k du_{[k]} \wedge dz,$$

where

$$u = \frac{w}{\langle w, z - a \rangle}. \tag{3.14}$$

Further, taking (3.8) into account, we find

$$\Omega(w, w, \ldots, w, z-a) = \frac{(n-1)!}{(2\pi i)^n} \sum_{k=1}^{n} \begin{vmatrix} u_1 u'_{11} & \cdots & u'_{1n} \\ \cdots & [k] & \cdots \\ u_n u'_{n1} & \cdots & u'_{nn} \end{vmatrix} d\bar{z}_{[k]} \wedge dz$$

$$= \frac{(n-1)!}{(2\pi i)^n \langle w, z-a\rangle^n} \sum_{k=1}^{n} \begin{vmatrix} w_1 w'_{11} & \cdots & w'_{1n} \\ \cdots & [k] & \cdots \\ w_n w'_{n1} & \cdots & w'_{nn} \end{vmatrix} d\bar{z}_{[k]} \wedge dz$$

$$= \omega(z-a, w). \quad \square$$

Note that the vector-valued function (3.14) satisfies the condition

$$\langle u, z-a\rangle = 1, \qquad z \in \partial D, a \in D. \tag{3.15}$$

We give (3.13) a more abstract form. Fix a point $a \in D$ and consider the surface $M_a = \{(z, u): \langle u, z-a\rangle = 1,\ z \in \partial D\}$ in the space $\mathbf{C}^{2n}$ of complex variables (z, u).

COROLLARY 3.7 (Leray). *Suppose that the vector-valued function $u(z, a)$, $z \in \partial D$, $a \in D$, belongs to the class $C^{(1)}(\partial D)$ with respect to z and satisfies condition* (3.15). *Let α be the cycle on M_a described by the point (z, u) as z runs along ∂D. This cycle is in some class $h \in H_{2n-1}(M_a)$. Then any $\varphi \in A_C(D)$ and any cycle $\beta \in h$ satisfy the Cauchy-Fantappiè formula*

$$\varphi(a) = \frac{(n-1)!}{(2\pi i)^n} \int_\beta \varphi(z) \sum_{k=1}^{n} (-1)^{k-1} u_k du_{[k]} \wedge dz. \tag{3.16}$$

PROOF. Observe that (3.16) is (3.13) for $\beta = \alpha$. To get (3.16) with any $\beta \in h$ from (3.13) it suffices to see that the form $\varphi(z) \cdot \omega(z-a, u)$ is closed on M_a. Indeed, only $2n-1$ of the differentials $dz_1, \ldots, dz_n, du_1, \ldots, du_n$ are independent on M_a, and the form under consideration has maximal dimension $2n-1$, so it is closed. $\square$

We point out that the Cauchy-Fantappiè formula (all the more so, the generalized Cauchy-Fantappiè formula) is not a single integral representation for functions $\varphi \in A_C(D)$, but a whole collection of such representations, depending on the choice of the vector-valued function w. For various classes of domains D it is possible to choose w in such a way that the conditions of Corollary 3.6 hold and w is holomorphic with respect to a in D. We thus obtain integral representations with holomorphic kernels (see §§4 and 8–11).

The classical Cauchy formula

$$f(a_1) = \frac{1}{2\pi i} \int_{\partial D} \frac{f(z_1)dz_1}{z_1 - a_1} \tag{3.17}$$

for holomorphic functions of a single complex variable has two remarkable properties which have been responsible for many of its applications:

1) It is universal, i.e., true for any domain D with a sufficiently nice boundary ∂D, and the Cauchy kernel $1/2\pi i(z_1 - a_1)$ does not depend on the shape of D.

2) The Cauchy kernel is holomorphic in $a_1 \in D$ for fixed $z_1 \in \partial D$.

There is no integral representation with two analogous properties for holomorphic functions of several complex variables; instead, there exist formulas that either are universal but have a nonholomorphic kernel (for example, the Martinelli-Bochner formula (1.1)) or have holomorphic kernels but are not universal (see §§4 and 8–11).

The following assertion can be obtained similarly to the way Corollary 3.6 was obtained.

COROLLARY 3.8. *If the domain D and the mapping f satisfy the conditions of Theorem 3.2 and the vector-valued function $w \in C^{(1)}(\partial D)$ is such that $\langle w, f\rangle \neq 0$ on ∂D, then for every $\varphi \in A_C(D)$*

$$\int_{\partial D} \varphi\omega(f, w) = \sum_{a \in E_f} \mu_a(f)\varphi(a), \tag{3.18}$$

where

$$\omega(f, w) = \frac{(n-1)!}{(2\pi i)^n} \frac{\sum_{k=1}^n (-1)^{k-1} w_k \, dw_{[k]} \wedge df}{\langle w, f\rangle^n}.$$

We remark in conclusion that (3.5), (3.12), and (3.18) can be given a more abstract form similarly to the way (3.13) was reduced to the form (3.16).

§4. The Cauchy formula. The logarithmic residue with respect to the skeleton*

1°. Let $D_i \subset \mathbf{C}^1_{z_i}$ be a domain with piecewise smooth boundary, $i = 1, \dots, n$. The domain $D = D_1 \times \cdots \times D_n \subset \mathbf{C}^n$ is called a *polycylinder domain* (in particular, a *polydisk* if all the D_i are disks). The set $\Gamma = \partial D_1 \times \cdots \times \partial D_n$, provided with the natural orientation induced by the orientation of D, is called the *skeleton* of D.

THEOREM 4.1. *Let D be a polycylinder domain and a any point of D, and suppose that $\varphi \in A_C(D)$. Then the Cauchy formula is valid:*

$$\varphi(a) = \frac{1}{(2\pi i)^n} \int_\Gamma \frac{\varphi(z)\, dz}{z - a}. \tag{4.1}$$

**Editor's note.* "Skeleton" is a literal rendering of "остров", but "distinguished boundary" is common in the English literature.

Note that the skeleton Γ plays the same role in the Cauchy formula (4.1) as the whole boundary of the domain plays in the classical Cauchy formula for $n = 1$. The existence of the integral representation over the skeleton implies that the functions in $A_C(D)$ satisfy the maximum modulus principle with respect to the skeleton (see Theorem 6.9).

The Cauchy formula (4.1) is an example of a formula with a holomorphic kernel that is not universal (it is true only for polycylinder domains).

PROOF. Formula (4.1) is easy to get by repeated application of the classical Cauchy formula for $n = 1$, but we give another proof by showing that (4.1) follows from the Cauchy-Fantappiè formula (3.16) for a suitable choice of the cycle of integration β. The same method is used later in deriving Weil's formula.

Let us represent ∂D in the form $\bigcup_1^n \gamma_i$, where $\gamma_i = \{z : z \in \overline{D}, z_i \in \partial D_i\}$. On each face γ_i we consider the vector-valued function $u^{(i)} = (u_1^{(i)}, \ldots, u_n^{(i)})$, where $u_k^{(i)} = \delta_{ik}(z_i - a_i)^{-1}$, $i, k = 1, \ldots, n$. Obviously, the $u^{(i)}$ satisfy condition (3.15). Now consider the oriented surface $\beta_i = \{(z, u^{(i)}), z \in \gamma_i\}$ in $\mathbf{C}^{2n}$. The union $\bigcup_1^n \beta_i$ is not a cycle, since each face γ_i has its own vector-valued function $u^{(i)}$. At the same time, ∂D is a cycle, and the set $\{(z, u), z \in \partial D\}$ also would be a cycle if u were continuous in $z \in \partial D$. The set $\bigcup_1^n \beta_i$ has "holes" located over the points where the faces γ_i intersect, i.e., over the edges. We "stop up" these "holes" as follows: if the k faces $\gamma_{i_1}, \ldots, \gamma_{i_k}$ meet at the edge $\gamma_{i_1 \cdots i_k}$, then we construct the set

$$\beta_{i_1 \cdots i_k} = \left\{(z, u) : z \in \gamma_{i_1 \cdots i_k}, u = \sum_{l=1}^{k} \lambda_l u^{(i_l)}, \lambda_l \geqslant 0, \sum_{l=1}^{k} \lambda_l = 1\right\}.$$

We add all the surfaces $\beta_{i_1 \cdots i_k}$ to $\bigcup_1^n \beta_i$; the resulting (oriented) surface β is a cycle. We apply (3.16) to β. All the $u^{(i)}$ are holomorphic in z; therefore, $du_{[k]}^{(i)} \wedge dz = 0$, $i, k = 1, \ldots, n$, and, hence, the integral in (3.16) over each β_i is equal to zero. If $k < n$, then all the $du_{[j]} \wedge dz$ are 0 on $\beta_{i_1 \cdots i_k}$, because u depends holomorphically on z and on the $k - 1$ independent parameters λ, i.e., there are $n + k - 1 < 2n - 1$ independent differentials in all. Therefore, the integral in (3.16) over $\beta_{i_1 \cdots i_k}$ is also equal to zero. It remains to compute this integral over $\beta_{i_1 \cdots i_n}$. On $\Gamma = \gamma_{1 \cdots n}$ the components u_k of the vector-valued function u have the form

$$u_k = \sum_{l=1}^{n} \lambda_l u_k^l = \frac{\lambda_k}{z_k - a_k}, \qquad k = 1, \ldots, n,$$

and so

$$\frac{(n-1)!}{(2\pi i)^n}\int_{\beta_{1,n}} \varphi(z)\sum_{k=1}^{n}(-1)^{k-1}u_k\,du_{[k]}\wedge dz$$
$$=\frac{(n-1)!}{(2\pi i)^n}\int_\Gamma \varphi(z)\frac{dz}{z-a}\int_{\substack{\lambda_1+\cdots+\lambda_n=1\\ \lambda_i\geq 0}}\sum_{k=1}^{n}(-1)^{k-1}\lambda_k\,d\lambda_{[k]}$$
$$=\frac{1}{(2\pi i)^n}\int_\Gamma \varphi(z)\frac{dz}{z-a},$$

where we have used the elementary equality

$$\int_{\substack{\lambda_1+\cdots+\lambda_n=1\\ \lambda_i\geq 0}}\sum_{k=1}^{n}(-1)^{k-1}\lambda_k\,d\lambda_{[k]}=\frac{1}{(n-1)!}. \tag{4.2}$$

To verify (4.2) it suffices to observe that, by Stokes' formula, the left-hand side of (4.2) is equal to

$$n\int_{\lambda_1+\cdots+\lambda_n\leq 1,\,\lambda_i\geq 0} d\lambda = n\frac{1}{n!}. \quad \square$$

COROLLARY 4.2. *Under the conditions of Theorem* 4.1,

$$D^\alpha\varphi(a)=\frac{\alpha!}{(2\pi i)^n}\int_\Gamma\frac{\varphi(z)\,dz}{(z-a)^{\alpha+I}}.$$

2°. The logarithmic residue formulas considered in §§2 and 3 for a holomorphic mapping involve integration over the whole boundary of a domain. We now investigate logarithmic residue formulas in which the integration is over an n-dimensional cycle. The logarithmic residue formulas in §§2 and 3 are essentially based on the Martinelli-Bochner formula (1.1), and the formulas of the present subsection are based on the Cauchy formula (4.1). The results here can, as a rule, be derived from the general formulas (3.5) or (3.17) in the same way as we just obtained (4.1) from (3.16), but we shall give simpler proofs based on the Cauchy formula (4.1). The form

$$\frac{df(z)}{f(z)}=\left(\frac{\partial f}{\partial z}\Big/f\right)dz=\frac{\partial \ln f(z)}{\partial z}dz$$

appears here as an analogue of the logarithmic differential for a holomorphic mapping.

We need the concept of an *analytic polyhedron*. This is defined to be an open set of the form $D=D_\rho=\{z: z\in G, |f_i(z)|<\rho_i, i=1,\dots,k\}$, where $f_i\in A(G)$ $(i=1,\dots,n)$ and G is a domain in $\mathbf{C}^n$, $D\Subset G$. If $k=n$, then D is called a *special analytic polyhedron*. The latter polyhedron play an important role in this subsection. We first mention the following facts about them.

LEMMA 4.3. *Every special analytic polyhedon D consists of finitely many connected components.*

PROOF. Every (bounded) domain $\tilde{D}$ can be approximated by domains with piecewise smooth boundaries in such a way that the conditions of Theorem 2.5 hold on the boundaries of the approximating domains if they hold on the boundary of $\tilde{D}$; therefore, the Rouché principle is applicable to $\tilde{D}$. It follows from the definition of D that for each connected component $D' \subset D$ there exists a point $a \in D'$ such that $f(a) = \zeta$, $|\zeta_i| < \rho_i$, $i = 1, \ldots, n$. Hence, the mapping $f(z) - \zeta$ has zeros in D'. On the other hand, by the Rouché principle (Theorem 2.5) the mappings $f(z)$ and $f(z) - \zeta$ have the same number of zeros in D', so $f(z)$ has zeros in D'. Thus, $f(z)$ has zeros in each connected component D' of the special analytic polyhedron D. We conclude from Theorem 2.4 that the number of connected components of D is finite. □

LEMMA 4.4. *The skeleton $\Gamma = \Gamma_\rho = \{z : z \in \overline{D}, |f_i(z)| = \rho_i, i = 1, \ldots, n\}$ of a special analytic polyhedron D is smooth for almost all $\rho = (\rho_1, \ldots, \rho_n)$, $\rho_i > 0$, $i = 1, \ldots, n$.*

A proof is obtained immediately by applying Sard's theorem (Theorem 0.11) to the mapping $z \to (|f_1(z)|^2, \ldots, |f_n(z)|^2)$. □

In all the results below concerning integration over the skeleton Γ_ρ it is assume that Γ_ρ is smooth. By Lemma 4.4, this situation can always be achieved by varying ρ by an arbitrarily small amount.

THEOREM 4.5 (Caccioppoli, Martinelli, and Sorani). *Let D be a special analytic polyhedron given with the help of a vector-valued function $f \in A^n(\overline{D})$. Then any $\varphi \in A_C(D)$ satisfies the formula*

$$\frac{1}{(2\pi i)^n} \int_\Gamma \varphi(z) \frac{df(z)}{f(z)} = \sum_{a \in E_f} \mu_a(f) \varphi(a), \tag{4.3}$$

where $E_f = \{z : z \in D, f(z) = 0\}$, and Γ is the skeleton of D.

PROOF. Since $\partial D = \bigcup_{j=1}^n \{z : z \in \overline{D}, |f_j(z)| = \rho_j\}$, f does not have zeros on ∂D, and, by Theorem 2.4, it has only isolated zeros in D. We surround each zero $a \in E_f$ with a neighborhood $U_a \Subset D$ in such a way that $\overline{U}_a \cap \overline{U}_b = \varnothing$ for $a \neq b$, $a, b \in E_f$. Let $\delta > 0$ be such that $\delta < \min\{\|f(z)\|, z \in \overline{D} \setminus \bigcup_{a \in E_f} U_a\}$. Then each connected component of the cycle $\gamma = \{z : z \in D, |f_1(z)| = \cdots = |f_n(z)| = \delta\}$ falls in one of the neighborhoods U_a, $a \in E_f$. Let $\gamma_a = \gamma \cap U_a$.

Note that $\Gamma - \gamma = \partial Q$, where $Q = \{z : z \in \bar{D}, |f_j(z)| = \delta t + \rho_j(1-t), j = 1, \ldots, n, 0 \leq t \leq 1\}$, i.e., $\Gamma \sim \gamma$ in $D_f = D \setminus \{z : f_1(z) \cdots f_n(z) = 0\}$. The form $\varphi(df/f)$ is holomorphic in D_f; therefore, by the Cauchy-Poincaré theorem,

$$\frac{1}{(2\pi i)^n} \int_\Gamma \varphi \frac{df}{f} = \sum_{a \in E_f} \frac{1}{(2\pi i)^n} \int_{\gamma_a} \varphi \frac{df}{f}. \tag{4.4}$$

If a is a simple zero of f, then U_a can be chosen small enough that f is biholomorphic in it. Making the change of variables $w = f(z)$ and applying the Cauchy formula (4.1), we get that

$$\frac{1}{(2\pi i)^n} \int_{\gamma_a} \varphi \frac{df}{f} = \frac{1}{(2\pi i)^n} \int_{\substack{|w_j| = \delta \\ j=1,\ldots,n}} \varphi\left(f|_{\bar{U}_a}^{-1}(w)\right) \frac{dw}{w} = \varphi\left(f|_{\bar{U}_a}^{-1}(0)\right) = \varphi(a). \tag{4.5}$$

Suppose now that a is a multiple zero of f. By Proposition 2.1, for almost all sufficiently small (in modulus) ζ the mapping $f(z) - \zeta$ has simple zeros $a^{(1)}, \ldots, a^{(k)}$ in U_a, where $k = \mu_a(f)$. For $\|\zeta\| < \min\{\delta, \min_{\partial U_a} \|f(z)\| - \delta\}$ the set $\gamma_{a,\zeta} = \{z : z \in U_a, |f_1(z) - \zeta_1| = \cdots = |f_n(z) - \zeta_n| = \delta\}$ forms a cycle in U_a, because $\gamma_{a,\zeta} \cap \partial U_a = \varnothing$. The cycle $\gamma_{a,\zeta}$ is obtained from γ_a by the homotopy $\gamma_{a,t\zeta} = \{z : z \in U_a, |f_j(z) - t\zeta_j| = \delta, j = 1, \ldots, n\}$, $0 \leq t \leq 1$, in the domain $U_a \setminus \{z : \prod_{j=1}^n (f_j(z) - \zeta_j) = 0\}$. Applying the Cauchy-Poincaré theorem once again along with formula (4.5) for the simple zeros, we have that

$$\frac{1}{(2\pi i)^n} \int_{\gamma_a} \varphi \frac{d(f-\zeta)}{f-\zeta} = \frac{1}{(2\pi i)^n} \int_{\gamma_{a,\zeta}} \varphi \frac{d(f-\zeta)}{f-\zeta} = \sum_{j=1}^k \varphi(a^{(j)}),$$

and, passing to the limit as $\zeta \to 0$,

$$\frac{1}{(2\pi i)^n} \int_{\gamma_a} \varphi \frac{df}{f} = \mu_a(f)\varphi(a). \tag{4.6}$$

Now (4.3) follows from (4.4)–(4.6). □

THEOREM 4.6. *If f and D satisfy the conditions of Theorem 2.4, then (4.3) holds for any $\varphi \in A_C(D)$, where $\Gamma = \{z : z \in \partial D, |f_2(z)| = \cdots = |f_n(z)| = \varepsilon\}$, and $\varepsilon < \min_{\partial D} \|f(z)\|$.*

PROOF. Since $\varepsilon < \min_{\partial D} \|f\|$, it follows that $|\Gamma| \subset D_f$. Note that if $D \cap E_f \neq \varnothing$, then $|\Gamma| \neq \varnothing$. Indeed, by the Rouché principle (Theorem 2.5), the mapping $f_1(z) - \varepsilon, \ldots, f_n(z) - \varepsilon$ has zeros in D. According to the second part of the proof of Theorem 2.4, this implies that the mapping $0, f_2(z) - \varepsilon, \ldots, f_n(z) - \varepsilon$ has zeros on ∂D, i.e., $|\Gamma|$ is not empty. The cycle Γ is homologous to $\gamma = \{z : z \in D, |f_1(z)| = \varepsilon, \ldots, |f_n(z)| = \varepsilon\}$ in $D_f = \bar{D} \setminus \{z : \prod_{j=1}^n f_j(z) = 0\}$, since $\Gamma - \gamma = \partial Q$, where $Q = \{z : z \in \bar{D}, |f_1(z)| \geqslant \varepsilon, |f_2(z)| = \cdots = |f_n(z)| = \varepsilon\}$, $Q \subset D_f$. By the Cauchy-Poincaré theorem and Theorem 4.5, applied to the domain $\tilde{D} = \{z : z \subset \tilde{D}, |f_j(z)| < \varepsilon, j = 1, \ldots, n\}$, we get that

$$\frac{1}{(2\pi i)^n} \int_\Gamma \varphi \frac{df}{f} = \frac{1}{(2\pi i)^n} \int_\gamma \varphi \frac{df}{f} = \sum_{a \in E_f} \mu_a(f) \varphi(a). \quad \square$$

COROLLARY 4.7. *Under the conditions of Theorems* 4.5 *and* 4.6, *the number N of zeros of f in D is expressed by the formula*

$$N = N(f, D) = \frac{1}{(2\pi i)^n} \int_\Gamma \frac{df}{f}.$$

The next theorem, which includes a variant of the Rouché principle and the logarithmic residue theorems, is useful for applications.

THEOREM 4.8 (Yuzhakov). *Let f, φ, D, and Γ be the same as in Theorem* 4.5, *and suppose that the mapping $g \in A^n(\bar{D})$ satisfies on Γ the inequality*

$$|g_j(z)| < |f_j(z)|, \qquad j = 1, \ldots, n. \tag{4.7}$$

Then: a) *the mappings f and $f + g$ have the same number of zeros in D*; *and* b)

$$\frac{1}{(2\pi i)^n} \int_\Gamma \varphi \frac{d(f+g)}{f+g} = \sum_{a \in E_{f+g}} \mu_a(f+g) \varphi(a). \tag{4.8}$$

LEMMA 4.9. *Under the conditions of Theorem* 4.8 *there exists a number $\delta_0 > 0$ such that for $0 < \delta < \delta_0$ the cycle Γ is homologous to the cycle $\gamma = \{z : z \in D, |f_j(z) + g_j(z)| = \delta, j = 1, \ldots, n\}$ in $D_{f+g} = \bar{D} \setminus \{z : \prod_{j=1}^{r} [f_j(z) + g_j(z)] = 0\}$.*

PROOF. Let $r_j = \max_{z \in \bar{D}} |g_j(z)|$, $j = 1, \ldots, n$. By the maximum modulus principle (see Theorem 6.9), it follows from (4.6) that $r_j < \rho_j$ $(j = 1, \ldots, n)$. Consider $\delta_0 = \min\{\rho_1 - r_1, \ldots, \rho_n - r_n\}$ and the family of cycles $\Gamma_p = \{z : z \in \bar{D}, |f_j(z) + g_j(z)| = \delta, j = 1, \ldots, p, |f_k(z)| = \rho_k, k = p+1, \ldots, n\}$, $0 < \delta < \delta_0$, $p = 0, 1, \ldots, n$. Obviously, $\Gamma_0 = \Gamma$ and $\Gamma_n = \gamma$. Further, $\Gamma_p \subset D_{f+g}$, since $f + g \neq 0$ on Γ_p by the choice of δ. Moreover, $\Gamma = \Gamma_0 \sim \Gamma_1 \sim \cdots \sim \Gamma_n = \gamma$ in D_{f+g}. Indeed, $\Gamma_p - \Gamma_{p+1} = \partial Q_p$, where $Q_p = \{z : z \in \bar{D}, |f_j(z) + g_j(z)| = \delta, j = 1, \ldots, p-1, |f_p(z) + g_p(z)| \geqslant \delta, |f_p(z)| \leqslant \rho_p, |f_k(z)| = \rho_k, k = p+1, \ldots, n\} \subset D_{f+g}$, $p = 1, \ldots, n$. $\square$

PROOF OF THEOREM 4.8. By the maximum modulus principle for a special analytic polyhedron (see Theorem 6.9), (4.7) gives us that $|g_j(z)| < \rho_j$ in $\bar{D}$ for $j = 1, \ldots, n$. Thus, the inequality $|g_j(z)| < |f_j(z)|$, $j = 1, \ldots, n$, holds on the face $\Delta_j = \{z : z \in \bar{D}, |f_j(z)| = \rho_j\}$. But since $\partial D = \bigcup_1^n \Delta_j$, we get assertion a) from Theorem 2.5. Further, by Lemma 4.9 and Theorem 4.5, applied to the mapping $f+g$ and the cycle γ,

$$\frac{1}{(2\pi i)^n}\int_\Gamma \varphi\frac{d(f+g)}{f+g} = \frac{1}{(2\pi i)^n}\int_\gamma \varphi\frac{d(f+g)}{f+g} = \sum_{a\in E_{f+g}} \mu_a(f+g)\varphi(a). \quad \square$$

EXAMPLE. Let m be the multiplicity of the zero of the mapping $w_1 = z_1 - z_2^2$, $w_2 = z_1^2 - z_2^3$ at the point (0, 1). Since the inequalities $|z_2|^2 < |z_1|$, and $|z_1|^2 < |z_2|^3$, hold on $\Gamma = \{z : |z_1| = \varepsilon^7, |z_2| = \varepsilon^4\}$, $0 < \varepsilon < 1$, it follows from Theorem 4.8 that

$$m = \frac{1}{(2\pi i)^2}\int_\Gamma \frac{d(z_1 - z_2^2)\wedge d(z_1^2 - z_2^3)}{(z_1 - z_2^2)(z_1^2 - z_2^3)}$$

$$= \frac{1}{(2\pi i)^2}\int_\Gamma \frac{(-3z_2^2 + 4z_1z_2)dz_1\wedge dz_2}{(z_1 - z_2^2)(z_1^2 - z_2^3)}$$

$$= \frac{1}{2\pi i}\int_{|z_2|=\varepsilon^4} \frac{(-3z_2^2 + 4z_2^3)dz_2}{z_2^4 - z_2^3} = 3.$$

THEOREM 4.10 (Caccioppoli, Martinelli, and Sorani). *Suppose that $f\in A^n(D)$, $\varphi\in A(D)$, and that the cycle Γ satisfies the weak homology relation*

$$\Gamma \approx \sum_{a\in E_f} n_a\gamma_a \tag{4.9}$$

$D_f = D\setminus\{z : f_1(z)\cdots f_n(z) = 0\}$, *where* $\gamma_a = \{z : |f_1(z)| = \cdots = |f_n(z)| = \varepsilon\}\cap U_a$; *here* U_a *is a neighborhood of a zero* $a\in E_f$ *and* ε *is sufficiently small. Then*

$$\frac{1}{(2\pi i)^n}\int_\Gamma \varphi\frac{df}{f} = \sum_{a\in E_f} n_a\mu_a(f)\varphi(a).$$

The proof follows at once from Corollary 0.5 and (4.6). $\square$

In connection with this theorem the problem arises of a topological description of cycles of the form (4.9). Let $F_j = \{z : z\in D, f_j(z) = 0\}$, $j = 1,\dots,n$ and $F = F_1\cup\cdots\cup F_n$. Then $D_f = D\setminus F$. The cycle γ_a is said to be *locally separating* for the analytic sets $F_1,\dots,F_n$ at the point a of their intersection. The subgroup of $H_n(D\setminus F)$ generated by the locally separating cycles is called the (globally) *separating subgroup*, and accordingly a cycle of the form (4.9) is called a *separating cycle*.

PROPOSITION 4.11. (Martinelli and Sorani). *A separating cycle satisfies the condition*

$$\Gamma \sim 0 \quad \text{in } D\setminus F_{[j]}, \qquad j = 1,\dots,n, \tag{4.10}$$

where $F_{[j]} = F_1\cup\cdots[j]\cdots\cup F_n$.

PROOF. $\gamma_a = \partial Q_a$, where $Q_a = \{z : z\in U_a, |f_1(z)| = \cdots[j]\cdots = |f_n(z)| = \varepsilon, |f_j(z)| \leqslant \varepsilon\}$, $Q_a \subset D\setminus F_{[j]}$. $\square$

The converse assertion is, in general, false.

EXAMPLE. $D = \mathbf{C}^2 \setminus \{z : |z_1| \leqslant 1, |z_2| \leqslant 1\}$, $f_1 = z_1$, $f_2 = z_2$, $\Gamma = \{z : |z_1| = |z_2| = 2\}$. Then the cycle Γ is not homologous to 0 in $D \setminus F$, and it satisfies condition (4.10), but is not separating, since $E_f = \varnothing$.

On the other hand, we have

THEOREM 4.12 (Tsikh). *If D is a domain of holomorphy, then a cycle $\Gamma \in Z_n(D \setminus F)$ is separating if it satisfies the conditions* (4.9).

Above we considered logarithmic residue formulas in $\mathbf{C}^n$ in which the integration was over cycles of dimension n or $2n - 1$. It is also possible to indicate formulas for cycles of intermediate dimension.

THEOREM 4.13 (Yuzhakov and Kuprikov). *Under the conditions of Theorem* 3.1

$$\int_{\Gamma_{2n-p-1}} \varphi K_{2n-p-1}(f) = \sum_{a \in E_f} \mu_a(f) \varphi(a),$$

where

$$K_{2n-p-1}(f) = \frac{(n-p-1)!}{(2\pi i)^n} (-1)^q$$

$$\times \sum_{\alpha_0 < \cdots < \alpha_p} \frac{(-1)^{\alpha_0 - 1} \bar{f}_{\alpha_0} \cdots \bar{f}_{\alpha_p} d\bar{f}_{[\alpha_0, \ldots, \alpha_p]} \wedge df}{\left(f_1 \bar{f}_1 + \cdots + f_{n-p-1} \bar{f}_{n-p-1} + f_{n-p} \bar{f}_{n-p} \cdot \cdots \cdot f_n \bar{f}_n\right)^{n-p}};$$

with $q = \frac{1}{2}(n-p)(n-p-1)$ and the summation over all ordered subsets $(\alpha_0, \ldots, \alpha_p)$ of $\{1, \ldots, n\}$ such that $1 \leqslant \alpha_0 \leqslant n-p$, $n-p \leqslant \alpha_1 < \cdots < \alpha_p \leqslant n$;

$$\Gamma_{2n-p-1} = \{z : z \in \partial D, |f_{n-p-1}(z)| = \cdots = |f_n(z)| = \varepsilon\},$$

with ε sufficiently small.

Theorem 3.1 and Theorem 4.6 are particular cases of this theorem (for $p = 0$ and $p = n - 1$ respectively).

COROLLARY 4.14 (Martinelli, Sommer, and Sorani). *Let $\varphi \in A_C(D)$, where D is a bounded domain with piecewise smooth boundary. Then for $a \in D$*

$$\int_{\Gamma_{2n-p-1}} \varphi K_{2n-p-1}(z - a) = \varphi(a). \tag{4.11}$$

This formula generalizes the Martinelli-Bochner formula (1.1) and the Cauchy formula (4.1) to the case of integration over cycles of intermediate dimension, but (4.11) can itself be obtained from the Cauchy-Fantappiè formula (3.16) by a suitable choice of the cycle β, similarly to the way we obtained the Cauchy formula in subsection 1° of this section.

In conclusion we prove

PROPOSITION 4.15 (Osgood). *If a holomorphic mapping $f: D \to \mathbf{C}^n$, $D \subset \mathbf{C}^n$, is one-to-one, then it is biholomorphic, i.e., f^{-1} is also holomorphic.*

PROOF. Take an arbitrary point $a \in D$ and a closed neighborhood $\bar{B}_a \subset D$ of it. Since f is one-to-one, it follows that a is an isolated zero of the mapping $f(z) - f(a)$, and the multiplicity is 1. By the Rouché principle (Theorem 2.5), for any $w \in U_\varepsilon = \{w \in \mathbf{C}^n : \|w - f(a)\| \leqslant \varepsilon\}$, where $\varepsilon = \min_{\partial B_a} \|f(z) - f(a)\|$, there exists a unique point $z = \varphi(w) \in B_a$ such that $f(z) = w$ (this implies that $f(D)$ is a domain), and, by (4.8),

$$z_j = \varphi_j(w) = \frac{1}{(2\pi i)^n} \int_\Gamma \frac{z_j df(z)}{[f(z) - w]^I}, \qquad j = 1, \ldots, n, \tag{4.12}$$

where $\Gamma = \{z \in B_a : |f_j(z) - f_j(a)| = \varepsilon, j = 1, \ldots, n\}$. According to Lemma 4.4, ε can be chosen in such a way that the set Γ is smooth and has real dimension n. Since the form under the integral sign in (4.12) is holomorphic with respect to w, the mapping $f^{-1}(w) = \varphi(w)$ is holomorphic in $\overline{U}_\varepsilon$. □

§5. The Andreotti-Norguet formula and generalizations of it

1°. Let $\alpha = (\alpha_1, \ldots, \alpha_n)$, where all the α_k are nonnegative integers. We use the notation $\langle w, f^{(\alpha)} \rangle = w_1 f_1^{\alpha_1} + \cdots + w_n f_n^{\alpha_n}$; here $w = (w_1, \ldots, w_n)$ and $f = (f_1, \ldots, f_n)$. Further,

$$\omega_\alpha(f, w) = \frac{(n-1)!\alpha!}{(2\pi i)^n} \frac{\sum_{k=1}^n (-1)^{k-1} w_k^{\alpha_k+1} dw_{[k]}^{(\alpha+I)} \wedge df}{\langle w, f^{(\alpha+I)} \rangle^n}.$$

THEOREM 5.1 (Andreotti and Norguet). *Suppose that $\varphi \in A_C(D)$, where D is a bounded domain with piecewise smooth boundary. Then for every point $z \in D$ and any α*

$$\int_{\partial D} \varphi(\zeta) \omega_\alpha(\zeta - z, \bar{\zeta} - \bar{z}) = D^\alpha \varphi(z). \tag{5.1}$$

The Andreotti-Norguet formula (5.1) generalizes the Martinelli-Bochner formula (1.1) to the case involving the computation not of the value of a holomorphic function at a given point but of the value of any derivative of it. For $\alpha = 0$ we get (1.1) from (5.1). We remark that for $n > 1$ the kernel ω_α is not a derivative of the kernel ω; it is only for $n = 1$ that (5.1) becomes the formula obtained by simple differentiation of the Cauchy formula.

PROOF. 1. It is easy to verify that the form $\omega_\alpha(\zeta - z, \bar{\zeta} - \bar{z})$ is closed. Fix a point $z \in D$; by Corollary 0.4, the integration over ∂D can be replaced by integration over $\partial B_r^{\alpha+I}(z)$.

2. In a sufficiently small neighborhood of z the holomorphic function $\varphi(\zeta)$ can be expanded in a multiple power series about z. Hence, to prove (5.1) it suffices to establish that

$$\frac{(n-1)!}{(2\pi i r^2)^n} \int_{\partial B_r^{(\alpha+I)}} (\zeta - z)^\beta \sum_{k=1}^n (-1)^{k-1} (\bar{\zeta}_k - \bar{z}_k)^{\alpha_k+1} d(\bar{\zeta} - \bar{z})_{[k]}^{(\alpha+I)} \wedge d\zeta$$

$$= \begin{cases} 0 & \text{if } \alpha \neq \beta, \\ 1 & \text{if } \alpha = \beta. \end{cases} \tag{5.2}$$

3. Note that under the change of variables $\zeta_k - z_k = e^{it_k}(\zeta'_k - z_k)$, $0 \leqslant t_k \leqslant 2\pi$, $k = 1,\dots,n$, the set $\partial B_r^{(\alpha+I)}$ is carried into itself, and the integrand form in (5.2) is multiplied by $e^{it_1(\beta_1-\alpha_1)}\cdots e^{it_n(\beta_n-\alpha_n)}$. Therefore, the part of (5.2) involving the case where $\alpha \neq \beta$ is true.

4. If $\alpha = \beta$, then, by Stokes' theorem, the integral in (5.2) is equal to

$$\frac{n!\,(\alpha+I)^I}{(2\pi i r^2)^n}\int_{B_r^{(\alpha+I)}}|\zeta_1 - z_1|^{2\alpha_1}\cdots|\zeta_n - z_n|^{2\alpha_n}d\bar{\zeta}\wedge d\zeta$$

$$= \frac{n!\,(\alpha+I)^I 2^n}{r^{2n}}\int_{\sum_1^n|\zeta_j - z_j|^{2(\alpha_j+1)}<r^2}|\zeta_1 - z_1|^{2\alpha_1+1}\cdots|\zeta_n - z_n|^{2\alpha_n+1}d|\zeta_1 - z_1|$$

$$\wedge\cdots\wedge d|\zeta_n - z_n|\,n!\int_{\substack{\sum_1^n\tau_j<1\\ \tau_j>0}}d\tau = 1. \quad \square$$

The Andreotti-Norquet formula can be generalized in the spirit of the Cauchy-Fantappiè formula.

THEOREM 5.2. *Suppose that the conditions of Theorem 5.1 hold and that the vector-valued function $w \in C_\zeta^{(1)}(\partial D)$ satisfies the condition*

$$\left\langle w(\zeta, z), (\zeta - z)^{\alpha+I}\right\rangle \neq 0, \qquad z \in D, \zeta \in \partial D.$$

Then

$$\int_{\partial D}\varphi(\zeta)\omega_\alpha(\zeta - z, w) = D^\alpha\varphi(z). \tag{5.3}$$

This can be proved either in the same way as Theorem 3.2 (Koppelman's way, which is analytic) or geometrically (Khenkin's way); we take the latter course. Each of the integrals (5.1) and (5.3) is an integral of one and the same exterior differential form

$$\varphi(\zeta)\sum_{k=1}^n(-1)^{k-1}v_k dv_{[k]}\wedge d\zeta \tag{5.4}$$

over the respective cycles

$$\gamma_0 = \Bigg\{(\zeta, v): \zeta \in \partial D, v_k = v_k^0$$

$$= \frac{(\bar{\zeta}_k - \bar{z}_k)^{\alpha_k+1}}{|\zeta_1 - z_1|^{2(\alpha_1+1)} + \cdots + |\zeta_n - z_n|^{2(\alpha_n+1)}}, k = 1,\dots,n\Bigg\}$$

and

$$\gamma = \left\{ (\zeta, v) : \zeta \in \partial D,\, v = v' = \frac{w}{\left\langle w, (\zeta - z)^{(\alpha+I)} \right\rangle} \right\}$$

lying on the surface

$$M_z^{(\alpha)} = \left\{ (\zeta, v) : \zeta \in \partial D,\, \left\langle v, (\zeta - z)^{(\alpha+I)} \right\rangle = 1 \right\}$$

in $\mathbf{C}^{2n}_{\zeta,v}$. The form (5.4) is closed on this surface, and to prove that the integral in (5.3) is equal to the integral in (5.1) it suffices to establish that the cycles γ_0 and γ are homologous on $M_z^{(\alpha)}$. Consider the set $Q = \{(\zeta, v) : \zeta \in \partial D,\, v = \lambda v^0 + (1 - \lambda) v',\, 0 \leqslant \lambda \leqslant 1\}$. It is clear that $\gamma - \gamma_0 = \partial Q$. Further,

$$\left\langle \lambda v^0 + (1-\lambda) v', (\zeta - z)^{(\alpha+I)} \right\rangle = \lambda \left\langle v^0, (\zeta - z)^{(\alpha+I)} \right\rangle + (1-\lambda) \left\langle v', (\zeta - z)^{(\alpha+I)} \right\rangle = 1,$$

i.e., $Q \subset M_z^{(\alpha)}$. Thus, the cycles γ_0 and γ are homologous on $M_z^{(\alpha)}$. □

Finally, we can make the following generalization of (5.1) in a way similar to what was done in the case of Theorem 3.2. Let

$$\Omega_\alpha(w^0, w^1, \ldots, w^{n-1}, f) = \frac{(-1)^{n(n-1)/2}}{(2\pi i)^n} \frac{\langle w^0, df \rangle}{\langle w^0, f^{(\alpha)} \rangle} \wedge d\frac{\langle w^1, df \rangle}{\langle w^2, f^{(\alpha)} \rangle} \wedge \cdots \wedge d\frac{\langle w^{n-1}, df \rangle}{\langle w^{n-1}, f^{(\alpha)} \rangle}.$$

THEOREM 5.3. *If the conditions of Theorem* 5.1 *hold and the vector-valued functions* $w^0 \in C_\zeta(\partial D)$ *and* $w^j \in C_\zeta^{(1)}(\partial D)$, $j = 1, \ldots, n-1$, *satisfy the condition*

$$\left\langle w^j(\zeta, z), (\zeta - z)^{(\alpha+I)} \right\rangle \neq 0, \qquad z \in D,\, \zeta \in \partial D,\, j = 0, 1, \ldots, n-1,$$

then

$$\int_{\partial D} \varphi(\zeta) \Omega_\alpha(w^0, w^1, \ldots, w^{n-1}, \zeta - z) = D^\alpha \varphi(z). \tag{5.5}$$

The proof is analogous to that of Theorem 3.2.

2°. The problem arises of whethere we can construct a multidimensional logarithmic residue formula based not on the Martinelli-Bochner integral representation (1.1) but on the more general Andreotti-Norguet integral representation (5.1), i.e., we pose the question of computing the integral

$$\int_{\partial D} \varphi \omega_\alpha(f, \bar f), \tag{5.6}$$

where f is in $A^n(\bar{D})$ and does not have zeros on ∂D, or, what is the same (compare (3.1) with (3.4), and (5.1) with (5.5)),

$$\int_{\partial D} \varphi \Omega_\alpha(w^0, w^1, \ldots, w^{n-1}, f). \tag{5.7}$$

We do not know the answer in the general case, so we list particular cases in which computation of the integrals (5.6) and (5.7) has been successful.

1. For $n = 1$ these integrals take the form

$$\frac{\alpha_1!}{2\pi i}\int_{\partial D} \frac{\varphi(\zeta_1)\,df(\zeta_1)}{f(\zeta_1)^{\alpha_1+1}} = \sum_{z_1 \in E_f} \mu_{z_1}(f) \frac{\alpha_1!}{[\mu_{z_1}(f)(\alpha_1+1)-1]!} \frac{d^{\mu_{z_1}(f)(\alpha_1+1)-1}}{dz_1^{\mu_{z_1}(f)(\alpha_1+1)-1}} \frac{\varphi(z_1)f'(z_1)}{\psi_{z_1}^{\alpha_1+1}(z_1)}.$$

Here $\mu_{z_1}(f)$, as usual, is the multiplicity of a zero z_1 of f; consequently, f has the form

$$f(z_1') = \psi_{z_1}(z_1')(z_1' - z_1)^{\mu_{z_1}(f)}$$

in a neighborhood of this zero where ψ_{z_1} is holomorphic in the given neighborhood and $\psi_{z_1}(z_1) \neq 0$.

2. If $\alpha = (0, \ldots, 0)$, see (3.1) and (3.4).

3. Suppose that all the roots of the system $f = 0$ in D are simple, i.e., in a neighborhood of each root the Jacobian $\partial f/\partial z$ is nonzero. Then the $\eta_k = f_k(z)$, $k = 1, \ldots, n$, form a system of local coordinates in this neighborhood. Since the forms ω_α and Ω_α are closed, the integral (5.6) (respectively, (5.7)) is equal to the sum of the integrals over the boundaries of sufficiently small neighborhoods of the roots of the system $f = 0$, and, by a change of variables in each of these neighborhoods the integral reduces to (5.1). Thus, in this case the integral (5.6) (respectively, (5.7)) is equal to

$$\sum_{a \in E_f} \left.\frac{\partial^{|\alpha|}\varphi(z)}{\partial f_1^{\alpha_1} \cdots \partial f_n^{\alpha_n}}\right|_{z=a}.$$

4. If the function φ can be divided by f^α, i.e., $\varphi f^{-\alpha} \in A_C(D)$, then the integral (5.6) can be written as

$$\frac{(n-1)!\alpha!}{(2\pi i)^n(\alpha+I)^I}\int_{\partial D} \frac{\varphi f^{-\alpha}\sum_{k=1}^n (-1)^{k-1} \bar{f}_k^{\alpha_k+1} d\bar{f}_{[k]}^{(\alpha+I)} \wedge df^{(\alpha+I)}}{[|f_1|^{2(\alpha_1+1)} + \cdots + |f_n|^{2(\alpha_n+1)}]^n},$$

i.e., an integral of the form (3.1) is obtained, though not for the mapping f but for the mapping $f^{(\alpha+I)} = (f_1^{\alpha_1+1}, \ldots, f_n^{\alpha_n+1})$, whose zeros differ from those of f only in their multiplicities. Therefore, the integral (5.6) (respectively, (5.7)) is equal to

$$\sum_{a \in E_{f^{(\alpha+I)}}} \left.\frac{\varphi(z)}{f(z)^\alpha(\alpha+I)^I}\right|_{z=a} = \sum_{a \in E_f} \left.\frac{\varphi(z)}{f(z)^\alpha}\right|_{z=a}.$$

5. If the conditions of Theorem 4.6 hold, then it can be shown as in the proof of that theorem that the integrals (5.6) and (5.7) are equal to

$$\frac{\alpha!}{(2\pi i)^n}\int_\Gamma \frac{df}{f^{(\alpha+I)}}. \tag{5.8}$$

If the conditions of Theorem 4.5 hold, then it can be shown (Γ' denotes the distinguished boundary of the special analytic polyhedron D) that the integrals (5.6) and (5.7) can be transformed to the form

$$\frac{\alpha!}{(2\pi i)^n}\int_{\Gamma'} \varphi \frac{df}{f^{(\alpha+I)}}.$$

We need a definition fow what follows: a point a is called an *ordinary point of an analytic set* S if for some neighborhood U of this point there exists a biholomorphic mapping F in U such that $F(S \cap U) = F(U) \cap L$, where L is an analytic plane in $\mathbf{C}^n$.

Assume that any point $a \in E_f$ is an ordinary point of the analytic sets $\{z : f_j(z) = 0\}$, $j = 1, \ldots, n$. Then in a neighborhood U_a of a there exist holomorphic functions $h_{ai}(z)$ such that $h_{ai}^{m_{ai}}(z) = f_i(z)$, where m_{ai} is the smallest degree of a homogeneous polynomial in the Taylor expansion of f_i in a neighborhood of a (see [148], p. 117), $i = 1, \ldots, n$. If it is also required that for every $a \in E_f$ the indicated homogeneous polynomials have an isolated zero at a, then $\mu_a(f) = m_{a1} \cdots m_{an}$ by Theorem 22.6. According to Corollary 4.7,

$$m_{a1} \cdots m_{an} = \mu_a(h_a^{(m)}) = \frac{1}{(2\pi i)^n} \int_\Gamma \frac{dh_a^{(m)}}{h_a^{(m)}}$$

$$= \frac{m_{a1} \cdots m_{an}}{(2\pi i)^n} \int_\Gamma \frac{dh_a}{h_a} = m_{a1} \cdots m_{an} \mu_a(h_a),$$

where $m = (m_{a1} \cdots m_{an})$. Hence, $\mu_a(h_a) = 1$; here $h_a = (h_{a1}, \ldots, h_{an})$. Therefore, the integral (5.8) is equal to

$$\frac{\alpha!}{(2\pi i)^n} \sum_{a \in E_f} m_{a1} \cdots m_{an} \int_{\gamma_a} \frac{dh_a}{h_a^{(\beta_a + I)}},$$

where $\gamma_a\{z : z \in U_a, |f_a(z)| = \cdots = |f_n(z)| = \delta\}$, with δ sufficiently small, and $\beta_a = (\beta_{a1}, \ldots, \beta_{an})$; with $\beta_{ai} = \alpha_i m_{ai}$, $i = 1, \ldots, n$. Since $\mu_{h_a}(a) = 1$, it follows that $(\partial h_a / \partial z)|_{z=0} \neq 0$, and so the functions $\eta_k = h_k(z)$, $k = 1, \ldots, n$, form a system of local coordinates in some neighborhood of the point a, and we obtain the following assertion.

PROPOSITION 5.4 (Bolotov). *Suppose that all the points $a \in E_f$ are ordinary points of the analytic sets $\{z : f_j(z) = 0\}$, $j = 1, \ldots, n$, and that the homogeneous polynomial of smallest degree in the Taylor expansions of the functions $f_1, \ldots, f_n$ in a neighborhood of each point $a \in E_f$ have an isolated zero there. Then the integrals* (5.6) *and* (5.7) *are equal to*

$$\sum_{a \in E_f} m_{a1} \cdots m_{an} \frac{\alpha!}{\beta_a!} \left. \frac{\partial^{|\beta_a|} \varphi(z)}{\partial h_{a1}^{\beta_{a1}} \cdots \partial h_{an}^{\beta_{an}}} \right|_{z=a}.$$

§6. The Bergman kernel function, the Szegö kernel, and integral representations with a holomorphic kernel on the Silov boundary

1°. Let Q be a compact subset of $\mathbf{C}^n$, and $\mathfrak{S}$ some collection of real-valued upper semicontinuous functions on Q. A set $S_{\mathfrak{S}}(Q) \subset Q$ is called a *Shilov boundary of Q with respect to* $\mathfrak{S}$ if 1) $S_{\mathfrak{S}}(Q)$ is a *determining set* for the collection on the compact set Q, i.e., it is closed, and for any $\varphi \in \mathfrak{S}$

$$\sup_{z \in Q} \varphi(z) = \sup_{z \in S_{\mathfrak{S}}(Q)} \varphi(z), \tag{6.1}$$

and 2) $S_{\mathfrak{S}}(Q)$ is minimal, i.e., no proper subset of it is determining.

The concept of a Shilov boundary is introduced similarly when $\mathfrak{S}$ is a collection of complex-valued functions, except that it is necessary to replace (6.1) by

$$\sup_{z \in Q} |\varphi(z)| = \sup_{z \in S_{\mathfrak{S}}(Q)} |\varphi(z)|.$$

THEOREM 6.1. *Suppose that a collection* $\mathfrak{S}$ *of real-valued upper semicontinuous functions on a compact set* Q *satisfies the condition that if* $\varphi \in \mathfrak{S}$, *then* $[\varphi(z) + \ln|z_i - c|] \in \mathfrak{S}$, $i = 1,\ldots,n$, *for any complex number* c. *Then the Shilov boundary* $S_{\mathfrak{S}}(Q)$ *exists and is unique.*

PROOF. The class of determining sets is not empty; for example, Q is determining. Let $\{\Gamma_k\}$ be a nested sequence of determining sets: $\Gamma_k \supset \Gamma_{k+1}$, $k = 1, 2, \ldots$. Then $\Gamma = \bigcap_1^\infty \Gamma_k$ is determining. This follows from the fact that for an upper semicontinuous function φ

$$\lim_{k\to\infty} \sup_{z\in\Gamma_k} \varphi(z) = \sup_{z\in\Gamma} \varphi(z).$$

It now follows from Zorn's lemma that there is a minimal determining set—a Shilov boundary.

We prove that $S_{\mathfrak{S}}(Q)$ is unique. Suppose that there exist two Shilov boundaries $S^1 = S^1_{\mathfrak{S}}(Q)$ and $S^2 = S^2_{\mathfrak{S}}(Q)$. Then there exist a point $z^0 \in S^1 \backslash S^2$ and an $r > 0$ such that the polydisk $U(z^0, r) = \{z : |z_i - z_i^0| < r, i = 1,\ldots,n\}$ does not intersect S^2. There exists a function $\varphi \in \mathfrak{S}$ that attains a maximal value M on S^1 in $U(z^0, r) \cap S^1$ and remains less than this M on $S^1 \backslash U(z^0, r)$ (otherwise S^1 would not be a minimal determining set). Let

$$m = \sup_{z\in S^1\backslash U(z^0, r)} \varphi(z);$$

then $m < M$.

On the other hand, there exists a point $z^1 \in S^2 \backslash S^1$ such that $\varphi(z^1) = M$. Since $z^1 \notin U(z^0, r)$, there exists an i such that $|z_i^1 - z_i^0| > r$. Let $R = \sup_{z\in S^1} |z_i - z_i^1|$ and choose $\lambda > 1$ in such a way that

$$m + \ln\left(R + \lambda|z_i^0 - z_i^1|\right) < M + \ln\left(\lambda|z_i^0 - z_i^1|\right).$$

Consider the function

$$\psi(z) = \left[\varphi(z) + \ln|z_i - z_i^1 - \lambda(z_i^0 - z_i^1)|\right] \in \mathfrak{S}.$$

Obviously, $\psi(z^1) = M + \ln(\lambda|z_i^0 - z_i^1|)$. If $z \in S^1 \backslash U(z^0, r)$, then

$$\begin{aligned}\psi(z) &\leq m + \ln|z_i - z_i^1 - \lambda(z_i^0 - z_i^1)| \\ &\leq m + \ln\left(R + \lambda|z_i^0 - z_i^1|\right) < M + \ln\left(\lambda|z_i^0 - z_i^1|\right);\end{aligned}$$

but if $z \in S^1 \cap U(z^0, r)$ then

$$\begin{aligned}\psi(z) &\leq M + \ln|z_i - z_i^0 + (1-\lambda)(z_i^0 - z_i^1)| \\ &< M + \ln\left(\lambda|z_i^0 - z_i^1| - |z_i^0 - z_i^1| + |z_i - z_i^0|\right) < M + \ln\left(\lambda|z_i^0 - z_i^1|\right).\end{aligned}$$

Thus, for $z \in S^1$ we have the inequality $\psi(z) < \psi(z^1)$, which is impossible, because S^1 is a Shilov boundary of Q with respect to $\mathfrak{S}$. □

COROLLARY 6.2. *Suppose that $\mathfrak{S}$ is a collection of complex-valued functions whose moduli are upper semicontinuous on a compact set Q, and that if $\varphi \in \mathfrak{S}$, then $\varphi(z)(z_i - c) \in \mathfrak{S}$, $i = 1,\ldots,n$, for any complex number c. Then $S_{\mathfrak{S}}(Q)$ exists and is unique.*

The proof reduces to a verification of the fact that the collection $\{\ln|\varphi(z)|\}$, $\varphi \in \mathfrak{S}$, satisfies the conditions of Theorem 6.1. □

If D is a bounded domain and $\mathfrak{S} = A_C(D)$, then $S_{\mathfrak{S}}(\overline{D})$ is simply called the *Shilov boundary of the domain* D and denoted by $S(D)$. If $\mathfrak{S} = A(\overline{D})$, then $S_{\mathfrak{S}}(\overline{D})$ is called the *Bergman boundary of the domain* D and denoted by $B(D)$. If $\mathfrak{S}$ is the set of all polynomials, then $S_{\mathfrak{S}}(\overline{D})$ is called the *polynomial boundary of the domain* D and denoted by $P(D)$.

COROLLARY 6.3. *Let D be a bounded domain in $\mathbf{C}^n$. Then the boundaries $S(D)$, $B(D)$, and $P(D)$ exist and are unique, and $P(D) \subset B(D) \subset S(D) \subset \partial D$.*

EXAMPLES. 1. If D is a ball, then $S(D) = B(D) = P(D) = \partial D$.

2. If D is a polydisk, then $S(D) = B(D) = P(D) = \Gamma$, where Γ is its skeleton.

3. *Example of a domain of holomorphy D such that the Shilov and Bergman boundaries have different dimension* (*Aĭzenberg*). Let $D = \{z \in \mathbf{C}^2 : |z_2| < |z_1|^{-\ln|z_1|},\ 0 < |z_1| < 1\}$. The functions $f_{m,l} = z_1^{-m} z_2^l$, m, $l = 1, 2, \ldots,$ belong to $A_C(D)$, and their moduli $|f_{m,l}|$ attain their respective maxima on $\overline{D}$ at a point $z = (z_1, z_2)$ with $z = (z_1, z_2)$, where $|z_1| = \exp(-m/2l)$, $|z_2| = |z_1|^{-\ln|z_1|}$. From this it is not hard to see that $S(D) = \{z \in \mathbf{C}^2 : |z_1| \leq 1, |z_2| = R(|z_1|)\}$, where

$$R(|z_1|) = \begin{cases} |z_1|^{-\ln|z_1|} & \text{if } 0 < |z_1| \leq 1, \\ 0 & \text{if } |z_1| = 0. \end{cases}$$

Every function in $A(\overline{D})$ is holomorphic also in the closed bidisk $\overline{U(0, 1)} = \{z \in \mathbf{C}^2 : |z_1| \leq 1,\ |z_2| \leq 1\}$; therefore $B(D) = S(U) = \Gamma = \{z \in \mathbf{C}^2 : |z_1| = |z_2| = 1\}$. Thus, the boundary $S(D)$ is three-dimensional, while $B(D)$ is two-dimensional.

2°. Let D be a bounded domain in $\mathbf{C}^n$, and suppose that $h(z, \bar{z})$ is in $C(\overline{D})$ and is positive in D. Denote by $L_h^2 = L_h^2(D)$ the collection of all functions $f \in A(D)$ with finite form

$$\|f\| = \|f\|_D^h = \left(\int_D |f|^2 h\, dv\right)^{1/2}, \tag{6.2}$$

where dv is the volume element, and the integral is understood as an improper integral; L_h^2 is a Hilbert space with the inner product $f(f, g) = \int_D f\bar{g}h\, dv$.

It is easy to see that if a polydisc satisfies $U(z^0, r) \Subset D$, then for every $f \in L_h^2$

$$|f(z^0)| \leq c\|f\|_D^h, \tag{6.3}$$

where the constant c depends on r and h but not on f nor on z^0.

It can be shown in the usual way that there are complete orthonormal systems in L_h^2. If $\{\varphi_m(z),\ m = 0, 1, 2, \dots\}$ is such a system, and a_m are the Fourier coefficients of an $f \in L_h^2$ with respect to this system, then the series

$$f(z) = \sum_{m=0}^{\infty} a_m \varphi_m(z) \tag{6.4}$$

converges in the norm and, hence (by (6.3)), absolutely and uniformly on compact subsets of D.

Let us next consider the series

$$\sum_{m=0}^{\infty} \varphi_m(z)\overline{\varphi_m(\zeta)} = K(z, \bar{\zeta}). \tag{6.5}$$

THEOREM 6.4 (Bergman). *The series* (6.5) *converges absolutely and uniformly on compact subsets of* $D \times D$. *For fixed* $\zeta \in D$ *this series converges in the* L_h^2*-norm with respect to* z.

PROOF. Let $a \in D$. Taking account of the orthonormality of the system $\{\varphi_m\}$ and inequality (6.3), we then get that

$$\sum_{m=0}^{M} |\varphi_m(a)|^2 = \int_D \left| \sum_{m=0}^{M} \varphi_m(z)\overline{\varphi_m(z)} \right|^2 h\, dv \geq \frac{1}{c^2} \left(\sum_{m=0}^{M} |\varphi_m(a)|^2 \right)^2,$$

and so

$$\sum_{m=0}^{M} |\varphi_m(a)|^2 \leq c^2. \tag{6.6}$$

Let $z \in D$ and $\zeta \in D$, and suppose that the polydisks $U(z, r)$ and $U(\zeta, \rho)$ are in D along with their closures. Then

$$\left(\sum_{m=0}^{M} |\varphi_m(z)\overline{\varphi_m(\zeta)}| \right)^2 \leq \sum_{m=0}^{M} |\varphi_m(z)|^2 \sum_{m=0}^{M} |\varphi_m(\zeta)|^2$$
$$\leq c^2(r) \cdot c^2(\rho),$$

which implies that (6.5) converges absolutely and uniformly on compact subsets of $D \times D$. Finally, it follows from (6.6) that for fixed $\zeta \in D$ the equality (6.5) is the expansion of $K(z, \bar{\zeta})$ in the complete orthonormal system $\{\varphi_m(z)\}$. □

The function $K(z, \bar{\zeta})$ is called the *Bergman-kernel function for D with respect to the weight h*, or simply the *kernel function*. Note that $K(z, \bar{\zeta}) = \overline{K(\zeta, \bar{z})}$ and that $K(z, \bar{\zeta})$ is holomorphic with respect to z and antiholomorphic with respect to ζ in $D \times D$.

THEOREM 6.5 (Bergman). *If $f \in L_h^2(D)$, then for $z \in D$*

$$f(z) = \int_D f(\zeta)K(z, \bar{\zeta})h(\zeta, \bar{\zeta})\, dv_\zeta. \tag{6.7}$$

PROOF. Let us represent f and K by (6.4) and (6.5). Then

$$\int_D f(\zeta)K(z, \bar{\zeta})h(\zeta, \bar{\zeta})\, dv_\zeta = \int_D \left(\sum_{m=0}^{\infty} a_m \varphi_m(\zeta) \right) \left(\sum_{l=0}^{\infty} \varphi_l(z) \overline{\varphi_l(\zeta)} \right) h(\zeta, \bar{\zeta})\, dv_\zeta$$

$$= \sum_{m=0}^{\infty} a_m \varphi_m(z) = f(z). \quad \square$$

We mention the following results of Bergman without proof.

1. The kernel function $K(z, \bar{\zeta})$ does not depend on the choice of the complete orthonormal system $\{\varphi_m\}$, but only on the domain D and the weight function h.

2. If $h \equiv 1$, then the positive-definite differential form

$$\sum_{\alpha,\beta=1}^{n} \frac{\partial^2 \ln K(z, \bar{z})}{\partial z_\alpha \partial \bar{z}_\beta} dz_\alpha d\bar{z}_\beta$$

determines a Hermitian metric in D that is invariant under biholomorphic mappings; it is called the *Bergman metric*.

We give examples of the computation of the kernel function, confining ourselves to complete n-circular domains D and to the weight function $h \equiv 1$. It is convenient to take the monomials $\{a_k z^k\}$, $k = (k_1, \ldots, k_n)$, $k_j = 0, 1, \ldots, j = 1, \ldots, n$, as a complete orthonormal system, where

$$a_k^2 = \left[\int_D |z^k|^2\, dv \right]^{-1}. \tag{6.8}$$

The completeness of this system follows from the fact that holomorphic functions can be expanded in a Taylor series for the class of domains under consideration, and the orthonormality follows from (6.8) and

$$\int_D z^k \bar{z}^{k'}\, dv = 0, \qquad k \neq k'.$$

1. *A polydisk*. For the polydisk $U(0, r)$, $r = (r_1, \ldots, r_n)$ we get from (6.8) that

$$a_k^2 = \frac{1}{\pi^n} \prod_{\nu=1}^{n} \frac{k_\nu + 1}{r_\nu^{2k_\nu+2}}.$$

Then, by (6.5) (here $x_\nu = z_\nu \bar{\zeta}_\nu r_\nu^{-2}$)

$$K(z, \bar{\zeta}) = \sum_k \frac{1}{\pi^n} \prod_{\nu=1}^n \frac{d_\nu + 1}{r_\nu^{2k_\nu+2}} z^k \bar{\zeta}^k = \frac{1}{\pi^n r_1^2 \cdots r_n^2} \sum_k \frac{\partial^n}{\partial x_1 \cdots \partial x_n} x^k$$

$$= \frac{1}{\pi^n r_1^2 \cdots r_n^2} \frac{\partial^n}{\partial x_1 \cdots \partial s_n} \frac{1}{(1 - x_1) \cdots (1 - x_n)}$$

$$= \frac{1}{\pi^n r_1^2 \cdots r_n^2} \frac{1}{(1 - x_1)^2 \cdots (1 - x_n)^2}$$

$$= \frac{1}{\pi^n} \prod_{\nu=1}^n \frac{r_\nu^2}{\left(r_\nu^2 - z_\nu \bar{\zeta}_\nu\right)^2}.$$

2. *A ball.* For the ball $B(O, R)$ formula (6.8) gives us that

$$a_k^2 = \frac{(k_1 + \cdots + k_n + n)!}{k_1! \cdots k_n! \pi^n R^{2(k_1 + \cdots + k_n + n)}},$$

and so (here $x = R^{-2} \sum_1^n z_j \bar{\zeta}_j$, $|x| < 1$)

$$K(z, \bar{\zeta}) = \frac{1}{\pi^n R^{2n}} \sum_k \frac{(k_1 + \cdots + k_n + n)!}{k_1! \cdots k_n! R^{2(k_1 + \cdots + k_n)}} z^k \bar{\zeta}^k$$

$$= \frac{1}{\pi^n R^{2n}} \sum_{m=0}^{\infty} \frac{(m+1) \cdots (m+n)}{R^{2n}} \sum_{k_1 + \cdots + k_n = m} \frac{m!}{k_1! \cdots k_n!} z^k \bar{\zeta}^k$$

$$= \frac{1}{\pi^n R^{2n}} \sum_{m=0}^{\infty} \frac{d^n}{dx^n} x^m = \frac{1}{\pi^n R^{2n}} \frac{d^n}{dx^n} \frac{1}{(1-x)} = \frac{n!}{\pi^n R^{2n} (1-x)^{n+1}}$$

$$= \frac{n! R^2}{\pi^n \left(R^2 - z_1 \bar{\zeta}_1 - \cdots - z_n \bar{\zeta}_n\right)^{n+1}}.$$

33. Let D be a bounded domain in $\mathbf{C}^n$ with a piecewise smooth boundary ∂D. For a function $f \in A_C(D)$ we write the Martinelli-Bochner formula (1.1) in the form

$$f(\zeta) = \int_{\partial D} f(\eta) M(\eta, \zeta)\, d\sigma_\eta,$$

where $\zeta \in D$, and $d\sigma$ is the surface element of ∂D. Further, by (6.7),

$$f(z) = \int_D \left[\int_{\partial D} f(\eta) M(\eta, \zeta)\, d\sigma_\eta \right] K(z, \bar{\zeta}) h(\zeta, \bar{\zeta})\, dv_\zeta$$

$$= \int_{\partial D} \left[\int_D M(\eta, \zeta) K(z, \bar{\zeta}) h(\zeta, \bar{\zeta})\, dv_\zeta \right] f(\eta)\, d\sigma_\eta$$

$$= \int_{\partial D} f(\eta) B(z, \eta)\, d\sigma_\eta.$$

Here the kernel

$$B(z,\eta)=\int_D M(\eta,\zeta)K(z,\bar\zeta)h(\zeta,\bar\zeta)\,dv_\zeta \tag{6.9}$$

is holomorphic in $z\in D$ and continuous in $\eta\in\partial D$ if the integral in (6.9) converges uniformly on each set of the form $\{(z,\eta): z\in Q,\ \eta\in\partial D\}$, where Q is an arbitrary compact subset of D. For this to be true it suffices that with respect to the variable ζ we have $\|M(\eta,\zeta)\|_D^h\leqslant c$, where c does not depend on η. This inequality is true when, for example, $h=[\rho(\zeta,\partial D)]^{4n-2}$. Thus, by combining the Martinelli-Bochner integral representation (1.1) and the Bergman formula (6.7) (with a specially chosen weight function), we have proved the following result by using an idea of Khenkin.

THEOREM 6.6 (Bungart). *For every bounded domain D with a piecewise smooth boundary ∂D there exists an integral representation*

$$f(z)=\int_{\partial D} f(\zeta)B(z,\zeta)\,d\sigma_\zeta, \tag{6.10}$$

$z\in D$, for functions $f\in A_C(D)$, where the kernel $B(z,\zeta)$ is holomorphic with respect to $z\in D$, and $B(z,\zeta)\in C\{(z,\zeta): z\in D,\ \zeta\in\partial D\}$.

The following assertion is also true.

THEOREM 6.7 (Bungart). *Under the conditions of Theorem 6.6 there exists an integral representation (6.10) with kernel $B(z,\bar\zeta)\in A\{(z,\bar\zeta): z\in D,\ \zeta\in D\}$ in the class $H^2_{d\sigma_\zeta}$ for every $z\in D$.*

A kernel such as the one in this theorem which is holomorphic in z and antiholomorphic in ζ is called a *Szegö kernel*. The question arises of whether we can get an analogue of formula (6.10) with the integration over the Shilov boundary $S(D)$ instead of over ∂D. A positive answer to this is contained in the next theorem, which we present without proof.

THEOREM 6.8 (Gleason). *On the Shilov boundary $S(D)$ of a domain D there exist a positive measure μ and a μ-measurable function $Q(z,\zeta)$, $z\in D$, $\zeta\in S(D)$, such that the following conditions hold:*

a) *For fixed $\zeta\in S(D)$ the function Q is in $A(D)$ with respect to z.*

b) *$Q(z,\zeta)$ is integrable with respect to the measure μ_ζ for a fixed $z\in D$.*

c) *The function $\int_{S(D)}|Q(z,\zeta)|\,d\mu_\zeta$ is continuous in D.*

d) *For every $f\in A_C(D)$ and $z\in D$*

$$f(z)=\int_{S(D)} f(\zeta)Q(z,\zeta)\,d\mu_\zeta. \tag{6.11}$$

Thus, the Shilov boundary plays the same role in the theory of functions of several complex variables as the whole boundary plays in the theory of functions of a single complex variable. Not only does every function in $A_C(D)$ attain a maximum on the Shilov boundary, but also the values of these functions inside the domain can be recovered by means of the integral representation (6.11) from the values on the Shilov boundary.

We complement Theorem 6.8 by a simple but often useful assertion that is in a certain sense a converse to this theorem.

THEOREM 6.9. *If M is a closed subset of the boundary ∂D of a bounded domain D, and for every $f \in A_C(D)$ and $z \in D$*

$$f(z) = \int_M f(\zeta) R(z, \zeta)\, d\mu_\zeta, \tag{6.12}$$

where μ is a positive measure on M and the function R is absolutely integrable with respect to the measure μ_ζ for fixed $z \in D$, then $M \supset S(D)$.

PROOF. If $f \in A_C(D)$, then for any positive integer m

$$|f(z)|^m = \left|\int_M f(\zeta)^m R(z, \zeta)\, d\mu_\zeta\right| \leqslant \left(\sup_M |f(\zeta)|\right)^m \int_M |R(z, \zeta)|\, d\mu_\zeta.$$

Taking the mth root and letting $m \to \infty$, we now get that M is a determining set. Hence, $M \supset S(D)$. □

§7. The Martinelli-Bochner-Koppelman integral representation for exterior differential forms

In this section formula (1.1) is generalized to exterior differential forms.

Let $w(\zeta, z) = (w_1, \ldots, w_n)$ be a vector-valued function defined on a set $U \subset \mathbf{C}^n_\zeta \times \mathbf{C}^n_\zeta \setminus \{(\zeta, z) : \zeta = z\}$, and such that $\langle w, \zeta - z \rangle = 1$. If $w_i \in C^{(1)}(U)$, then we write

$$W_{p,q}(w, \zeta, z) = \frac{(-1)^{q+p(n-q-1)}\binom{n-1}{q}}{(2\pi i)^n p!\,(n-p)!} D_{1,q,n-q-1}\left(w, \bar\partial_z w, \bar\partial_\zeta w\right) \wedge D_{p,n-p}(\partial z, \partial\zeta),$$

$$W_{p,-1} \equiv W_{p,n} \equiv 0.$$

The forms $W_{p,q}(w, \zeta, z)$ with the special choice $w = t(\zeta, z) = (\bar\zeta - \bar z)/|\zeta - z|^2$ are integral representation kernels for exterior differential forms; therefore, we introduce the extra notation

$$U_{p,q}(\zeta, z) = W_{p,q}(t(\zeta, z), \zeta, z).$$

Observe that $W_{0,0}(w, \zeta, z) = \omega(\zeta - z, w)$, since

$$W_{0,0}(w, \zeta, z) = \frac{1}{(2\pi i)^n n!} D_{1,n-1}(w, \bar{\partial}_\zeta w) \wedge D_n(\partial \zeta)$$
$$= \frac{(n-1)!}{(2\pi i)^n} \sum_{k=1}^{n} (-1)^{k-1} w_k d_\zeta w_{[k]} \wedge d\zeta.$$

THEOREM 7.1 (Koppelman). *Let D be a bounded domain with piecewise smooth boundary ∂D, and let $\gamma \in C^{(1)}_{p,q}(\bar{D})$. Then the Martinelli-Bochner-Koppelman formula is valid*:

$$J^1_{p,q}(D, \gamma) - J^2_{p,q}(D, \bar{\partial}_\zeta \gamma) - \bar{\partial}_z J^2_{p,q-1}(D, \gamma) = \begin{cases} \gamma(z), & z \in D \\ 0, & z \notin \bar{D}, \end{cases} \tag{7.1}$$

where

$$J^1_{p,q}(D, \gamma)(z) = \int_{(\partial D)_\zeta} \gamma(\zeta) \wedge U_{p,q}(\zeta, z),$$

$$J^2_{p,q}(D, \gamma)(z) = \int_{D_\zeta} \gamma(\zeta) \wedge U_{p,q-1}(\zeta, z).$$

Note that $U_{p,q}(\zeta, z)$ has type (p, q) with respect to z and type $(n-p, n-q-1)$ with respect to ζ. For $n = 1$ in the case $p = q = 0$ the third term is not present in (7.1) by definition, and

$$U_{0,0}(\zeta, z) = W_{0,0}\left(\frac{1}{\zeta_1 - z_1}, \zeta_1, z_1\right)$$
$$= \frac{1}{2\pi i} D_{1,0,0}\left(\frac{1}{\zeta_1 - z_1}, \dots\right) \wedge D_{0,1}(dz_1, d\zeta_1) = \frac{1}{2\pi i} \frac{d\zeta_1}{\zeta_1 - z_1}$$

so that the classical *Pompeiu formula*, sometimes called the *Cauchy-Green formula*,

$$\frac{1}{2\pi i} \int_{\partial D} \frac{\varphi(\zeta_1)\, d\zeta_1}{\zeta_1 - z_1} - \frac{1}{2\pi i} \int_D \frac{\varphi'_{\bar{\zeta}_1}(\zeta_1)\, d\bar{\zeta}_z \wedge d\zeta_1}{\zeta_1 - z_1} = \begin{cases} \varphi(z), & z \in D, \\ 0, & z \notin \bar{D}, \end{cases}$$

follows from (7.1), where $\varphi \in C^{(1)}(\bar{D})$. In the particular case where $p = q = 0$ and $n > 1$ we have $U_{0,0}(\zeta, z) = \omega(\zeta - z, \bar{\zeta} - z)$, and we arrive at the Martinelli-Bochner formula (1.4) for smooth functions.

Formula (7.1) is true also under somewhat more general requirements: for example, it holds when D is the union of finitely many domains of the type indicated in Theorem 7.1 whose closures are disjoint. Further, we note that

$$\bar{\partial}_z t = (\bar{\zeta} - \bar{z}) \bar{\partial}_z \frac{1}{|\zeta - z|^2} - \frac{1}{|\zeta - z|^2} \bar{\partial}_z \bar{z},$$

and so, for $q > 1$

$$D_{1,q,n-q-1}(t, \bar\partial_z t, \bar\partial_\zeta t) = \frac{1}{|\zeta - z|^2}\bar\partial_z \frac{1}{|\zeta - z|^2} \wedge D_{2,q-1,n-q-1}(\bar\zeta - \bar z, \bar\partial_z t, \bar\partial_\zeta t)$$
$$- \frac{1}{|\zeta - z|^4} D_{1,1,q-1,n-q-1}(\bar\zeta - \bar z, \bar\partial_z \bar z, \bar\partial_z t, \bar\partial_\zeta t).$$

The first determinant on the right-hand side has two identical columns of functions; consequently, it equals zero by Property 0.12. Application of this device $q - 1$ more times gives us that

$$D_{1,q,n-1}(t, \bar\partial_z t, \bar\partial_\zeta t) = \frac{(-1)^q}{|\zeta - z|^{2q+2}} D_{1,q,n-q-1}(\bar\zeta - \bar z, \bar\partial_z \bar z, \bar\partial_\zeta t).$$

We proceed similarly with the last $n - q - 1$ columns, and get that

$$D_{1,q,n-q-1}(t, \bar\partial_z t, \bar\partial_\zeta t)$$
$$= \frac{(-1)^q}{|\zeta - z|^{2n}} D_{1,q,n-q-1}(\bar\zeta - \bar z, \bar\partial_z z, \bar\partial_\zeta \bar\zeta)$$
$$= \frac{(-1)^q q!(n-q-1)!}{|\zeta - z|^{2n}} \sum_J{}' \sum_{\substack{k=1\\ k\notin J}}^{n} \sigma(J,k)(\bar\zeta_k - \bar z_k)\, d\bar z_J d\bar\zeta_{[J,k]}, \tag{7.2}$$

where $J = (i_1, \ldots, i_q)$ is a multi-index, the prime after the summation sign indicates summation over J with $1 \leqslant i_1 < \cdots < i_q \leqslant n$, the constant $\sigma(J, k)$ is determined by the equality $\sigma(J,k)\, d\bar z = d\bar z_k \wedge d\bar z_J \wedge d\bar z_{[J,k]}$, and the factors $q!$ and $(n - q - 1)!$ are due to the permutations of the q columns consisting of differentials of $\bar z$ and the $n - q - 1$ columns consising of differentials of $\bar\zeta$, respectively. Using the equality

$$D_{p,n-p}(\partial z, \partial\zeta) = p!\,(n-p)! \sum_{\mathcal{J}}{}' \sigma(\mathcal{J})\, dz_{\mathcal{J}}\, d\zeta_{[\mathcal{J}]},$$

where $\mathcal{J} = (j_1, \ldots, j_p)$ and $\sigma(\mathcal{J}) dz = dz_{\mathcal{J}} \wedge dz_{[\mathcal{J}]}$, we get the following expression for the Martinelli-Bochner-Koppelman kernel:

$$U_{p,q}(\zeta, z) = \frac{(-1)^{p(n-q-1)}(n-1)!}{(2\pi i)^n} \sum_{J,\mathcal{J}}{}' \sigma(J,k)\sigma(\mathcal{J})$$
$$\times \frac{\bar\zeta_k - \bar z_k}{|\zeta - z|^{2n}} d\bar\zeta_{[J,k]} \wedge d\zeta_{[\mathcal{J}]} d\bar z_J \wedge dz_{\mathcal{J}}. \tag{7.3}$$

To prove Theorem 7.1 we need one more form of expression for the kernel $U_{0,q}$. If

$$v_q = \frac{(-n)^n(n-1)!}{(2\pi i)^n} \sum_J{}' \sum_{\substack{k=1 \\ k \notin J}}^{n} \sigma(J,k)(-1)^{k-1} d\bar{\zeta}_{[J,k]} \wedge d\zeta_{[k]}\, d\bar{z}_J,$$

then

$$U_{0,q} = \frac{(-1)^{q+1}}{1-n} \partial_\zeta \frac{1}{|\zeta - z|^{2n-2}} \wedge v_q. \tag{7.4}$$

LEMMA 7.2. *Let $w(\zeta, z)$ be a vector-valued function of class $C^{(2)}$ on the set $U \subset (\mathbf{C}^n_\zeta \times \mathbf{C}^n_z) \setminus \{(\zeta, z) : \zeta = z\}$ such that $\langle w, \zeta - z \rangle = 1$ and $0 \leqslant q \leqslant n$. Then*

$$\bar{\partial}_\zeta W_{p,q}(w, \zeta, z) = (-1)^{p+q} \bar{\partial}_z W_{p,q-1}(w, \zeta, z). \tag{7.5}$$

In particular, $\bar{\partial}_\zeta W_{0,0} = \bar{\partial}_z W_{p,n-1} = 0.$

PROOF. For $1 \leqslant q \leqslant n$ Property 0.14 gives us that

$$\begin{aligned} \bar{\partial}_z D_{1,q-1,n-q}(w, \bar{d}_z w, \bar{\partial}_\zeta w) &= D_{q,n-q}(\bar{\partial}_z w, \bar{\partial}_\zeta w) \\ &\quad + (-1)^{q-1}(n-q) D_{1,q-1,1,n-q-1}(w, \bar{\partial}_z w, \bar{\partial}_z \bar{\partial}_\zeta w, \bar{\partial}_\zeta w), \end{aligned} \tag{7.6}$$

and for $0 \leqslant q \leqslant n-1$

$$\begin{aligned} \bar{\partial}_\zeta D_{1,q,n-q-1}(w, \bar{\partial}_z w, \bar{\partial}_\zeta w) &= (-1)^q D_{q,n-q}(\bar{\partial}_z w, \bar{\partial}_\zeta w) \\ &\quad + q D_{1,q-1,1,n-q-1}(w, \bar{\partial}_z w, \bar{\partial}_\zeta \bar{\partial}_z w, \bar{\partial}_\zeta w). \end{aligned} \tag{7.7}$$

According to Property 0.13, the first determinants in (7.6) and (7.7) are equal to zero. The second terms are equal to zero for $q = n$ in (7.6) and for $q = 0$ in (7.7); therefore, (7.5) holds for $q = 0, n$.

If $1 \leqslant q \leqslant n-1$, then the second determinants in (7.6) and (7.7) are equal, since the double forms $\bar{\partial}_z \bar{\partial}_\zeta w$ and $\bar{\partial}_\zeta \bar{\partial}_z$ are equal. Hence,

$$\bar{\partial}_\zeta\left[(n-q) D_{1,q,n-q-1}(w, \bar{\partial}_z w, \bar{\partial}_\zeta w)\right] = \bar{\partial}_z\left[(-1)^{q-1} q D_{1,q-1,n-q}(w, \bar{\partial}_z w, \bar{\partial}_\zeta w)\right].$$

Multiplying the last equality by

$$\frac{(n-1)!\,(-1)^{p(n-q-1)}}{(2\pi i)^n q!\,(n-q)!\,p!\,(n-p)!} D_{p,n-p}(\partial z, \partial \zeta),$$

we arrive at (7.5). □

LEMMA 7.3. *If $f \in C^{(1)}$ in a neighborhood of the point $0 \in \mathbf{C}^n$, then*

$$\lim_{\varepsilon \to 0} \int_{|\zeta| = \varepsilon} \frac{f(\zeta)\, d\bar{\zeta} \wedge d\zeta_{[k]}}{|\zeta|^{2n-m}} = \begin{cases} (-1)^{n-k-1} \dfrac{(2\pi i)^n}{n!} \dfrac{\partial f}{\partial \bar{\zeta}_k}(0), & m = 0, \\ 0, & m > 0; \end{cases} \tag{7.8}$$

in particular,

$$\lim_{\varepsilon\to 0}\int_{|\zeta|=\varepsilon}\frac{f(\zeta)\zeta_j d\bar\zeta\wedge d\zeta_{[k]}}{|\zeta|^{2n}}=(-1)^{n-k-1}\frac{(2\pi i)^n}{n!}f(0)\delta_{jk}. \tag{7.9}$$

PROOF. In fact,

$$\int_{|\zeta|=\varepsilon}\frac{f(\zeta)d\bar\zeta\wedge d\zeta_{[k]}}{|\zeta|^{2n-m}}=\frac{1}{\varepsilon^{2n-m}}\int_{|\zeta|=\varepsilon}f(\zeta)\,d\bar\zeta\wedge d\zeta_{[k]}$$

$$=(-1)^{n-k-1}\varepsilon^{m-2n}\int_{|\zeta|<\varepsilon}\frac{\partial f}{\partial\zeta_k}d\bar\zeta\wedge d\zeta$$

$$=(-1)^{n-k-1}\varepsilon^{m-2n}\left[\frac{\partial f}{\partial\zeta_k}(0)\int_{|\zeta|<\varepsilon}d\bar\zeta\wedge d\zeta+\int_{|\zeta|<\varepsilon}\left(\frac{\partial f}{\partial\zeta_k}(\zeta)-\frac{\partial f}{\partial\zeta_k}(0)\right)d\bar\zeta\wedge d\zeta\right]$$

$$=(-1)^{n-k-1}\varepsilon^{m-2n}\left[\frac{(2\pi i)^n}{n!}\varepsilon^{2n}\frac{\partial f}{\partial\zeta_k}(0)+O(\varepsilon^n)\right],$$

and, passing to the limit as $\varepsilon\to 0$, we get (7.8). To prove (6.9) it now suffices to note that $(\partial/\partial\zeta_k)f(\zeta)|_{\zeta=0}=f(0)\delta_{jk}$. □

PROOF OF THEOREM 7.1. Because the operations performed on the form γ on the left-hand side of (7.1) are additive, it suffices to carry out the proof for forms like

$$\gamma=f(z)\,d\bar z_J\wedge dz_{\mathcal{J}},$$

where $f\in C^{(1)}(\bar D)$, $J=(i_1,\dots,i_q)$ and $\mathcal{J}=(j_1,\dots,j_p)$. Further,

$$\begin{aligned}\gamma(\zeta)\wedge U_{p,q}(\zeta,z)&=\frac{(-1)^{q+p(n-q-1)}\binom{n-1}{q}}{(2\pi i)^n p!\,(n-p)!}f(\zeta)\,d\bar\zeta_J\wedge d\zeta_{\mathcal{J}}\\&\quad\wedge D_{1,q,n-q-1}(t,\bar\partial_z t,\bar\partial_\zeta t)\wedge D_{p,n-p}(\partial z,\partial\zeta)\\&=f(\zeta)\,d\bar\zeta_J\wedge\frac{(-1)^q\binom{n-1}{q}}{(2\pi i)^n}D_{1,q,n-q-1}(t,\bar\zeta_z t,\bar\partial_\zeta t)\wedge d\zeta\,dz_{\mathcal{J}}\\&=f(\zeta)\,d\bar\zeta_J\wedge U_{0,q}(\zeta,z)\wedge d\zeta_{\mathcal{J}}.\end{aligned}$$

Similarly,

$$\begin{aligned}\partial\bar\gamma\wedge U_{p,q}(\zeta,z)&=\bar\partial(f\,d\bar\zeta_J)\wedge U_{0,q}\,dz_{\mathcal{J}},\\ \gamma\wedge U_{p,q-1}(\zeta,z)&=f\,\bar d\zeta_J\wedge U_{0,q-1}\,dz_{\mathcal{J}},\end{aligned}$$

i.e., it suffices to obtain (7.1) for forms of type $(0,q)$—more precisely, for forms like $\gamma=f(z)\,d\bar z_J$.

Finally, the kernels $U_{0,q}$ are invariant under the transformations $\zeta_j \to \zeta_{\tau(j)}$, $z_j \to z_{\tau(j)}$, $j = 1,\dots,n$, where τ is an arbitrary permutation, since the same rows are thereby interchanged in the two determinants of $U_{0,q}$. Hence, it suffices to consider forms like

$$\gamma = f(z)d\bar{z}_1 \wedge \cdots \wedge d\bar{z}_q.$$

For $z \notin \bar{D}$, (7.1) is a corollary of Stokes' formula and Lemma 7.2.

Let $z \in D$. We write

$$\theta_q^* = U_{0,q} - \bar{\partial}_z g \wedge v_{q-1},$$

where g was introduced in (1.7). Since $\gamma \wedge U_{0,q}$ has type $(n, n-1)$ with respect to ζ and the form $\gamma \wedge \bar{\partial}_z g \wedge v_{q-1}$ has type $(n-1, n)$, it follows that

$$\begin{aligned} d_\zeta(\gamma \wedge \theta_q^*) &= \bar{\partial}_\zeta(\gamma \wedge U_{0,q}) - \partial_\zeta(\gamma \wedge \bar{\partial}_z g \wedge v_{q-1}) \\ &= \bar{\partial}_\zeta \gamma \wedge U_{0,q} + (-1)^q \gamma \wedge \bar{\partial}_\zeta U_{0,q} - \partial_\zeta \gamma \wedge \bar{\partial}_z g \wedge v_{q-1} \\ &\quad - (-1)^q \gamma \wedge \partial_\zeta \bar{\partial}_z g \wedge v_{q-1}. \end{aligned}$$

With the help of (7.4) and Lemma 7.2 we now get that

$$\begin{aligned} (-1)^q \gamma \wedge \bar{\partial}_\zeta \bar{\partial}_z g \wedge v_{q-1} &= (-1)^q \gamma \wedge \bar{\partial}_z(\partial_\zeta g \wedge v_{q-1}) \\ &= \gamma \wedge \bar{\partial}_z U_{0,q-1} = (-1)^q \gamma \wedge \bar{\partial}_\zeta U_{0,q}; \end{aligned}$$

consequently,

$$d_\zeta(\gamma \wedge \theta_q^*) = \bar{\partial}_\zeta \gamma \wedge U_{0,q} - \bar{\partial}_\zeta \gamma \wedge \bar{\partial}_z g \wedge v_{q-1}, \tag{7.10}$$

and so the second integral in

$$i(z) = \int_{\partial D_\zeta} \gamma \wedge \theta_q^* - \int_{D_\zeta} d_\zeta(\gamma(\zeta) \wedge \theta_q^*) \tag{7.11}$$

converges absolutely.

Let us now show that $i(z)$ coincides with the left-hand side of (7.1). To do this we consider the form $\gamma \wedge g v_{q-1}$, which has type $(n-1, n)$ with respect to ζ.

By Stokes' formula,

$$\begin{aligned} \int_{\partial D_\zeta} \gamma \wedge g v_{q-1} - \int_{\partial B_\varepsilon(z)} \gamma \wedge g v_{q-1} &= \int_{D \setminus B_\varepsilon(z)} \bar{\partial} \gamma \wedge g v_{q-1} \\ &\quad + (-1)^q \int_{D \setminus B_\varepsilon(z)} \gamma \wedge \partial_\zeta g \wedge v_{q-1}. \end{aligned} \tag{7.12}$$

According to Lemma 7.3, the limit of the second integral on the left-hand side of (7.12) as $\varepsilon \to 0$ is equal to zero. Consequently, (7.4) implies that

$$\int_{D_\zeta} \gamma \wedge U_{0,q-1} = \int_{\partial D_\zeta} \gamma \wedge g v_{q-1} - \int_{D_\zeta} \bar{\partial}\gamma(\zeta) \wedge g v_{q-1}. \tag{7.13}$$

We can differentiate under the integral sign in the integrals on the right-hand side of (7.13): after differentiation the first integral will be a proper integral as before, but the second will have a singularity of the form $|\zeta - z|^{1-2n}$, and therefore converges absolutely. Hence, the left-hand side of (7.13) is also differentiable, and

$$\int_{\partial D_\zeta} \gamma(\zeta) \wedge \bar{\partial}_z g \wedge v_{q-1} - \int_{D_\zeta} \partial\gamma(\zeta) \wedge \bar{\partial}_z g \wedge v_{q-1} - \bar{\partial}_z \int_{D_\zeta} \gamma \wedge U_{0,q-1} = 0. \tag{7.14}$$

Adding (7.11) and (7.14) and using the definition of $U_{0,q}$ and (7.10), we get that $i(z)$ is equal to the left-hand side of (7.1).

We show that $i(z) = \gamma(z)$. From Stokes' formula, applied to the domain $D \setminus \overline{B_\varepsilon(z)}$, and (7.11) it follows that

$$i(z) = \lim_{\varepsilon \to 0} \int_{\partial B_\varepsilon(z)} \gamma(\zeta) \wedge \theta_q^*.$$

Let us compute this limit; to do this we pass from the form θ_q^* to the form

$$\theta_q = (-1)^q \bar{\partial}_\zeta g \wedge v_q - \partial_z g \wedge v_{q-1}.$$

From the invariance of the first differential we have that $d_\zeta g|_{\partial B_\varepsilon(z)} = 0$, or $\bar{\partial}_\zeta g|_{\partial B_\varepsilon(z)} = \partial_\zeta g|_{\partial B_\varepsilon(z)}$; therefore, (7.4) gives us that $\theta_q^*|_{\partial B_\varepsilon(z)} = \theta_q|_{\partial B_\varepsilon(z)}$.

Let $g_k = (\zeta_k - z_k)|\zeta - z|^{-2n}$, $k = 1, \ldots, n$. We have that

$$\gamma(\zeta) \wedge \theta_q = (-1)\frac{(n-1)!}{(2\pi i)^n} f(\zeta)$$

$$\times \Big\{ (-1)^q d\bar{\zeta}_1 \wedge \cdots \wedge d\bar{\zeta}_q \wedge \sum_{k=1}^n g_k d\zeta_k \wedge \sum_J{}' \sum_{j \notin J} \sigma(J, j)(-1)^{j-1} d\bar{\zeta}_{[J,j]}$$

$$\wedge d\zeta_{[j]} d\bar{z}_J - d\bar{\zeta}_1 \wedge \cdots \wedge d\bar{\zeta}_q \wedge \sum_{j=1}^n (-g_k) d\bar{z}_k$$

$$\wedge \sum_{\mathfrak{J}}{}' \sum_{k \notin \mathfrak{J}} \sigma(\mathfrak{J}, k)(-1)^{k-1} d\bar{\zeta}_{[\mathfrak{J},k]} \wedge d\zeta_{[k]} d\bar{z}_{\mathfrak{J}} \Big\}$$

$$= \frac{(-1)^n(n-1)!}{(2\pi i)^n} f(\zeta)$$

$$\times \Bigg\{ (-1)^q \sum_{k=q+1}^{n} g_k d\bar{\zeta}_1 \wedge \cdots \wedge d\bar{\zeta}_q \wedge d\bar{\zeta}_k \wedge d\bar{\zeta}_{[1,\ldots,q,k]}$$

$$\wedge \Bigg[\sum_{j=1}^{q} \sigma((1,\ldots[j]\ldots,q,k),j)(-1)^{j-1} d\zeta_{[j]} d\bar{z}_1 \wedge \cdots [j] \cdots \wedge d\bar{z}_q \wedge d\bar{z}_k$$

$$+ \sigma((1,\ldots,q),k)(-1)^{k-1} d\zeta_{[k]} d\bar{z}_1 \wedge \cdots \wedge d\bar{z}_q \Bigg]$$

$$+ d\bar{\zeta}_1 \wedge \cdots \wedge d\bar{\zeta}_q \wedge d\bar{\zeta}_{[1,\ldots,q]} \wedge \sum_{k=1}^{q} \sigma((1,\ldots[k]\ldots,q),k)$$

$$\times (-1)^{k-1} \wedge d\zeta_{[k]} \sum_{j=1}^{n} g_j d\bar{z}_j \wedge d\bar{z}_1 \wedge \cdots [k] \cdots \wedge d\bar{z}_q \Bigg\}.$$

Note that

$$\begin{aligned}
(-1)^q \sigma((1,\ldots,q),k) d\bar{\zeta}_1 \wedge \cdots \wedge d\bar{\zeta}_q \wedge d\bar{\zeta}_k \wedge d\bar{\zeta}_{[1,\ldots,q,k]} \\
&= \sigma((1,\ldots,q),k) d\bar{\zeta}_k \wedge d\bar{\zeta}_1 \wedge \cdots \wedge d\bar{\zeta}_q \wedge d\bar{\zeta}_{[1,\ldots,q,k]} \\
&= d\bar{\zeta}, \sigma((1,\ldots[k]\ldots,q),k)\, d\zeta_1 \wedge \cdots \wedge d\bar{\zeta}_q \wedge d\bar{\zeta}_{[1,\ldots,q]} d\bar{z}_k \\
&\quad \wedge d\bar{z}_1 \wedge \cdots [k] \cdots \wedge d\bar{z}_q \\
&= \sigma((1,\ldots[k]\ldots,q),k) d\bar{\zeta}_k \wedge d\bar{\zeta}_1 \wedge \cdots [k] \cdots \wedge d\bar{\zeta}_q \wedge d\bar{\zeta}_q \\
&\quad \wedge d\bar{\zeta}_{[1,\ldots,q]} d\bar{z}_1 \wedge \cdots \wedge d\bar{z}_q \\
&= d\bar{\zeta} d\bar{z}_1 \wedge \cdots \wedge d\bar{z}_q.
\end{aligned}$$

Consequently,

$$\gamma \wedge \theta_q = \frac{(-1)^n(n-1)!}{(2\pi i)^n} f \sum_{k=1}^{n} (-1)^{k-1} g_k d\bar{\zeta} \wedge d\zeta_{[k]} \wedge d\bar{z}_1 \wedge \cdots \wedge d\bar{z}_q + \chi,$$

where χ consists of terms like

$$cf(\zeta) g_k d\bar{\zeta} \wedge d\zeta_{[j]} d\bar{z}_J, \qquad j \neq k, J = (i_1,\ldots,i_q).$$

By Lemma 7.3,

$$i(z) = \lim_{\varepsilon\to 0}\int_{\partial B_\varepsilon(z)} \gamma(\zeta)\wedge\theta_q = \frac{(n-1)!}{(2\pi i)^n}\sum_{k=1}^{n}\lim_{\varepsilon\to 0}(-1)^{n+k-1}$$

$$\times\int_{\partial B_\varepsilon(z)}\frac{f(\zeta)(\bar\zeta_k - \bar z_k)}{|\zeta - z|^{2n}}d\bar\zeta\wedge d\zeta_{[k]}d\bar z_1\wedge\cdots\wedge d\bar z_q + \lim_{\varepsilon\to 0}\int_{\partial B_\varepsilon(z)}\chi$$

$$= \frac{(n-1)!}{(2\pi i)^n}\sum_{k=1}^{n}\frac{(2\pi i)^n}{n!}f(z)\,d\bar z_1\wedge\cdots\wedge d\bar z_q = \gamma(z). \quad \square$$

Theorem 7.1 can be strengthened somewhat by relaxing the requirements on the form γ. Here and below, the derivative of a continuous function is understood to be the derivative in the sense of the theory of generalized functions.

THEOREM 7.4. *If D is a bounded domain with a piecewise smooth boundary ∂D, and $\gamma \in C_{p,q}(\bar D)$, while $\bar\partial\gamma \in C_{p,q+1}(\bar D)$, then formula* (7.1) *is valid.*

Finally, the following generalization of (7.1) (which can be obtained without difficulty) will be useful in applications of it.

THEOREM 7.5. *Suppose that the conditions of Theorem 7.4 hold and the function $\varphi(\zeta, z) \in C^{(1)}(\bar D \times D)$ is such that $\varphi(z, z) \equiv 1$ for $z \in D$. Then the integral representation*

$$\gamma(z) = J^1_{p,q}(D, \gamma\varphi) - J^2_{p,q}(D, \varphi\bar\partial_\zeta\gamma) - \bar\partial_z J^2_{p,q-1}(D, \gamma\varphi)$$
$$-J^2_{p,q}(D, \bar\partial_\zeta\varphi\wedge\gamma) + J^2_{p,q-1}(D, \bar\partial_z\varphi\wedge\gamma). \tag{7.15}$$

CHAPTER II

Integral Representations of Special Form for Holomorphic Functions

§8. Modifications and the simplest form for holomorphic functions

1°. For $n = 1$ the Cauchy-Fantappiè formula (3.13) or the generalized Cauchy-Fantappiè formula (3.12) becomes the classical Cauchy formula. However, for holomorphic functions of a single complex variable it is possible to get other integral representations that recover the values of a function at points $z \in D$ from its values on ∂D and that differ from the Cauchy formula, for example, by means of a conformal mapping.

Let $w_1 = \psi(z_1)$ be a conformal mapping of a closed domain $\overline{D}$ onto a closed domain $\overline{G}$, and consider the inverse mapping $z_1 = \theta(w_1)$, $z_1 \in \overline{D}$, $w_1 \in \overline{G}$. From the classical Cauchy formula (3.17) we now get that

$$f(\theta(w_1)) = \frac{1}{2\pi i}\int_{\partial G} f(\theta(\eta_1))\frac{\theta'(\eta_1)d\eta_1}{\theta(\eta_1) - \theta(w_1)}$$
$$= \frac{1}{2\pi i}\int_{\partial G} f(\theta(\eta_1))\frac{\theta'(\eta_1)(\eta_1 - w_1)}{\theta(\eta_1) - \theta(w_1)}\frac{d\eta_1}{\eta_1 - w_1},$$

or, in other notation,

$$f(z_1) = \frac{1}{2\pi i}\int_{\partial D} f(\zeta_1)\varphi(\zeta_1, z_1)\frac{d\zeta_1}{\zeta_1 - z_1}, \tag{8.1}$$

where the function φ is jointly continuous for $z_1 \in D$, $\zeta \in \overline{D}$ and holomorphic with respect to ζ_1 in $\overline{D}$, with $\varphi(z_1, z_1) \equiv 1$. Formula (8.1) does not follow from the general formulas (3.12) and (3.13).

It is natural to generalize (3.12) and (3.13) in the corresponding way. The next result is easy to obtain.

PROPOSITION 8.1. *Suppose that* $w^j(\zeta, z)$ *satisfies the conditions* $w^0 \in C_\zeta(\partial D)$, $w \in C_\zeta^{(1)}(\partial D)$, $j = 1,\dots,n-1$ *and* $\langle w^j, \zeta - z\rangle \neq 0$, $\zeta \in \partial D$, $z \in D$, $j = 0, 1,\dots,n-1$, *the function* $\varphi(\zeta, z)$ *belongs to* $A_C(D)$ *with respect to* ζ *for fixed* $z \in D$, *and* $\varphi(z, z) \equiv 1$. *Then for every* $f \in A_C(D)$ *and any* $z \in D$ *the following modifications of* (3.12) *and* (3.13) *are valid*:

$$\int_{\partial D} f\varphi\Omega(w^0, w^1,\dots,w^{n-1}, \zeta - z) = f(z), \tag{8.2}$$

$$\int_{\partial D} f\varphi\omega(\zeta - z, w^1) = f(z). \tag{8.3}$$

PROOF. We apply (3.12) to the function $f(\zeta)\varphi(\zeta, z)$ for a fixed $z \in D$:

$$\int_{\partial D} f\varphi\Omega = f(z)\varphi(z, z) = f(z). \quad \square$$

For $n = 1$ formulas (8.2) and (8.3) contain all possible integral representations that are analogues of the Cauchy integral (see §26). Even when $n > 1$ formula (8.3) has a certain completeness.

PROPOSITION 8.2. 1. *The kernel* $\varphi\omega(z - a, w^1)$ *is transformed into a kernel of the same form under a biholomorphic mapping of the closed domain* $\overline{D}$.

2. *If* $n = 2$, *the functions* $F_1(z)$ *and* $F_2(z)$ *belong to* $A_C(D)$, *and* $F_1 + F_2 \neq 0$ *in* D, *then the reproducing kernel*

$$\frac{F_1\varphi_1\omega(z - a, w^1) + F_2\varphi_2\omega(z - a, w^2)}{F_1 + F_2}$$

is again a kernel of the same form as in (8.3).

We mention one more formal generalization of the Cauchy-Fantappiè formula (3.13) consisting in replacement of the inner product $\langle w, \zeta - z\rangle$ by $\langle w, Q(\zeta) - Q(z)\rangle$ in the denominator of the kernel $\omega(z - a, w)$, where $Q \in A^n(\overline{D})$. By Theorem 25.2 (Hefer),

$$Q_i(\zeta) - Q_i(z) = \sum_{j=1}^{n} q_{ij}(\zeta_j - z_j), \qquad i = 1,\dots,n,$$

where $q_{ij}(\zeta, z) \in A^n(\overline{D} \times \overline{D})$. Let $\Omega = \det|q_{ij}|$. Then (3.13) gives us

PROPOSITION 8.3 (Aĭzenberg). *For every* $f \in A_C(D)$ *and any point* $z \in D$

$$\int_{\partial D} f\varphi\Omega\tilde{\omega}(Q(\zeta) - Q(z), w) = f(z), \tag{8.4}$$

where

$$\tilde{\omega}(F, w) = \frac{(n-1)!}{(2\pi i)^n} \frac{\Sigma_{k=1}^{n}(-1)^{k-1} w_k dw_{[k]} \wedge d\zeta}{\langle w, F\rangle^n}.$$

Although (8.4) is formally more general than the Cauchy-Fantappiè formula (3.13), each is a simple consequence of the other.

2°. The Martinelli-Bochner formula (1.1) can be obtained from the Cauchy-Fantappiè formula (3.13) by taking $w = \bar{\zeta} - \bar{z}$.

3°. A domain $D = \{z\colon \rho(z, \bar{z}) < 0\}$ is said to be a *regular linearly convex* domain if the real function ρ belongs to the class $C^{(2)}$ in a neighborhood of $\bar{D}$, $\operatorname{grad}\rho \neq 0$ on ∂D, and for every $\zeta \in \partial D$ the tangential analytic plane $\{z\colon \langle z - \zeta, \operatorname{grad}\rho(\zeta)\rangle = 0\}$ does not intersect D. For such a domain we can regard $\operatorname{grad}\rho$ as the w in (3.13), in which case we get

$$f(z) = \frac{(n-1)!}{(2\pi i)^n} \int_{\partial D} \frac{f(\zeta)\Sigma_{k=1}^{n}\delta_k d\bar{\zeta}_{[k]} \wedge d\zeta}{\left[\rho'_{\zeta_1}(\zeta_1 - z_1) + \cdots + \rho'_{z_n}(\zeta_n - z_n)\right]^n} \tag{8.5}$$

for $f \in A_C(D)$, where

$$\delta_k = \begin{vmatrix} \rho'_{\zeta_1} & \cdots & \rho'_{\zeta_n} \\ \rho''_{\zeta_1\bar{\zeta}_1} & \cdots & \rho''_{\zeta_n\bar{\zeta}_1} \\ \cdots & [k] & \cdots \\ \rho''_{\zeta_1\bar{\zeta}_n} & \cdots & \rho''_{\zeta_n\bar{\zeta}_n} \end{vmatrix}, \qquad k = 1, \ldots, n.$$

In (8.5), as in Cauchy's formula for $n = 1$, the outside variable z appears only in the denominator of the kernel, and the denominator is a linear function of z, taken to the power n. This nice structure for the kernel results in a number of beautiful applications of (8.5).

We represent ∂D as a union of disjoint sets Γ_m, $m = 1, \ldots, n$, in such a way that $\rho'_{\bar{\zeta}_n} \neq 0$ on Γ_m. Moreover, on Γ_m

$$d\bar{\zeta}_{[k]} \wedge d\zeta = (-1)^{m-k} \frac{\rho'_{\bar{\zeta}_k}}{\rho'_{\bar{\zeta}_m}} d\bar{\zeta}_{[m]} \wedge d\zeta.$$

Therefore, (8.5) can be written in the form

$$f(z) = \frac{(n-1)!}{(2\pi i)^n} \sum_{m=1}^{n} (-1)^{m-1} \int_{\Gamma_m} \frac{f(\zeta)L(\rho)d\bar{\zeta}_{[m]} \wedge \zeta}{\rho'_{\bar{\zeta}_m}\langle \operatorname{grad}\rho(\zeta), \zeta - z\rangle^n}, \tag{8.6}$$

where $L(\rho)$ is the Levi determinant

$$L(\rho) = -\begin{vmatrix} 0 & \rho'_{\zeta_1} & \cdots & \rho'_{\zeta_n} \\ \rho'_{\bar{\zeta}_1} & \rho''_{\zeta_1\bar{\zeta}_1} & \cdots & \rho''_{\zeta_n\bar{\zeta}_1} \\ \vdots & \vdots & & \vdots \\ \rho'_{\bar{\zeta}_n} & \rho''_{\zeta_1\bar{\zeta}_n} & \cdots & \rho''_{\zeta_n\bar{\zeta}_n} \end{vmatrix}.$$

A point $\zeta \in \partial D$ is called a *point of strict pseudoconvexity* if the *Levi form*

$$\sum_{j,k=1}^{n} \frac{\partial^2\rho(\zeta)}{\partial\zeta_j\partial\bar{\zeta}_k}\eta_j\bar{\eta}_k$$

is positive for ann η such that

$$\left\{\eta\colon \sum_{j=1}^{n} \frac{\partial\rho}{\partial\zeta_j}\eta_j = 0\right\}. \tag{8.7}$$

PROPOSITION 8.4. *The Levi determinant $L(\rho)$ is equal to the product of the eigenvalues of the Levi form on the hyperplane* (8.7), *multiplied by* $|\operatorname{grad}\rho(\zeta)|^2$.

It follows from this proposition that the integration in (8.5) is over the closure of the set of points of strict pseudoconvexity in ∂D. From Theorem 6.9 it now follows that this closure contains the Shilov boundary $S(D)$ in the case of a regular linearly convex domain D. The requirement of linear convexity for D can be discarded in this assertion.

THEOREM 8.5 (Rossi and Basener). *Let D be a bounded domain with $\partial D \in C^{(2)}$. Then $S(D)$ is contained in the closure of the set of points of strict pseudoconvexity in ∂D.*

Note that every regular convex domain is also a regular linearly convex domain; therefore, the integral representation (8.5) is valid for it. In particular, for the ellipsoid $D = \{z\colon \alpha_1|z_1|^2 + \cdots + \alpha_n|z_n|^2 < R^2\}$, $\alpha_i > 0$, $i = 1, \ldots, n$, we have

$$f(z) = \frac{\alpha_1\cdots\alpha_n(n-1)!}{(2\pi i)^n}\int_{\partial D}\frac{f(\zeta)\sum_{k=1}^{n}(-1)^{k-1}\bar{\zeta}_k d\bar{\zeta}_{[k]}\wedge d\zeta}{\left[R^2 - \alpha_1\bar{\zeta}_1 z_1 - \cdots - \alpha_n\bar{\zeta}_n z_n\right]^n}. \tag{8.8}$$

For $\alpha_1 = \cdots = \alpha_n = R = 1$ this formula is called the *Bochner formula* for functions holomorphic in the unit ball. The kernel in (8.8) is holomorphic in z and antiholomorphic in ζ (i.e., holomorphic in $\bar{\zeta}$). Such reproducing kernels for integration over ∂D are called *Szegö kernels* (see §6).

4°. Let D be a bounded domain with piecewise smooth boundary ∂D, and for an arbitrary point $z \in D$ choose a $v \in D$ such that the ball $B_r(v)$ is in D and $z \in B_r(v)$. If $\zeta \in \partial D$, then

$$\begin{aligned}|\langle \bar\zeta - \bar v, \zeta - z\rangle| &= |\langle \bar\zeta - \bar v, \zeta - v\rangle - \langle \bar\zeta - \bar v, z - v\rangle| \\ &\geqslant |\zeta - v^2|^2 - |\langle \bar\zeta - \bar v, \zeta - v\rangle| \\ &\geqslant |\zeta - v|(|\zeta - v| - |z - v|) > 0;\end{aligned}$$

therefore, we can take $\bar\zeta - \bar v$ as the w in the Cauchy-Fantappiè formula, and for $z \in B_r(v)$ we get the *Bochner-Ono formula*

$$f(z) = \frac{(n-1)!}{(2\pi i)^n} \int_{\partial D} \frac{f(\zeta)\Sigma_{k=1}^n (-1)^{k-1}(\bar\zeta_k - \bar v_k)\, d\bar\zeta_{[k]} \wedge d\zeta}{\langle \bar\zeta - \bar v, \zeta - z\rangle^n}. \tag{8.9}$$

It is possible to choose an at most countable set of points $v \in D$ such that the corresponding balls $B_{r(v)}(v)$ cover the whole of D. The kernel in (8.9) is holomorphic with respect to $z \in B_r(v)$, but is not even continuous with respect to z in the whole domain D. If D is a ball, then (8.9) becomes the Bochner formula.

5°. We consider a bounded domain of holomorphy Q and a function $\Psi(z, \bar\zeta)$ on $Q \times Q$ that is holomorphic in z and antiholomorphic in ζ. Suppose that $\Psi(z, \bar z)$ is real; the domain $D = \{z\colon \Psi(z, \bar z) < 0\}$ is bounded and has a piecewise smooth boundary, $\bar D \subset Q$, and for $\zeta \in \partial D$ and $z \in D$

$$\Psi(z, \bar z) - \Psi(z, \bar\zeta) - \overline{\Psi(z, \bar\zeta)} + \Psi(\zeta, \bar\zeta) \geqslant 0, \tag{8.10}$$

whence $\operatorname{Re} \Psi(z, \bar\zeta) < 0$. Under these assumptions we can take the inner product $\langle w, \zeta - z\rangle$ in (3.13) to be the difference $\Psi(z, \bar\zeta) - \Psi(\zeta, \bar\zeta) \equiv \Psi(z, \bar\zeta) \neq 0$ for $z \in D$ and $\zeta \in \partial D$.

The integral representation so obtained is interesting in that its kernel is holomorphic in z, and the "equation" of the boundary ∂D of D is in the denominator of the kernel. For example, for the domain $D = B_r^\alpha(0) = \{z\colon |z_1|^{2\alpha_1} + \cdots + |z_n|^{2\alpha_n} < 1\}$ we have the integral formula

$$f(z) = \frac{(n-1)!}{(2\pi i)^n} \times \int_{\partial D} \frac{f(\zeta)\Pi_{k=1}^n\left(\alpha_k \bar\zeta_k^{\alpha_k - 1}\Sigma_{i+j=\alpha_k-1} z_k^i \zeta_k^j\right)\Sigma_{k=1}^n(-1)^{k-1}\alpha_k^{-1}\,\bar\zeta_k\, d\bar\zeta_{[k]} \wedge d\zeta}{\left(1 - z_1^{\alpha_1}\bar\zeta_1^{\alpha_1} - \cdots - z_n^{\alpha_n}\bar\zeta_n^{\alpha_n}\right)^n}.$$

An example of a function satisfying condition (8.10) is a function of the form

$$\Psi(z,\bar{\zeta}) = F(z) + F(\bar{\zeta}) + \sum_{k=1}^{n} a_k z_k^{m_k} \bar{\zeta}_k^{m_k},$$

where the m_k are positive integers, $a_k > 0$, $k = 1,\dots,n$, and $\overline{F(\bar{\zeta})} = F(\zeta)$.

Another example is based on the following theorem.

THEOREM 8.6 (Bishop). *Suppose that D is a domain of holomorphy in $\mathbf{C}^n$. Then there exist functions $\varphi_1(z),\dots,\varphi_{n+1}(z)$ holomorphic in D such that for any $R > 0$ the set $D_R = \{z\colon \sum_{k=1}^{n+1} |\varphi_k(z)|^2 < R\}$ has the property that $D_R \Subset D$ and, moreover, the boundary ∂D_R is smooth for almost all $R > 0$.*

Let us consider the function

$$K(z,\bar{\zeta}) = \sum_{m=1}^{n+1} \varphi_m(z)\overline{\varphi_m(\zeta)},$$

which is holomorphic with respect to z and antiholomorphic with respect to ζ in $D \times D$. For each connected component of D_R inequality (8.10) is valid if we set $\Psi(z,\bar{\zeta}) = K(z,\bar{\zeta}) - R$, since

$$\Psi(z,\bar{z}) - \Psi(z,\bar{\zeta}) - \overline{\Psi(z,\bar{\zeta})} + \Psi(\zeta,\bar{\zeta}) = \sum_{m=1}^{n+1} |\varphi_m(z) - \varphi_m(\zeta)|^2.$$

For $R > 0$ such that ∂D_R is smooth we now get that for $f \in A_C(D)$ and $z \in D_R$

$$f(z) = \int_{\partial D_R} f(\zeta)\mu, \tag{8.11}$$

where the kernel μ is an exterior differential form like the Cauchy-Fantappiè form (see (3.13)), and the inner product $\langle w, \zeta - z\rangle$ is taken to be the function $K(z,\bar{\zeta}) - R = K(z,\bar{\zeta}) - K(\zeta,\bar{\zeta})$, to which the Hefer theorem is applicable. We remark that in (8.11) we need not integrate over the whole boundary ∂D_R, but only over the boundary of the connected component of D_R containing z. The kernel μ in (8.11) is holomorphic in z and does not depend on R, and $\bigcup_R D_R = D$. It is natural to say that the representation (8.11) is *universal with respect to the given domain D*.

We remark in conclusion that the Bergman kernel function could be taken as $K(z,\bar{\zeta})$ (see §6).

In the next three sections we illuminate some other consequences of the Cauchy-Fantappiè integral representation.

§9. The Bergman-Weil formula

Suppose that the functions $\chi_1(z),\ldots,\chi_N(z)$, $N \geqslant n$, are holomorphic in the domain $D \subset \mathbf{C}^n$, and that the planar domains $D_1,\ldots,D_N$ are such that $D_i \subset \chi_i(D)$, $i = 1,\ldots,N$. The set $\Delta = \{z\colon z \in D,\ \chi_i(z) \in D_i,\ i = 1,\ldots,N\}$ is called an *analytic polyhedron*. A connected component of an analytic polyhedron is called a *Weil polyhedron* if the boundaries ∂D_i are piecewise smooth and the intersection of k of the faces $\gamma_i = \{z\colon z \in D,\ \chi_i \in \partial D_i,\ \chi_j \in D_j,\ i \neq j\}$ has dimension at most $2n - k$ (so that the faces intersect in general position). Let $\gamma_{i_1\cdots i_n} = \bigcap_{k=1}^n \gamma_{i_k}$. We give these n-dimensional edges the natural orientation determined by the order of the faces $\gamma_{i_1},\ldots,\gamma_{i_n}$. The union of all such n-dimensional edges is called the *skeleton* of the polyhedron Δ and denoted by σ.

By Hefer's Theorem 25.2,

$$\chi_i(\zeta) - \chi_i(z) = \sum_{j=1}^n q_{ij}(\zeta_j - z_j), \qquad i = 1,\ldots,N.$$

Let $\Omega_{i_1\cdots i_n}$ denote the determinant $\det\|q_{ij}\|$, $i = i_1,\ldots,i_n$, $j = 1,\ldots,n$.

THEOREM 9.1 (Weil). *Let Δ be a Weil polyhedron, and let $f \in A_C(\Delta)$. Then for $z \in \Delta$*

$$f(z) = \frac{1}{(2\pi i)^n} \sum_{i_1<\cdots<i_n} \int_{\gamma_{i_1\cdots i_n}} \frac{f(\zeta)\Omega_{i_1\cdots i_n}(\zeta,z)\,d\zeta}{\prod_{k=1}^n\left[\chi_{i_k}(\zeta) - \chi_{i_k}(z)\right]}. \tag{9.1}$$

This is called the *Bergman-Weil integral representation*. It generalizes the Cauchy integral formula for polycylinder domains and becomes that formula when $N = n$ and $\chi_i = z_i$, $i = 1,\ldots,n$. Formula (9.1) recovers the value of the holomorphic function f in Δ from its values on the n-dimensional skeleton σ and has a kernel that is holomorphic in z; moreover, the form of the kernel depends on the shape of the domain.

On the whole, the proof repeats that of Theorem 4.1. On each face γ_i we consider the vector-valued function

$$u^i = (u_1^i,\ldots,u_n^i), \quad \text{where } u_k^i = q_{ik}\left[\sum_{j=1}^n q_{ij}(\zeta_j - z_j)\right]^{-1},$$

and over each edge $\gamma_{\iota_1\cdots\iota_n}$ we consider the set $\beta_{\iota_1\cdots\iota_n}=\{(\zeta,u)\colon\ \zeta\in\gamma_{\iota_1\cdots\iota_n},\ u=\Sigma_1^n\lambda_l u^{\iota_l},\ \lambda_l\geqslant 0,\ \Sigma_1^n\lambda_l=1\}$. As in the proof of Theorem 4.1, it is established that

$$f(z)=\frac{(n-1)!}{(2\pi i)^n}\sum_{\iota_1<\cdots<\iota_n}\int_{\beta_{\iota_1\cdots\iota_n}}f(\zeta)\sum_{k=1}^{n}(-1)^{k-1}u_k\,du_{[k]}\wedge d\zeta$$

$$=\frac{(n-1)!}{(2\pi i)^n}\sum_{\iota_1<\cdots<\iota_n}\int_{\gamma_{\iota_1\cdots\iota_n}}\frac{f(\zeta)\Omega_{\iota_1\cdots\iota_n}(\zeta,z)\,d\zeta}{\prod_{k=1}^n\left[\chi_{\iota_k}(\zeta)-\chi_{\iota_k}(z)\right]}$$

$$\times\int_{\substack{\lambda_1+\cdots+\lambda_{n+1}=1\\ \lambda_l\geqslant 0}}\sum_{k=1}^{n}(-1)^{k-1}\lambda_k\,d\lambda_{[k]}$$

$$=\frac{1}{(2\pi i)^n}\sum_{\iota_1<\cdots<\iota_n}\int_{\gamma_{\iota_1\cdots\iota_n}}\frac{f(\zeta)\Omega_{\iota_1\cdots\iota_n}(\zeta,z)\,d\zeta}{\prod_{k=1}^n\left[\chi_{\iota_k}(\zeta)-\chi_{\iota_k}(z)\right]}.$$

For simplicity we perform the calculations for $n=2$:

$$\int_{\beta_{\iota_1\iota_2}}f(\zeta)(u_1\,du_2-u_2\,du_1)\wedge d\zeta$$

$$=\int_0^1 d\lambda_1\int_{\gamma_{\iota_1\iota_2}}f(\zeta)\big[\big(\lambda_1u_1^{\iota_1}+(1-\lambda_1)u_1^{\iota_2}\big)\big(u_2^{\iota_1}-u_2^{\iota_2}\big)$$

$$-\big(\lambda_1u_2^{\iota_1}+(1-\lambda_1)u_2^{\iota_2}\big)\big(u_1^{\iota_1}-u_1^{\iota_2}\big)\big]\,d\zeta$$

$$=\int_0^1 d\lambda\int_{\gamma_{\iota_1\iota_2}}f(\zeta)\big(-u_1^{\iota_1}u_2^{\iota_2}+u_1^{\iota_2}u_2^{\iota_1}\big)\,d\zeta=\int_{\gamma_{\iota_1\iota_2}}\frac{f(\zeta)\Omega_{\iota_1\iota_2}(\zeta,z)\,d\zeta}{\prod_{k=1}^2\big(\chi_{\iota_k}(\zeta)-\chi_{\iota_k}(z)\big)}.\qquad\square$$

§10. Integral representation for strictly pseudoconvex domains

A bounded domain D is said to be *strictly pseudoconvex* if there exist a neighborhood $U\supset\overline{D}$ and a function $\rho\in C^{(2)}(U)$ such that $D=\{z\colon z\in U,\ \rho(z)<0\}$, where $\operatorname{grad}\rho\neq 0$ on ∂D and the function ρ is *strictly plurisubharmonic* in U, i.e.,

$$\sum_{j,k=1}^{n}\frac{\partial^2\rho(z)}{\partial z_j\,\partial\bar z_k}w_j\bar w_k>0$$

for $z\in U$ and all $w\in\mathbf{C}^n$, $w\neq 0$.

Strictly pseudoconvex domains play an important role in multidimensional complex analysis. Any domain of holomorphy can be approximated from the inside by these domains (as well as by Weil polyhedra; see §25). We need some auxiliary propositions in order to get an integral representation of functions holomorphic in strictly pseudoconvex domains from the Cauchy-Fantappiè formula. Let

$$D_\delta = \{z: z \in U, \rho(z) < \delta\}, \qquad V_\delta = \{z: z \in U, |\rho(z)| < \delta\},$$
$$U_{\varepsilon,\delta} = \{(\zeta, z): \zeta \in V_\delta, z \in D_\delta, |\zeta - z| < \varepsilon\}$$

where $\varepsilon, \delta > 0$. The space of $C^{(1)}$-functions in V_δ with values in a space H is denoted by $C^{(1)}(V_\delta, H)$.

LEMMA 10.1. *For every strictly pseudoconvex domain D there exist positive constants* $\varepsilon, \delta, \sigma$ *and functions*

$$F \in C^{(1)}(U_{\varepsilon,\delta}), \qquad G \in C^{(1)}(U_{\varepsilon,\delta}), \qquad \Phi \in C^{(1)}(V_\delta, A(D_\delta))$$

that satisfy the following conditions:

1) $\Phi = FG$ *on* $U_{\varepsilon,\delta}$, $F(z, z) = 0$, $|G| > \delta$ *on* $U_{\varepsilon,\delta}$, *and* $|\Phi| > \delta$ *outside* $U_{\varepsilon,\delta}$.

2) $2\operatorname{Re} F(\zeta, z) \geqslant \rho(\zeta) - \rho(\zeta) + \sigma|\zeta - z|^2$ *on* $U_{\varepsilon,\delta}$.

3) $d_\zeta F(\zeta, z)|_{\zeta=z} = -d_z F(\zeta, z)|_{\zeta=z} = \partial\rho(z)$.

PROOF. Suppose first that D is strictly convex; more precisely, suppose that ρ is strictly convex in U, and then let

$$\Phi = F = \sum_{i=1}^{n} \frac{\partial\rho(\zeta)}{\partial\zeta_i}(\zeta_i - z_i).$$

Taylor's formula implies the existence of a $\sigma > 0$ such that

$$2\operatorname{Re} F(\zeta, z) \geqslant \rho(\zeta) - \rho(z) + |\zeta - z|^2$$

for $\zeta \in D_\delta$, and $z \in D_\delta$, and the assertions of the lemma are valid in this case.

In the general case there exists (see Lemma 10.2) for every $\tilde{\zeta} \in \partial D$ a biholomorphic mapping $w(z, \tilde{\zeta})$ of some neighborhood $\tilde{U}_{\tilde{\zeta}}$ of $\tilde{\zeta}$ onto some neighborhood B of zero in $\mathbf{C}^n_w$ such that the inverse mapping $z(w, \tilde{\zeta})$ carries the strictly plurisubharmonic function $\rho(z)$ into the strictly convex function $\rho_{\tilde{\zeta}}(w) = \rho(z(w, \tilde{\zeta}))$ of the variable w in B. By Taylor's formula, there exist $\tilde{\varepsilon}', \tilde{\delta}' > 0$ such that for $|w'| < \tilde{\varepsilon}'$ and $|w - w'| < \tilde{\varepsilon}'$ we have

$$2\operatorname{Re} \sum_{j=1}^{n} \frac{\partial\rho_{\tilde{\zeta}}(w')}{\partial w_j}(w'_j - w_j) \geqslant \rho_{\tilde{\zeta}}(w') - \rho_{\tilde{\zeta}}(w) + \tilde{\sigma}'|w' - w|^2.$$

Therefore, there exist $\varepsilon', \sigma' > 0$ such that

$$2\operatorname{Re} \sum_{j=1}^{n} \frac{\partial\rho_{\tilde{\zeta}}(w(\zeta, \tilde{\zeta}))}{\partial w_j}\left[w_j(\zeta, \tilde{\zeta}) - w_j(z, \tilde{\zeta})\right] \geqslant \rho(\zeta) - \rho(z) + \sigma'|\zeta - z|^2 \tag{10.1}$$

if $|\zeta - \tilde{\zeta}| < \varepsilon'$ and $|\zeta - z| < \varepsilon'$. Let

$$F_{\tilde{\zeta}} = \sum_{j=1}^{n} \frac{\partial \rho_{\tilde{\zeta}}(w_j(\zeta, \tilde{\zeta}))}{\partial w_j}[w_j(\zeta, \tilde{\zeta}) - w_j(z, \tilde{\zeta})]$$

$$= \sum_{j=1}^{n} \frac{\partial \rho(\zeta)}{\partial \zeta_j}\left[\sum_{i=1}^{n} \frac{\partial z_j(w(\zeta,\tilde{\zeta}))}{\partial w_i}[w_i(\zeta, \tilde{\zeta}) - w_i(z, \tilde{\zeta})]\right]. \qquad (10.2)$$

Here again $|\zeta - \tilde{\zeta}| < \varepsilon'$ and $|\zeta - z| < \varepsilon'$. We consider also points $\tilde{\zeta}^1, \ldots, \tilde{\zeta}^N$ on ∂D such that

$$\partial D = \bigcup_{j=1}^{N} B_{\varepsilon_j'}(\tilde{\zeta}^j).$$

Suppose that the functions $\{\varphi_i\}_{i=1,\ldots,N}$ form a parition of unity in $V_{\delta'}$, with the support of φ_j in $B_{\varepsilon_j'}(\tilde{\zeta}^j)$. Let

$$F(\zeta, z) = \sum_{j=1}^{n} F_{\tilde{\zeta}^j}(\zeta, z)\varphi_j(\zeta), \qquad (10.3)$$

where $\zeta \in V_{\delta'}$ and $|\zeta - z| < 2\varepsilon = \min_j \varepsilon_j'$.

We now get from (10.1)–(10.3) that

$$2\operatorname{Re} F(\zeta, z) \geqslant \rho(\zeta) - \rho(z) + \sigma|\zeta - z|^2, \qquad (\zeta, z) \in U_{2\varepsilon,\delta}. \qquad (10.4)$$

Here $\sigma = \min_j \sigma_j'$. This proves assertions 2); 3) can be verified directly.

For small $\delta' > 0$ it follows from (10.4) that for $\zeta \in V_{\delta'}$, $z \in D_{\delta'}$ and $\varepsilon < |\zeta - z| < 2\varepsilon$

$$\operatorname{Re} F(\zeta, z) > 0. \qquad (10.5)$$

Fix a continuous single-valued branch in $\ln F$ in the domain $\{F\colon \operatorname{Re} F > 0\}$ and consider a real-valued $C^{(\infty)}$-function $\chi(z)$ with compact support in the ball $B_{2\varepsilon}(0)$ such that $\chi(z) = 1$ in $\overline{B_\varepsilon(0)}$. We introduce an exterior differential form $A(\zeta, z)$ of type (0, 1) that depends smoothly on the parameter $\zeta \in V_{\delta'}$:

$$A(\zeta, z) = \begin{cases} \bar{\partial}_z[(\ln F)\chi(\zeta - z)] & \text{if } z \in \operatorname{supp}\operatorname{grad}_z \chi(\zeta - z), \\ 0 & \text{if } z \notin \operatorname{supp}\operatorname{grad}_z \chi(\zeta - z). \end{cases} \qquad (10.6)$$

Since $A(\zeta, z) \in C^{(1)}(V_\delta, C(\overline{D}_\delta))$ for $0 < \delta < \delta'$, it follows (see Theorem 25.3) that there is a function $B(\zeta, z) \in C^{(1)}(V_\delta, C(\overline{D}_\delta))$ such that

$$\bar{\partial}_z B(\zeta, z) = A(\zeta, z). \qquad (10.7)$$

Let

$$\Phi(\zeta, z) = \begin{cases} F(\zeta, z)\exp[-B(\zeta, z)] & \text{for } \zeta \in V_\delta,\ z \in D_\delta,\ |\zeta - z| \leq \varepsilon, \\ \exp[\ln F\chi(\zeta - z) - B(\zeta, z)] & \text{for } \zeta \in V_\delta,\ z \in D_\delta,\ |\zeta - z| \geq \varepsilon. \end{cases} \tag{10.8}$$

This function is well-defined, because $F\exp[-B(\zeta, z)] = \exp[\ln F - B]$ for $|\zeta - z| = \varepsilon$. Assertion 1) of the lemma now follows from (10.6)–(10.8). □

LEMMA 10.2. *Let $D = \{z\colon z \in U, \rho(z) < 0\}$ be a strictly pseudoconvex domain. For any point $\tilde{\zeta} \in \partial D$ there exists a biholomorphic mapping of some neighborhood $\tilde{U_{\zeta}}$ of $\tilde{\zeta}$ onto some neighborhood B of zero in $\mathbf{C}^n_z$ under which the domain $D \cap \tilde{U_{\zeta}}$ is biholomorphically equivalent to a convex domain in $\mathbf{C}^n_w$, and the inverse mapping carries the strictly plurisubharmonic function $\rho(z)$ into a strictly convex function.*

PROOF. By a parallel translation we can take the point $\tilde{\zeta}$ to zero, and then in some neighborhood of zero

$$\rho(z) = 2\mathrm{Re} \sum_{j=1}^{n} \frac{\partial \rho(0)}{\partial z_j} z_j + \frac{1}{2} \sum_{j,k=1}^{n} \frac{\partial^2 \rho(0)}{\partial z_j \partial \bar{z}_k}$$
$$+\mathrm{Re} \sum_{j,k=1}^{n} \frac{\partial^2 \rho(0)}{\partial z_j \partial z_k} z_j z_k + o\left(|z|^2\right).$$

By the implicit function theorem, we can take the local coordinates in a smaller neighborhood of zero to be the function

$$w_1(z) = 2 \sum_{j=1}^{n} \frac{\partial \rho(0)}{\partial z_j} + \sum_{j,k=1}^{n} \frac{\partial^2 \rho(0)}{\partial z_j \partial z_k} z_j z_k$$

together with any linear functions $w_2, \ldots, w_n$ that are coordinates in the hyperplane

$$\left\{z\colon \frac{\partial \rho(0)}{\partial z_1} z_1 + \cdots + \frac{\partial \rho(0)}{\partial z_n} z_n = 0\right\}.$$

Strict pseudoconvexity at 0 now means that

$$\sum_{j,k=1}^{n} a_{jk} \zeta_j \bar{\zeta}_k \geq \gamma |\zeta|^2,$$

where $\gamma > 0$ and $a_{jk} = \partial^2 \rho(0)/\partial w_j \partial \bar{w}_k$. In the new local coordinates w we have

$$\rho = x_1 + \sum_{j,k=1}^{n} a_{jk} w_j \bar{w}_k + o\left(|w|^2\right). \tag{10.9}$$

Here $x_1 = \operatorname{Re} w_1$. The function (10.9) is strictly convex in the coordinates of w. If $\varepsilon > 0$ is sufficiently small, then $B_\varepsilon(0) \cap D$ is a convex domain in the coordinates of w, where $B_\varepsilon(0) = \{w: |w| < \varepsilon\}$. □

LEMMA 10.3. *Under the conditions of Lemma* 10.1

$$\Phi(\zeta, z) = \sum_{i=1}^{n} P_i(\zeta, z)(\zeta_i, z_i),$$

where $P_i(\zeta, z) \in C^{(1)}(V_\delta, A(D_\delta))$, $i = 1, \ldots, n$, $\zeta \in V_\delta$ *and* $z \in D_\delta$.

PROOF. Let $C(\zeta, z, w) = \Phi(\zeta, z) - \Phi(\zeta, w)$; then

$$C(\zeta, z, w) \in C^{(1)}(V_\delta, A(D_\delta \times D_\delta)).$$

For fixed $\zeta \in V_\delta$ the function C is in the ideal J of holomorphic functions that are zero on the set $\{(z, w): z \in D_\delta, w \in D_\delta, z = w\}$. By Hefer's theorem (see §25), this ideal has generators $w_1 - z_1, \ldots, w_n - z_n$, and the expansion of $C(\zeta, z, w)$ in these generators can be made continuously differentiable with respect to the parameter ζ (see Theorem 25.2′), i.e., there exist $Q_i(\zeta, z, w) \in C^{(1)}(V_\delta, A(D_\delta \times D_\delta))$, $i = 1, \ldots, n$, such that

$$C(\zeta, z, w) = \sum_{i=1}^{n} Q_i(\zeta, z, w)(w_i - z_i).$$

We remark that $F(\zeta, \zeta) = 0$; hence, also $\Phi(\zeta, \zeta) = 0$ (see the proof of Lemma 10.1), and so

$$\Phi(\zeta, z) = C(\zeta, z, \zeta) = \sum_{i=1}^{n} P_i(\zeta, z)(\zeta_i - z_i),$$

where $P_i(\zeta, z) = Q_i(\zeta, z, \zeta)$, $i = 1, \ldots, n$. □

Using the vector-valued function

$$P = P(\zeta, z) = (P_1(\zeta, z), \ldots, P_n(\zeta, z)),$$

whose existence was proved in Lemma 10.3, we can now get the following assertion from the Cauchy-Fantappiè formula.

THEOREM 10.4 (Khenkin and Ramírez de Arellano). *Under the condition of Lemma* 10.1 *the formula*

$$f(z) = \int_{\partial D} f(\zeta)\omega(\zeta - z, P(\zeta, z)) \tag{10.10}$$

holds for every $f \in A_C(D)$ *and any point* $z \in D$.

The kernel in the integral representation (10.10) is not written explicitly; however, the estimates in Lemma 10.1 enable us to get information about the singularity of this kernel, and that proves to be sufficient for important applications.

THEOREM 10.5 (Khenkin). *For every point $z \in \overline{D}$ and any $\delta, 0 < \delta < \gamma_0$,*

$$\int_{\partial D \setminus B(z,\delta)} \frac{d\sigma_\zeta}{|\Phi(\zeta, z)|^n} \leqslant \gamma_1 \ln \frac{\gamma_1}{\delta + |\rho(z)|}, \tag{10.11}$$

where $d\sigma_\zeta$ is the surface element of ∂D, and the positive constants γ_0 and γ_1 do not depend on z nor δ.

To prove this theorem we need some auxiliary facts.

LEMMA 10.6. *There exist $\sigma > 0$, $\delta > 0$, and $\gamma > 0$ such that for every $z \in V_\sigma$ it is possible to make a smooth change of variables $\eta = \eta(\zeta)$ in $B_\delta(z)$ with the properties that*

1) $\eta_1(\zeta) = \rho(\zeta) - \rho(z) + i \operatorname{Im} F(\zeta, z)$.
2) $\eta(z) = 0$,
3) $|\zeta - z| \geqslant \gamma|\eta(\zeta) - \eta(z)| \geqslant \gamma^2|\zeta - z|$ *for* $\zeta \in B_{2\delta}(z)$, *and*
4) $1/\gamma \geqslant |\partial\eta/\partial\zeta| \geqslant \gamma$ *for* $\zeta \in B_{2\delta}(z)$,

where $F(\zeta, z)$ is the function constructed in Lemma 10.1.

PROOF. There exists a $\sigma > 0$ such that $d_\zeta F(\zeta, z)|_{\zeta=z} \neq 0$ for every $z \in V_\sigma$. Consequently, there exist $\tilde{\gamma} > 0$ and $\tilde{\delta} > 0$ such that for $z \in V_\sigma$ it is possible to make a smooth change of variables $\tilde{\eta} = \tilde{\eta}(\zeta)$ in $B_{\tilde{\delta}}(z)$ under which we have properties 2)–4) for $\delta = \tilde{\delta}$ and $\gamma = \tilde{\gamma}$ and the property 1′) expressed by $\tilde{\eta}_1(\zeta) = F(\zeta, z)$. Further, $\operatorname{Im} \tilde{\eta}_1(\zeta) = \operatorname{Im} F(\zeta, z) = \operatorname{Im} \eta_1(\zeta)$ and, according to assertion 3) in Lemma 10.1, we get that

$$d_\zeta \operatorname{Re} \tilde{\eta}_1(\zeta)|_{\zeta=z} = d_\zeta \operatorname{Re} F(\zeta, z)|_{\zeta=z} = \tfrac{1}{2} d_\zeta \rho(\zeta)|_{\zeta=z} = \tfrac{1}{2} d_\zeta \operatorname{Re} \eta_1(\zeta)|_{\zeta=z},$$

so there exist a γ and a δ as claimed in our lemma. □

LEMMA 10.7. *For every ε and δ such that $0 < \varepsilon < 1$ and $0 < \delta < 1$*

$$J = \int_{\delta^2 \leqslant t_2^2 + t_3^2 + \cdots + t_{2n}^2 \leqslant 1} \frac{dt_2 \wedge \cdots \wedge dt_{2n}}{\left(t_2^2 + \cdots + t_{2n}^2 + \varepsilon^2 + |t_2|\right)^n}$$
$$\leqslant \gamma_0 |\ln(\varepsilon + \delta)| + \gamma_0, \tag{10.12}$$

where $\gamma_0 > 0$ does not depend on ε nor δ.

PROOF. We consider the case $n > 1$ of interest to us. Let $r^2 = t_2^2 + \cdots + t_{2n}^2$. Passing to a spherical system of coordinates, we find that

$$J \leqslant 2\int_\delta^1 dr \int_0^{\pi/2} \frac{r^{2n-2}\sin^{2n-3}\varphi\, d\varphi}{(r^2+\varepsilon^2+r\cos\varphi)^n} \leqslant 2\int_\delta^1 dr \int_0^1 \frac{r^{2n-2}ds}{(r^2+\varepsilon^2+rs)^n}$$

$$\leqslant \frac{2}{n-1}\int_\delta^1 \frac{r^{2n-3}dr}{(r^2+\varepsilon^2)^{n-1}} \leqslant \frac{2}{n-1}\int_\delta^1 \frac{dr}{r+\varepsilon} = \frac{2}{n-1}\ln\frac{1+\varepsilon}{\varepsilon+\delta}$$

$$\leqslant \gamma_0|\ln(\varepsilon+\delta)| + \gamma_0. \quad \square$$

PROOF OF THEOREM 10.5. According to Lemma 10.1,

$$\int_{\partial D\setminus B_\delta(z)} \frac{d\sigma_\zeta}{|\Phi(\zeta,z)|^n} \leqslant \int_{(B_{\delta'}(z)\setminus B_\delta(z))\cap\partial D} \frac{d\sigma_\zeta}{|F(\zeta,z)|^n|G(\zeta,z)|^n}$$

$$+\int_{\partial D\setminus B_{\delta'}(z)} \frac{d\sigma_\zeta}{|\Phi(\zeta,z)|^n}$$

$$\leqslant \frac{1}{\sigma^n}\left[\int_{(B_{\delta_1}(z)\setminus B_\delta(z))\cap\partial D} \frac{d\sigma_\zeta}{|F(\zeta,z)|^n} + \int_{\partial D} d\sigma_\zeta\right], \tag{10.13}$$

where σ, F, and G are from Lemma 10.1, and the sufficiently small δ' satisfies the conditions of Lemmas 10.1 and 10.6. In the ball $B_{\delta'}(z)$ we make a change of variables $\eta = \eta(\zeta)$ as in Lemma 10.6. Let $t_{2t-1} = \operatorname{Re}\eta_t$ and $t_{2t} = \operatorname{Im}\eta_t$. From 2) in Lemma 10.1 and 1) is Lemma 10.6 we get

$$|F(\zeta,z)| \geqslant \left[\left(\frac{t_1+\sigma|\zeta-z|^2}{2}\right)^2 + t_2^2\right]^{1/2} \geqslant \frac{1}{\sqrt{2}}\left(\frac{|t_1|+\sigma|\zeta-z|^2}{2} + |t_2|\right)$$

for $\zeta \in B_{\delta'}(z)$. Since $t_1 = -\rho(z)$ for $\zeta \in (\partial D)\cap B_{\delta'}(z)$, we now find from 3) and 4) of Lemma 10.6 that

$$\int_{(B_{\delta'}(z)\setminus B_\delta(z))\cap\partial D} \frac{d\sigma_\zeta}{|F(\zeta,z)|^n}$$

$$\leqslant \gamma_2 \int_{\delta^2 \leqslant t_2^2+\cdots+t_{2n}^2 \leqslant 1} \frac{dt_2\wedge\cdots\wedge dt_{2n}}{\left(t_2^2+\cdots+t_{2n}^2+|\rho(z)|+|t_2|\right)^n}, \tag{10.14}$$

where $\gamma_2 > 0$ does not depend on z or δ. Then (10.11) follows from (10.13), (10.14), and (10.12). $\square$

In conclusion we give a result about the connection between the Khenkin-Ramírez de Arellano formula (10.10), written in the form

$$f(z) = \int_{\partial D} f(\zeta)H(z,\zeta)\,d\sigma, \tag{10.15}$$

and the integral representation with Szegö kernel

$$f(z) = \int_{\partial D} f(\zeta) S(z, \bar{\zeta})\, d\sigma_\zeta. \tag{10.16}$$

Let $A(z, \zeta) = \overline{H(\zeta, z)} - H(z, \zeta)$. The integrals on the right-hand sides of (10.15) and (10.16) give rise to operators in $L^2_{d\sigma}(\partial D)$ which we denote by $\tilde{H}$ and $\tilde{S}$. The operator $\tilde{A}$ is introduced similarly. The following result is based on Theorem 10.5.

THEOREM 10.8 (Kerzman and Stein). *Let D be a strictly pseudoconvex domain with $C^{(\infty)}$-boundary. The kernel $A(z, \zeta)$ is absolutely integrable with respect to $\zeta \in \partial D$ for fixed $z \in \partial D$. The equality*

$$\tilde{S} = \tilde{H} + \tilde{H}\tilde{A} + \cdots + \tilde{H}\tilde{A}^k + \cdots$$

is valid in the asymptotic sense, i.e.,

$$\tilde{S} - \tilde{H} - \sum_{j=1}^{k} \tilde{H}\tilde{A}^j$$

is a bounded operator from $L^2_{d\sigma}(\partial D)$ to $C^{(\alpha(k))}(\partial D)$.d Here $\alpha(k) \to \infty$ as $k \to \infty$.

§11. Integral formulas for functions holomorphic in n-circular domains

1°. Let D be a bounded convex n-circular domain centered at zero and such that the function $\psi(|\zeta|)$ determining the boundary $\partial D = \{z\colon |\zeta_1| = \psi('|\zeta|),\ '|\zeta| \in \overline{G}\}$ is twice continuously differentiable; here $'|\zeta| = (|\zeta_2|, \ldots, |\zeta_n|)$, and G is the projection of the image of D in the absolute octant of moduli $(|\zeta_1|, \ldots, |\zeta_n|)$ on the hyperplane $\{|\zeta_1| = 0\}$. In this case the integral representation (8.5) holds for functions holomorphic in D and of class $C(\overline{D})$, where $\rho(z) = |\zeta_1| - \psi('|\zeta|)$. If we set

$$\varphi_j('|\zeta|) = \frac{|\zeta_j|\psi'_{|\zeta_j|}}{|\zeta_2|\psi'_{|\zeta_2|} + \cdots + |\zeta_n|\psi'_{|\zeta_n|} - \psi^2}, \qquad j = 2, \ldots, n,$$

then we get the formula

$$f(z) = \frac{(n-1)!}{(2\pi i)^n} \int_G d\varphi('|\zeta|)_{[1]} \int_{\Delta_\zeta} \frac{f(\zeta)}{(1 - y_1 - \cdots - y_n)^n} \frac{d\zeta}{\zeta}, \tag{11.1}$$

where

$$d\varphi('|\zeta|)_{[1]} = d\varphi_2('|\zeta|) \wedge \cdots \wedge d\varphi_n('|\zeta|),$$
$$y_j = z_j \bar{\zeta}_j('|\zeta|)|\zeta_j|^{-2}, \qquad j = 1, \ldots, n,$$

$$\varphi_1('|\zeta|) = 1 - \varphi_2 - \cdots - \varphi_n = -\psi\left[|\zeta_2|\psi'_{|\zeta_2|} + \cdots + |\zeta_n|\psi'_{|\zeta_n|} - \psi\right]^{-1},$$

and $\Delta_\zeta = \{\zeta': |\zeta_i| = |\zeta_i|, i = 1,\ldots,n\}$. In the integral representation (11.1) the integral over ∂D is written as a repeated integral: the integration is first over the arguments of the variables, and then over their moduli. We form the following change of variables in (11.1): $\varphi_j('|\zeta|) = \tau_j$, $|\zeta|_1 = r_1(\tau)$, $\zeta_1 = r_1(\tau)\xi$, $\zeta_j = r_j(\tau)\xi e^{i\theta_j}$ $\theta_j = \arg\zeta_j - \arg\zeta_1$, $j = 2,\ldots,n$, $|\xi| = 1$, $\tau = (\tau_1,\ldots,\tau_n)$, $\tau_1 = 1 - \tau_2 - \cdots - \tau_n$, $r = (r_1(\tau),\ldots,r_n(\tau))$, $\theta = (\theta_2,\ldots,\theta_n)$. The function f on ∂D can be written as a function of (ξ, r, θ), where ξ varies on the unit circle, $0 \leq \theta_j \leq 2\pi, j = 2,\ldots,n$, and

$$(\tau_2,\ldots,\tau_n) \in \Delta^* = \{(\tau_2,\ldots,\tau_n): \tau_2 + \cdots + \tau_n < 1, \tau_i > 0, i = 2,\ldots,n\}.$$

Now (10.11) is transformed as follows:

$$(f(z) = \frac{(n-1)!}{(2\pi i)^n}\int_{\Delta_*} d\tau_{[1]}\int_{\Delta_*} d\theta_{[1]}\int_{|\xi|=1}\frac{f(\xi,r,\theta)}{(1-u\bar{\xi})^n}\frac{d\xi}{\xi}, \tag{11.2}$$

where $\Delta_* = \{(\theta_2,\ldots,\theta_n): 0 \leq \theta_j \leq 2\pi, j = 2,\ldots,n\}$ and

$$u = \frac{\tau_1}{r_1(\tau)}z_1 + \frac{\tau_2 e^{-i\theta_2}}{r_2(\tau)}z_2 + \cdots + \frac{\tau_n e^{-i\theta_n}}{r_n(\tau)}z_n.$$

Formula (11.2) can be rewritten in a more convenient form for what follows; namely,

$$f(z) = \frac{(n-1)!}{(2\pi i)^n}\int_{\Delta^*} d\tau_{[1]}\int_{\Delta_*} d\theta_{[1]}\int_{|\xi|=1}\frac{\xi^{n-1}fj(\xi,r,\theta)}{(\xi-u)^n}d\xi. \tag{11.3}$$

We now consider the linear differential operators

$$L_0(f) = f, \qquad L_k(f) = (n-k)L_{k-1}(f) + \sum_{\nu=1}^{n} z_\nu\frac{\partial}{\partial z_\nu}L_{k-1}(f),$$

$$k = 1,2,\ldots,n-1.$$

Let $M_k(\xi, u) = (n-k-1)!\xi^{n-k-1}(\xi, u)^{k-n}$. Then

$$L_k(M_k) = \frac{(n-1)!\xi^{n-1}}{(\xi-u)^n}, \qquad k = 0,1,\ldots,n-1. \tag{11.4}$$

By (11.4), formula (11.3) can be written as

$$f(z) = \frac{1}{(2\pi i)^n}\int_{\Delta^*} d\tau_{[1]}\int_{\Delta_*} d\theta_{[1]}\int_{|\xi|=1} f(\xi,r,\theta)L_k(M_k)\,d\xi. \tag{11.5}$$

Applying the inverse operators L_k^{-1} (whose existence is easily ascertained) to both sides of (11.5) and changing the notation for the functions, we get the next assertion.

THEOREM 11.1 (Temlyakov, Opial, and Siciak). *Let $k \in \{0,\dots,n-1\}$ be fixed. Then the integral representation*

$$f(z) = \frac{(n-k-1)!}{(2\pi i)^n} \int_{\Delta^*} d\tau_{[1]} \int_{\Delta_*} d\theta_{[1]} \int_{|\xi|=1} \frac{\xi^{n-k-1} L_k(f)\, d\xi}{(\xi, u)^{n-k}} \tag{11.6}$$

is valid for every $f \in A(D)$ such that $L_k(f) \in A_C(D)$.

For $n = 2$ the formulas (11.6) give the *Temlyakov integral of the first kind* ($k = 1$)

$$f(z) = \frac{1}{4\pi^2 i} \int_0^{2\pi} dt \int_0^1 d\tau \int_{|\xi|=1} \frac{F(r_1(\tau)\xi, r_2(\tau)\xi e^{it})}{\xi - u}\, d\xi$$

and the *Temlyakov integral of the second kind* ($k = 0$)

$$f(z) = \frac{1}{4\pi^2 i} \int_0^{2\pi} dt \int_0^1 d\tau \int_{|\xi|=1} \frac{\xi f(r_1(\tau)\xi, r_2(\tau)\xi e^{it})}{(\xi - u)^2}\, d\xi,$$

where $F = L_1(f)$ and

$$u = \frac{z_1 \tau}{r_1(\tau)} + \frac{z_1(1-\tau)e^{-it}}{r_2(\tau)}.$$

The Temlyakov integral of the first kind is convenient in that the inside integral in it has a Cauchy kernel. This circumstance helps in applications.

2°. Despite the great generality of the Cauchy-Fantappiè formula, the question of finding integral representations with holomorphic kernel for concrete domains $D \subset \mathbf{C}^n$ still remains. Therefore, it is of interest to determine ways of constructing integral formulas with holomorphic kernels for specific domains or classes of domains. One such way will be considered in this and the next subsection.

Let D be a bounded complete n-circular domain with center at 0, and let λ be a finite measure on $|\partial D|$. The measure λ is said to be *admissible* if for any nonnegative integers $(k_1,\dots,k_n) = k$

$$\operatorname*{Ess\,Sup}_{|\partial D|} |z^k| = \max_{|\partial D|} |z^k|.$$

We present without proof the following result on the existence of integral representations with Szegö kernel.

THEOREM 11.2 (Aĭzenberg). *Let λ_0 be a finite measure on $|\partial D|$. Then an integral representation ($z \in D$)*

$$f(z) = \frac{1}{(2\pi i)^n} \int_{|\partial D|} d\lambda_0 \int_{\Delta_\zeta} f(\zeta) h(z\bar{\zeta}) \frac{d\zeta}{\zeta} \tag{11.7}$$

exists for any $f \in A_C(D)$, where the Szegö kernel $h(z\bar{\zeta}) = h(z_1\bar{\zeta}_1,\ldots,z_n\bar{\zeta}_n)$ is in $A(\bar{D})$ with respect to $\bar{\zeta}$ for fixed $z \in D$ and in $A(D)$ with respect to z for fixed $\zeta \in \partial D$ and does not depend on f, if and only if λ_0 is admissible. The Szegö kernel can be expanded for $z \in D$ and $\zeta \in \partial D$ in a multiple power series

$$h(z\bar{\zeta}) = \sum_{\substack{\alpha=(\alpha_1,\ldots,\alpha_n)\\ \alpha_i \geq 0}} a_\alpha z^\alpha \bar{\zeta}^\alpha, \tag{11.8}$$

where

$$a_\alpha = \left[\int_{|\partial D|} |\zeta^{2\alpha}| d\lambda_0\right]^{-1}. \tag{11.9}$$

We remark that admissible measures are measures which are *massive on the Shilov boundary* of D (i.e., $\overline{|\partial D| \setminus E} \supset |S(D)|$ for every $E \supset |\partial D|$ of zero λ-measure).

Let us consider the computation of some Szegö kernels, restricting ourselves to the case $n = 2$. If the boundary of D has the form $\partial D = \{z\colon |z_2| = \Phi(|z_1|),\ 0 \leq |z_1| < r\}$, where Φ is a decreasing function, then the integration with respect to λ over $|\partial D|$ can be replaced by integration over $[0, r]$ with respect to some measure λ' that can be given, in particular, by means of an increasing continuous function.

1. Consider the sequence of domains $D_p = \{z\colon |z_2|^2 + |z_1|^{2/p} < 1,\ 0 \leq |z_1| < 1\}$, $p = 1, 2, \ldots$, and take the measure λ' on $[0, 1]$ to be the measure generated by the function $|z_1|^2$. The Szegö kernel h_p for D_p will have the form

$$h_p(z_1\bar{\zeta}_1, z_1\bar{\zeta}_2) = \sum_{k_1 \geq 0, k_2 \geq 0} a_k^{(p)} (z_1\bar{\zeta}_1)^{k_1} (z_2\bar{\zeta}_2)^{k_2},$$

where

$$a_k^{(p)} = \left[\int_0^1 |\zeta_1|^{2k_1} \left(1 - |\zeta_1|^{2/p}\right)^{k_2} d|\zeta_1|^2\right]^{-1} = \frac{[p(k_1+1)+k_2]!}{p[p(k_1+1)-1]!\,k_2!}.$$

It remains to determine the function

$$h_p(x_1, x_2) = \frac{1}{p} \sum_{k_1 \geq 0, k_2 \geq 0} \frac{[p(k_1+1)+k_2]!}{[p(k_1+1)-1]!\,k_2!} x_1^{k_1} x_2^{k_2}$$

in finite form where $x_1 = z_1\bar{\zeta}_1$ and $x_2 = z_2\bar{\zeta}_2$.

Note that $z \in D_p$ and $\zeta \in \partial D_p$; therefore, $|x_2| < 1$ and $|x_1(1-x_2)^{-p}| < 1$. Then

$$\frac{1}{(1-x_2)^p - x_1} = \frac{1}{(1-x_2)^p}\frac{1}{1-x_1/(1-x_2)^p} = \sum_{m=0}^{\infty} x_1^m\left(\frac{1}{1-x_2}\right)^{p(m+1)}$$
$$= \sum_{m=0}^{\infty}\frac{x_1^m}{[p(m+1)-1]!}\frac{d^{p(m+1)-1}}{dx_2^{p(m+1)-1}}\frac{1}{1-x_2}$$
$$= \sum_{k_1,k_2=0}^{\infty}\frac{[p(k_1+1)+k_2-1]!}{[p(k_1+1)-1]!k_2!}x^{k_1}x^{k_2}.$$

We now get that

$$h_p(x_1,x_2) = \frac{1}{p}\frac{d}{dx_2}\frac{1}{(1-x_2)^p - x_1} = \frac{(1-x_2)^{p-1}}{[(1-x_2)^p - x_1]^2}.$$

Thus, for $f \in A_C(D_p)$ we have the integral representation ($z \in D_p$)

$$f(z) = \frac{1}{(2\pi i)^2}\int_0^1 d|\zeta_1|^2\int_{\Delta_\zeta}\frac{f(\zeta)(1-z_2\bar{\zeta}_2)^{p-1}}{[(1-z_2\bar{\zeta}_2)^p - z_1\bar{\zeta}_1]^2}\frac{d\zeta}{\zeta}. \tag{11.10}$$

Here $|\zeta_2|^2 = 1 - |\zeta_1|^{2/p}$. In the case $p = 1$ this gives us

$$f(z) = \frac{1}{(2\pi i)^2}\int_0^1 d|\zeta_1|^2\int_{\Delta_\zeta}\frac{f(\zeta)}{(1-z_1\bar{\zeta}_1 - z_2\bar{\zeta}_2)^2}\frac{d\zeta}{\zeta} \tag{11.11}$$

for the ball $D_1 = B_1(0)$. This is the Bochner integral representation (see (8.8)) and is, moreover, a particular case of (11.1). Thus, in the case of a ball the Temlyakov integral representation is an integral representation with Szegö kernel.

2. Suppose that $D = \{z\colon |z_1| + |z_2| < 1,\ 0 \leqslant |z_1| < 1\}$, and take the measure λ' on $[0, 1]$ to be ordinary Lebesque measure. Then

$$a_k = \left[\int_0^1 |\zeta_1|^{2k_1}(1-|\zeta_1|)^{2k_2}\, d|\zeta_1|\right]^{-1} = \frac{(2k_1+2k_2+1)!}{(2k_1)!\,(2k_2)!},$$

and the Szegö kernel for the given case has the form

$$h(x_1,x_2) = \sum_{\substack{k_1 \geqslant 0\\ k_2 \geqslant 0}}\frac{(2k_1+2k_2+1)!}{(2k_1)!\,(2k_2)!}x_1^{k_1}x_2^{k_2},$$

where again $x_i = z_i\bar{\zeta}_i$, $i = 1, 2$. Since $|z_1| + |z_2| < 1$ and $|\zeta_1| + |\zeta_2| = 1$, it follows that $\sqrt{|x_1|} + \sqrt{|x_2|} < 1$, so that

$$\begin{aligned}
\sum_k \frac{(2k_1 + 2k_2 + 1)!}{(2k_1)!\,(2k_2)!} x_1^{k_1} x_2^{k_2}
&= \frac{1}{4}\Bigg[\sum_{m_1,m_2=0}^{\infty} \frac{(m_1+m_2+1)!}{m_1!m_2!}\left(\sqrt{x_1}\right)^{m_1}\left(\sqrt{x_2}\right)^{m_2} \\
&\quad + \sum_{m_1,m_2=0}^{\infty} \frac{(m_1+m_2+1)!}{m_1!m_2!}\left(\sqrt{x_1}\right)^{m_1}\left(-\sqrt{x_2}\right)^{m_2} \\
&\quad + \sum_{m_1,m_2=0}^{\infty} \frac{(m_1+m_2+1)!}{m_1!m_2!}\left(-\sqrt{x_1}\right)^{m_1}\left(\sqrt{x_2}\right)^{m_2} \\
&\quad + \sum_{m_1,m_2=0}^{\infty} \frac{(m_1+m_2+1)!}{m_1!m_2!}\left(-\sqrt{x_1}\right)^{m_1}\left(-\sqrt{x_2}\right)^{m_2}\Bigg] \\
&= \frac{1}{4}\Bigg[\frac{1}{\left(1-\sqrt{x_1}-\sqrt{x_2}\right)^2} + \frac{1}{\left(1-\sqrt{x_1}+\sqrt{x_2}\right)^2} \\
&\quad + \frac{1}{\left(1+\sqrt{x_1}-\sqrt{x_2}\right)^2} + \frac{1}{\left(1+\sqrt{x_1}+\sqrt{x_2}\right)^2}\Bigg] \\
&= \frac{(1-x_1-x_2)(1+2x_1x_2-x_1^2-x_2^2)+8x_1x_2}{\left[(1-x_1-x_2)^2-4x_1x_2\right]^2}.
\end{aligned}$$

3. The domains $D_{p,q} = \{z\colon |z_1|^{2/p} + |z_2|^{2/q} < 1\}$, $p, q = 1, 2, \ldots$, represent a more general case. With the same measure as in the first example it can be shown that.

$$h_{p,q}(x_1, x_2) = \frac{1}{pq}\frac{\partial}{\partial x_1}\sum_{j_1=1}^{p}\sum_{j_2=1}^{q}\frac{1}{1 - y_{j_1 1} - y_{j_2 2}},$$

where $y_{j_1 1} = x_1^{1/p}\varepsilon_{j_1}$, $y_{j_2 2} = x_2^{1/q}\varepsilon^1_{j_2}$, the ε_{j_1} are all the pth roots of unity ($j_1 = 1, \ldots, p$) and the $\varepsilon^1_{j_2}$ are all the qth roots of unity ($j_2 = 1, \ldots, q$). The Szegö kernels $h_{p,q}$ are rational.

3°. We fix a domain D_0 and an admissible measure λ_0 on $|\partial D_0|$. By Theorem 11.2, the integral representation (11.7) with Szegö kernel is valid. We consider an arbitrary domain D and pose the following problem: if possible, choose functions $\varphi_1(|\zeta_1|,|\zeta_n|),\dots,\varphi_n(|\zeta_1|,\dots,|\zeta_n|)$ and a measure λ on $|\partial D|$ such that the integral representation

$$f(z)=\frac{1}{(2\pi i)^n}\int_{|\partial D|}d\lambda\int_{\Delta_\zeta}f(\zeta)h\left(z\bar{\zeta}\frac{\varphi}{|\zeta|^2}\right)\frac{d\zeta}{\zeta} \tag{11.12}$$

holds for D, where

$$h\left(z\bar{\zeta}\frac{\varphi}{|\zeta|^2}\right)=h\left(z_1\bar{\zeta}_1\frac{\varphi_1}{|\zeta_1|^2},\dots,z_n\bar{\zeta}_n\frac{\varphi_n}{|\zeta_n|^2}\right)=\sum_k a_k z^k\bar{\zeta}^k\frac{\varphi^k}{|\zeta^{2k}|} \tag{11.13}$$

(h is the same function as in the integral representation (11.7)), with

$$a_k=\left[\int_{|\partial D|}\varphi^k\,d\lambda\right]^{-1}. \tag{11.14}$$

It is always possible to choose $\varphi_1,\dots,\varphi_n$ and the measure λ on $|\partial D|$ in such a way that the a_k in (11.14) equal the corresponding a_k in (11.9). (This can be realized by a change of variables under which the integral (11.9) over $|\partial D_0|$ becomes the integral (11.14) over $|\partial D|$). If the functions $\varphi_1,\dots,\varphi_n$ so chosen satisfy the condition that the series (11.13) converges for $z\in D$ and $\zeta\in\partial D$, then the desired integral representation (11.12) exists. It turns out that our problem has a solution for a certain class $\mathfrak{M}$ of n-circular domains that depends only on the domain D_0 and not on the choice of the measure λ_0 (or, what is the same, not on the choice of the Szegö kernel h). The kernels (11.13) are called *extended Szegö kernels*. In particular, if D_0 is a ball, then $\mathfrak{M}$ is the class of bounded convex n-circular domains with certain natural restrictions on their boundaries. But the Temlyakov integral representation (11.1) is an extension to convex n-circular domains of an integral representation with Szegö kernel for the ball. We shall not give a description of the class $\mathfrak{M}$ in the general case (see [10] or [17]), remarking only that for any complete n-circular domain of holomorphy D with $C^{(2)}$-boundary there exists an integral representation (11.12) with kernel that is rational in z and is an extension of a Szegö kernel for the domain $D_p=\{z\colon |z_1|^{2/p}+\cdots+|z_n|^{2/p}<1\}$ for some positive integer p (see [47] or [91]).

Results analogous to those in 2° and 3° can be obtained also for tubular domains, where instead of a general basis of monomials we use a continuous basis of exponentials, and instead of power series we use Fourier integrals (see [60]).

4°. Let D be a bounded complete n-circular domain with center at zero (no particular smoothness of ∂D is assumed), and M the set of vectors k whose coordinates are relatively prime nonnegative integers. For each $k\in M$ let $\zeta(k)$

be a point at which the monomial $|\zeta^k|$ attains a maximum on ∂D (such a point may not be unique, in which case we choose one of them). The corresponding skeleton $\Delta_{\zeta(k)}$ is denoted simply by $\Delta(k)$. Finally, consider any skeleton in ∂D and denote it by $\Delta(0)$.

THEOREM 11.3 (de Leeuw, Aĭzenberg, and Gindikin). *For any $f \in A_C(D)$ the integral representation*

$$f(z) = \frac{1}{(2\pi i)^n}\left[\int_{\Delta(0)} f(\zeta)\frac{d\zeta}{\zeta} + \sum_{k\in M} z^k \int_{\Delta(k)} \frac{f(\zeta)}{\zeta^k - z^k}\frac{d\zeta}{\zeta}\right] \tag{11.15}$$

holds for $z \in D$, where the series converges absolutely and uniformly on compact subsets of D.

PROOF. It can be assumed at once that $f \in A(\overline{D})$, since we can consider a homothety D_r $(r < 1)$ in place of D and then pass to the limit as $r \to 1$. We represent $f(z)$ in D by the power series

$$\begin{aligned} f(z) &= \sum_k c_k z^k = c_0 + \sum_{k\in M}\sum_{l=1}^{\infty} c_{kl} z^{kl} \\ &= \frac{1}{(2\pi i)^n}\left[\int_{\Delta(0)} f(\zeta)\frac{d\zeta}{\zeta} + \sum_{k\in M}\sum_{l=1}^{\infty} z^{kl}\int_{\Delta(k)}\frac{f(\zeta)}{\zeta^{kl}}\frac{d\zeta}{\zeta}\right] \\ &= \frac{1}{(2\pi i)^n}\left[\int_{\Delta(0)} f(\zeta)\frac{d\zeta}{\zeta} + \sum_{k\in M}\int_{\Delta(k)}\frac{f(\zeta)z^k}{\zeta^k - z^k}\frac{d\zeta}{\zeta}\right], \end{aligned}$$

where $kl = (k_1 l, \ldots, k_n l)$ and

$$\sum_{l=1}^{\infty}\left(\frac{z^k}{\zeta^k}\right)^l = \frac{z^k}{\zeta^k - z^k},$$

since $|z^k| < |\zeta^k|$ for $z \in D$ and $\zeta \in \Delta(k)$. □

We remark that formula (11.15) is universal for the class of domains under consideration.

§12. The Schwarz kernel and integral representation of holomorphic functions with nonnegative real part

The classical *Riez-Herglotz theorem* (see, for example, [134], Chapter VII, §2) asserts that a function $f(z_1)$ holomorphic in the unit disk $\{z_1: |z_1| < 1\}$ has a nonnegative real part in it if and only if there exists a nonnegative measure ν on the circle $\{z_1: |z_1| = 1\}$ such that for $|z_1| < 1$

$$f(z_1) = \frac{1}{2\pi} = \int_{|\zeta_1|=1}\frac{\zeta_1 + z_1}{\zeta_1 - z_1}\,d\nu_{\zeta_1} + i\,\mathrm{Im}\, f(0). \tag{12.1}$$

In this section the *Riez-Herglotz representation* (12.1) is generalized to the multidimensional case.

Let D be a bounded domain in $\mathbf{C}^n$. Denote by $R(D)$ the class of functions in $A(D)$ with nonnegative real part, and let $R_C(D) = R(D) \cap C(\overline{D})$. We consider the following two conditions on D.

A. *For every $f \in R(D)$ there exist a point $z' \in D$, a neighborhood $U(z') \subset D$ of it, and a sequence $f_m \in R_C(D)$, $m = 1, 2, \ldots$, such that $f(z) = \lim_{m\to\infty} f_m(z)$ for $z \in U(z')$.*

This condition is satisfied, for example, by *strongly starlike domains* D (i.e., domains such that $D_r = rD \supset \overline{D}$ for $r > 1$), or, more generally, by domains D for which there exist biholomorphic mappings g_m with $g_m(D) \supset \overline{D}$, $m = 1, 2, \ldots$, and $\lim_{m\to\infty} g_m(z) = z$ for $z \in U(z')$ (for example, a topological product of annuli), or, finally, by domains that are biholomorphically equivalent to the indicated domains, provided that the corresponding biholomorphic mapping extends to a homeomorphism of the closed domains. It can be shown that condition A holds for strictly pseudoconvex domains (see the theorem of Cole and Range in [245] or [186], Chapter VI, Theorem 8). At the same time, we do not know any examples of domains that are the interiors of their closures and do not satisfy condition A.

B. *There exist a closed set $M \supset \partial D$, a nonnegative measure μ on subsets of M, and a function $h(\bar{\zeta}, z) \in C\{(\zeta, z)\colon \zeta \in \overline{D}, z \in D\}$ holomorphic with respect to z in D and having the property that $h(\bar{z}, \zeta) = \overline{h(\bar{\zeta}, z)}$ on $D \times D$, such that for every $f \in A_C(D)$ and $z \in D$ the integral representation with Szegö kernel h*

$$f(z) = \int_M f(\zeta) h(\bar{\zeta}, z)\, d\mu_\zeta \tag{12.2}$$

is valid and, moreover, μ is a representing measure for a point $z^0 \in D$, i.e., for any $f \in A_C(D)$

$$f(z^0) = \int_M f(\zeta)\, d\mu. \tag{12.3}$$

If we do not require that h be continuous with respect to ζ in the closed domain $\overline{D}$ but instead restrict ourselves to the condition $h \in H^2_\mu(D)$, then the Szegö kernel exists in a very general situation (see Theorem 6.7). Our more stringent requirement of continuity is realized under these conditions if D is circular and strongly starlike, and certainly always when the Szegö kernel is found in finite form (see [10], [83], [84], [92] and [191]).

In the case of an n-circular domain D it is convenient to consider a measure that is the product of a measure depending only on $(|\zeta_1|, \ldots, |\zeta_n|)$ and the measure corresponding to the exterior differential form $d \arg \zeta_1 \wedge \cdots \wedge d \arg \zeta_n$

(see §11). Then a Szegö kernel with the properties we want exists if and only if $M \supset S(D)$ and the measure μ is massive on $S(D)$ (i.e., $\overline{\partial D \setminus E} \supset S(D)$ for every $E \subset \partial D$ of zero μ-measure; see §11). Here the Szegö kernel is $h = h(z_1\bar{\zeta}_1, \ldots, z_n\bar{\zeta}_n)$. For a number of n-circular domains the Szegö kernel has been found in finite form (see §11, as well as [10], [83], [84] and [92]).

Finally, these kernels have been explicitly written out also for symmetric domains (see [191]).

The relation (12.3) is valid for $z^0 = 0$ in the cases mentioned where the Szegö kernel is written out explicitly.

The requirement B implies the existence of a function

$$Q(\zeta, z) \in C\{(\zeta, z) : \zeta \in \bar{D}, z \in D\}$$

that is holomorphic with respect to z in D for fixed $\zeta \in \bar{D}$, has the property that $Q(\zeta, z^0) \equiv 1$, and is such that the *Schwarz integral representation*

$$f(z) = \int_M [\operatorname{Re} f(\zeta)] Q(\zeta, z)\, d\mu_\zeta + i \operatorname{Im} f(z^0) \tag{12.4}$$

holds for $f \in A_C(D)$ and $z \in D$.

PROPOSITION 12.1 *Relation* (12.3) *is equivalent to the equality*

$$h(\bar{\zeta}, z^0) \equiv 1. \tag{12.5}$$

The Schwarz kernel can be given by the formula

$$Q(\zeta, z) = 2h(\bar{\zeta}, z) - 1. \tag{12.6}$$

PROOF. Obviously, (12.3) follows from (12.2) and (12.5). Suppose now that (12.3) holds, and consider

$$\int_M h(\bar{\zeta}, z)\, d\mu_\zeta. \tag{12.7}$$

Since μ is a real measure, it has the reproducing property also for antiholomorphic functions. Therefore, the integral (12.7) is equal to $h(\bar{z}^0, z)$; but, (12.2), the same integral is equal to 1. Hence, $\overline{h(\bar{z}^0, z)} \equiv 1$, which leads to (12.5) if we take into account the property $h(\bar{z}, \zeta) = \overline{(\bar{\zeta}, z)}$ of the Szegö kernel.

Further, it follows from (12.5) and (12.6) that $Q(\zeta, z^\circ) \equiv 1$. It remains to show that (12.4) holds. Indeed, (12.2), (12.3), and (12.5) give us that

$$\begin{aligned}
&\int_M \frac{f(\zeta) + \overline{f(\zeta)}}{2} [2h(\bar{\zeta}, z) - 1]\, d\mu_\zeta \\
&\qquad = f(z) + \overline{f(\bar{z}^0) h(\bar{z}^0, z)} - \frac{f(z^0) + \overline{f(z^0)}}{2} \\
&\qquad = f(z) - i \operatorname{Im} f(z^0). \quad \square
\end{aligned}$$

We remark that the Schwarz formula can also be constructed in other ways, as can a formula recovering f from $\operatorname{Im} f$. For example,

$$f(z) = 2i\int_M [\operatorname{Im} f(\zeta)]\, h(\bar{\zeta}, z)\, d\mu_\zeta + \overline{f(z^0)}\,.$$

In the cases where it is not possible to find the Szegö kernel in finite form for an n-circular domain, we can consider the generalized Szegö kernel (11.13), which can be written out explicitly for a large class of n-circular domains (see [10], [17], [47] or [91]). And in this case we can find the Schwarz kernel from the generalized Szegö kernel by using a formula analogous to (12.6).

Observe that, for example, circular strongly starlike domains satisfy requirements A and B.

An important role in what follows is played by the class $[\operatorname{Re} A_C(D)]_\mu^\perp$. If E is such that for every real $u \in Ph_C(D)$ there is a sequence u_m, $m = 1, 2, \ldots$, $u_m \in \operatorname{Re} A_C(D)$, that converges uniformly to u on M, then $[\operatorname{Re} A_C(D)]_\mu^\perp = [Ph_C(D)]_\mu^\perp$ (for example, D strongly starlike).

THEOREM 12.2 (Aĭzenberg and Dautov). *$f \in R(D)$ if and only if the integral representation*

$$f(z) = \int_M Q(\zeta, z)\, d\nu_\zeta + i \operatorname{Im} f(z^0) \tag{12.8}$$

holds for $z \in D$ with a nonnegative measure ν such that

$$\int_M \varphi\, d\nu = 0 \tag{12.9}$$

for all $\varphi \in [\operatorname{Re} A_C(D)]_\mu^\perp$.

In the case where $n = 1$, D is the unit disk, and (12.4) is the classical Schwarz formula, Theorem 12.2 gives us the Riesz-Herglotz theorem, since $[\operatorname{Re} A_C(D)]_\mu^\perp$ contains only the zero function under these conditions.

PROOF. *Necessity*. Suppose that $f \in R(D)$ and $f_m \in R_C(D)$, $m = 1, 2, \ldots$, is a sequence of functions as in condition A. This sequence forms a normal family of holomorphic functions (see [131], §119); therefore, it contains a subsequence converging uniformly to f on compact subsets of D. Let us denote this subsequence again by f_m $(m = 1, 2, \ldots)$. We represent each f_m by formula (12.4),

$$f_m(z) = \int_M Q(\zeta, z)\, d\nu_{m,\zeta} + i \operatorname{Im} f_m(z^0), \tag{12.10}$$

where $d\nu_m = \operatorname{Re} f_m\, d\mu$, and then we get from the condition $q(\zeta, z^0) \equiv 1$ that

$$\operatorname{Re} f_m(z^0) = \int_M d\nu_m. \tag{12.11}$$

The measures ν_m are nonnegative. From (12.11) and the equality $\lim_{m\to\infty}\operatorname{Re} f_m(z^0) = \operatorname{Re} f(z^0)$ it follows that the total variations of these measures are bounded; therefore, $\{\nu_m\}$ contains a subsequence that converges to some nonnegative measure ν in the weak topology (i.e., in the C-topology of the space $C^*(M)$; see [70], Chapter V, §4). Denote such a subsequence again by ν_m ($m = 1, 2, \ldots$). We now pass to the limit in (12.10) as $m \to \infty$ and get (12.8).

From the way in which the measures ν_m are specified it follows that each of them satisfies condition (12.9); hence, the limit mesure also satisfies this condition.

Sufficiency. Consider the function f representable in D by (12.8) under condition (12.9). Obviously, $f \in A(D)$. For $z \in D$

$$\operatorname{Re} f(z) = \int_M [\operatorname{Re} Q(\zeta, z)]\, d\nu_\zeta. \tag{12.12}$$

Further, for every $\psi \in \operatorname{Re} A_C(D)$ we have the Poisson formula (see Proposition 12.3)

$$\psi(z) = \int_M \psi(\zeta)\mathcal{P}(\zeta, z)\, d\mu_\zeta,$$

and this together with (12.4) implies that the difference $\operatorname{Re} Q(\zeta, z) - \mathcal{P}(\zeta, z)$ is in $[\operatorname{Re} A_C(D)]_\mu^\perp$; therefore, (12.12) and (12.9) give us that $z \in D$

$$\operatorname{Re} f(z) = \int_M \mathcal{P}(\zeta, z)\, d\nu_\zeta.$$

Since the Poisson kernel $\mathcal{P}$ and the measure ν are nonnegative, we have that $f \in R(D)$. □

PROPOSITION 12.3. *For $f \in A_C(D)$ and $z \in D$ the Poisson integral representation*

$$f(z) = \int_M f(\zeta)\mathcal{P}(\zeta, z)\, d\mu_\zeta \tag{12.13}$$

is valid, where the Poisson kernel is

$$\mathcal{P}(\zeta, z) = \frac{|h(\bar{\zeta}, z)|^2}{H(\bar{z}, z)}.$$

Here for $z \in D$

$$h(\bar{z}, z) > 0. \tag{12.14}$$

PROOF. Application of (12.2) gives

$$h(\bar{z}, z) = \int_M h(\bar{z}, \zeta)h(\bar{\zeta}, z)\, d\mu_\zeta = \int_M |h(\bar{\zeta}, z)|^2\, d\mu_\zeta. \tag{12.15}$$

For every $z \in D$ we have $\operatorname{Ess\,sup}_M |h(\bar{\zeta}, z)| > 0$, for otherwise (12.2) would imply that $f(z) = 0$ for all $f \in A_C(D)$. Therefore, (12.15) gives us (12.14). Moreover,

$$\int_M f(\zeta)\mathcal{P}(\zeta, z)\, d\mu_\zeta = \frac{1}{h(\bar{z}, z)} \int_M f(\zeta) h(\bar{z}, z)\, d\mu_\zeta$$
$$= \frac{1}{h(\bar{z}, z)} f(z) h(\bar{z}, z) = f(z). \quad \square$$

In order to make the generalization of the Riesz-Herglotz theorem given in Theorem 12.2 clearer for a concrete strongly starlike domain D, we must find a family of functions (for example, polynomials) in $[Ph_C(D)]_\mu^\perp$ such that 1) they can be written out explicitly, and 2) their linear span is dense in $[Ph_C(D)]_\mu^\perp$ in the $C(M)$-metric; and then we must replace condition (12.9) by the requirement that ν is orthogonal to the elements of this family.

Let D_0 be a bounded circular strongly starlike domain, and μ a nonnegative measure on M that is massive on the Shilov boundary $S(D)$, with M and μ invariant with respect to the *rotations* $z \to e^{i\theta}z$, $0 \leqslant \theta \leqslant 2\pi$. Then we can choose a complete orthonormal system $\{\varphi_l^k(z)\}$ of homogeneous polynomials in $H_\mu^2(D)$ (with respect to the measure μ), where l is the degree of the polynomial and k is its index among the homogeneous polynomials of given degree l. Let $\varphi_{-l}^k(z) = \overline{\varphi_l^k(z)}$, $l = 1, 2, \ldots$, and let

$$P_{J\mathfrak{J}}(z, \bar{z}) = \sum_k a_{J\mathfrak{J}k}\varphi^k_{|J|-|\mathfrak{J}|}(z),$$

where $J = (i_1, \ldots, i_n)$, $\mathfrak{J} = (j_1, \ldots, j_n)$, i_l, with nonnegative integers j_l ($l = 1, \ldots, n$), and $|J| = i_1 + \cdots + i_n$. Define

$$a_{J\mathfrak{J}k} = \int_M z^J \bar{z}^{\mathfrak{J}} \varphi^k_{|J|-|\mathfrak{J}|}(z)\, d\mu.$$

PROPOSITION 12.4. *Any polynomial $P \in [Ph_C(D_0)]_\mu^\perp$ can be represented as a linear combination of the polynomials*

$$z^J\bar{z}^{\mathfrak{J}} - P_{J\mathfrak{J}}(z, \bar{z}), \tag{12.16}$$

each of which is in $[Ph_C(D_0)]_\mu^\perp$.

LEMMA 12.5. *Any polynomial of the form* (12.16) *is in $[Ph_C(D_0)]_\mu^\perp$.*

PROOF. Since D_0 is strongly starlike, every $f \in Ph_C(D_0)$ can be approximated uniformly on D_0 by functions pluriharmonic in a neighborhood of $\overline{D}_0$, and each function holomorphic on $\overline{D}_0$ can be approximated uniformly by linear combinations of the homogeneous polynomials $\varphi_l^k(z)$, $l = 0, 1, 2, \ldots$, (since D_0 is circular); therefore, f can be approximated uniformly on $\overline{D}_0$ by linear combinations of the polynomials $\varphi_l^k(z)$, $l = 0, \pm 1, \pm 2, \ldots$. Thus, it suffices to show that

$$\int_M \left[z^J \bar{z}^{\mathcal{J}} - P_{J\mathcal{J}}(z, \bar{z})\right] \varphi_l^k(z)\, d\mu = 0, \qquad l = 0, \pm 1, \pm 2, \ldots \tag{12.17}$$

The invariance of μ with respect to rotations implies (as in the proof of Theorem 1.1) that for $|N| \neq |L|$

$$\int_M z^N \bar{z}^L\, d\mu = 0, \tag{12.18}$$

Hence, (12.17) is valid for $l \neq |J| - |\mathcal{J}|$. But if $l = |J| - |\mathcal{J}|$, then (12.17) holds because of the form of the coefficients $a_{J\mathcal{J}k}$. □

LEMMA 12.6. *Any polynomial $P(z, \bar{z})$ can be represented in the form*

$$P(z, \bar{z}) = P_1(z, \bar{z}) + P_2(z, \bar{z}), \tag{12.19}$$

where P_1 is a linear combination of the polynomials (12.16), *and P_2 is pluriharmonic.*

PROOF. Write P in the form

$$P(z, \bar{z}) = \sum_{J, \mathcal{J}} b_{J\mathcal{J}} z^J \bar{z}^{\mathcal{J}},$$

where only finitely many of the coefficients $b_{J\mathcal{J}}$ are nonzero. Further,

$$P(z, \bar{z}) = \sum_{J, \mathcal{J}} b_{J\mathcal{J}}\left[z^J \bar{z}^{\mathcal{J}} - P_{J\mathcal{J}}(z, \bar{z})\right] + \sum_{J, \mathcal{J}} P_{J\mathcal{J}}(z, \bar{z}).$$

It remains to observe that $P_{J\mathcal{J}}$ is a holomorphic polynomial for $|J| \geq |\mathcal{J}|$ and antiholomorphic for $|J| \leq |\mathcal{J}|$. □

PROOF OF PROPOSITION 12.4. We represent the polynomial $P \in [Ph_C(D_0)]_\mu^\perp$ by (12.19). It must be shown that $P_2 \equiv 0$. Note that $\bar{P}_2 \in Ph_C(D_0)$ and $P_2 \in [Ph_C(D_0)]_\mu^\perp$; therefore,

$$\int_M |P_2|^2\, d\mu = 0,$$

so that $P_2 = 0$ on $S(D)$, i.e., $P_2 \equiv 0$. □

We next consider an n-circular strongly starlike domain D_1 and assume that M and μ are invariant with respect to the *n-rotations* $z \to (z_1 e^{i\theta_1}, \ldots, z_n e^{i\theta_n})$, $0 \leq \theta_j \leq 2\pi$, and that the nonnegative measure μ is massive on $S(D)$. Under these conditions we can take the system of monomials $\{\alpha_J z^J\}$ as a complete orthonormal system, where

$$\alpha_J = \left[\int_M z^J \bar{z}^J\, d\mu\right]^{-1/2}.$$

The next assertion can be proved similarly to the way Proposition 12.4 was proved.

PROPOSITION 12.7. *Every polynomial $P \in [Ph_C(D_1)]_\mu^\perp$ can be represented as a linear combination of the polynomials*

$$z^J \bar{z}^{\mathcal{J}}, \quad J \nleq \mathcal{J}, \quad J \ngeq \mathcal{J}, \tag{12.20}$$
$$z^J \bar{z}^{J+\mathcal{J}}\left(\alpha^2_{J+\mathcal{J}+I_k} - \alpha^2_{J+\mathcal{J}}|z_k|^2\right), \qquad z^{J+\mathcal{J}} \bar{z}^J\left(\alpha^2_{J+\mathcal{J}+I_k} - \alpha^2_{J+\mathcal{J}}|z_k|^2\right),$$

each of which is in $[Ph_C(D_1)]_\mu^\perp$.

We again return to a circular domain D_0 and consider an arbitrary measure ν on M. For $z \in D_0$ let

$$F_\nu(z) = \int_M \mathcal{P}(\zeta, z)\, d\nu_\zeta.$$

LEMMA 12.8. *If ν is orthogonal to the polynomials* (12.16), *then $F_\nu(z) \in Ph(D_0)$.*

PROOF. Formula (12.3) is valid for $f \in A_C(D)$ and $z \in D_0$. By using the facts that D_0 is strongly starlike and the Poisson kernel $\mathcal{P}(\zeta, z)$ is real, it is not hard to show that (12.13) holds also for $f \in Ph_C(D_0)$. Let

$$H_a(f) = \sum_{i,j=1}^n \frac{\partial^2 f}{\partial z_i \partial \bar{z}_j} a_i a_j, \qquad a \in \mathbf{C}^n,$$

and let T be any differential operator with constant coefficients. From (12.13) it follows that for every $f \in Ph_C(D_0)$

$$\int_M f(\zeta) T_z[H_{a,z}(\mathcal{P}(\zeta, z))]\, d\mu_\zeta = 0,$$

where the index z on the operators T and H_a means that they must be used with respect to z. In particular,

$$T_z[H_{a,z}(\mathcal{P}(\zeta, z))]\,|_{z=0} \in [Ph_C(D_0)]_\mu^\perp.$$

We observe that in our case

$$h(\bar{\zeta}, z) = \sum_{l=0}^\infty \sum_k \varphi_l^k(z) \overline{\varphi_l^k(\zeta)},$$

since, by (12.2),

$$\int_M \varphi_l^k(\zeta) h(\bar{\zeta}, z)\, d\mu_\zeta = \varphi_l^k(z);$$

from this and the form of $\mathcal{P}(\zeta, z)$ it follows that $T_z[H_{a,z}(\mathcal{P}(\zeta, a))]|_{z=0}$ is a polynomial in ζ and $\bar{\zeta}$. Therefore,

$$T[H_a(F_\nu(z))]|_{z=0} = \int_M T_z[H_{a,z}(\mathcal{P}(\zeta, z))]|_{z=0}\, d\nu_\zeta = 0,$$

i.e., $H_a(F_\nu(z))$ and all its derivatives at 0 vanish. The function $F_\nu(z)$ is real-analytic; consequently, $H_a(f_\nu(z))$ is also real-analytic in D_0, and so it is identically equal to zero, which means that $F_\nu(z) \in Ph(D_0)$. □

Suppose now that D is a bounded symmetric domain (see [174], §24). Such domains can be described in a nice way. For simplicity we consider classical symmetric domains or topological products of them. These domains are circular and strongly starlike (see [191]), and on their Shilov boundaries there exist nonnegative rotation-invariant measures μ that are massive on $S(D)$ and such that the Szegö kernels can be explicitly written out. (For example, this kernel can be given by (12.6).) If D is also n-circular, then μ is invariant with respect to n-rotations.

THEOREM 12.9 (Aĭzenberg and Dautov). *If D is a classical symmetric domain or a topological product of such domains, then $f \in R(D)$ if and only if there exists a representation*

$$f(z) = \int_{S(D)} q(\zeta, z)\, d\nu_\zeta + i \operatorname{Im} f(0),$$

where $z \in D$, ν is a nonnegative measure, and

$$\int_{S(D)} P d\nu = 0 \tag{12.21}$$

for all polynomials of the form (12.16). *If D is also n-circular, then* (12.16) *can be replaced by* (12.20).

PROOF. By Theorem 12.2, it suffices to show that (12.21) implies (12.9). By Lemma 12.6, (12.21) gives us that $F_\nu(z) \in Ph(D)$. Moreover (see [291]), $f_\nu(rz)\, d\mu$ converges weakly to ν as $r \to 1-0$, i.e., for any $\varphi \in C(S(D))$

$$\lim_{r \to 1-0} \int_{S(D)} \varphi(z) F_\nu(rz)\, d\mu = \int_{S(D)} \varphi(z)\, d\nu. \tag{12.22}$$

If $\varphi \in [Ph_C(D)]_\mu^\perp$, then the integral in (12.22) is equal to zero, since $F_\nu(rz) \in PH_C(D)$, $0 < r < 1$. □

We give three typical examples.

EXAMPLE 1 (a polydisk). Take

$$D = U(0,1) = \{z : |z_k| < 1,\ k = 1, \dots, n\},$$

$$S(D) = \{z : |z_k| = 1,\ k = 1, \dots, n\};$$

the measure μ is generated by the differential form

$$d\arg\zeta_1 \wedge \cdots \wedge d\arg\zeta_n,$$

and the Szegö kernel has the form

$$h(\bar\zeta, z) = (1 - z_1\bar\zeta_1)^{-1} \cdots (1 - z_n\bar\zeta_n)^{-1}.$$

In this case the polynomials of the second and third type in (12.20) are equal to zero, and we get the *Korányi-Punkánszky formula* for a characterization of $f \in R(D)$:

$$f(z) = \int_{S(D)} \left[\frac{2}{(1 - z_1\bar\zeta_1)\cdots(1 - z_n\bar\zeta_n)} - 1 \right] d\nu_\zeta + i \operatorname{Im} f(0),$$

where $z \in D$, the measure ν is nonnegative, and

$$\int_{S(D)} z^J \bar\zeta^{\mathfrak{J}}\, d\nu = 0$$

if $J \nleqslant \mathfrak{J}$ and $J \ngeqslant \mathfrak{J}$.

EXAMPLE 2 (a ball). Take $D = B_1(0) = \{z\colon |z| < 1\}$, $S(D) = \partial D$, $d\mu = (n-1)!(2\pi)^{-n} ds$, where ds is the surface element of the sphere ∂D, and the Szegö kernel has the form (see (8.7))

$$h(\bar\zeta, z) = (1 - z_1\bar\zeta_1 - \cdots - z_n\bar\zeta_n)^{-n}.$$

We arrive at the following characteristic formula for functions $f \in R(D)$:

$$f(z) = \int_{\partial D} \left[\frac{2}{(1 - z_1\bar\zeta_1 - \cdots - z_n\bar\zeta_n)^n} - 1 \right] d\nu_\zeta + i \operatorname{Im} f(0),$$

where $z \in D$, the measure ν is nonnegative, and $\int_{\partial D} P\, d\nu = 0$ for polynomials $P(z, \bar z)$ of the form

$$z^J \bar z^{\mathfrak{J}}, \qquad J \nleqslant \mathfrak{J}, \qquad J \ngeqslant \mathfrak{J},$$
$$z^J \bar z^{J+\mathfrak{J}}\left[{}_k + j_k = 1 - (|J| + |\mathfrak{J}| + n)|z_k|^2\right],$$
$$z^{J+\mathfrak{J}} \bar z^{J}\left[i_k + j_k + 1 - (|J| + |\mathfrak{J}| + n)|z_k|^2\right].$$

EXAMPLE 3 ("generalized unit disk"). Let $\tau_0 = \{w\colon ww^* < \mathrm{I}^{(2)}\} \subset \mathbf{C}^4$, where w is a 2×2 matrix, $w^* = \bar w'$ is the adjoint of w, and $\mathrm{I}^{(2)}$ is the 2×2 identity matrix. The Shilov boundary of τ_0 is the set $S(\tau_0) = U(2) = \{X\colon XX^* = \mathrm{I}^{(2)}\}$ of all 2×2 unitary matrices (see [52]–[54]); $S(\tau_0)$ admits the parametrization $X = e^{i\varphi/2}u$, $X \in S(\tau_0)$, $0 \leqslant \varphi \leqslant 2\pi$, $u \in SU(2)$ and the group $SU(2)$ is isomorphic to the three-dimensional unit sphere S_3. Let $d\mu = (4\pi^3)^{-1} d\varphi \wedge du$, where du is Lebesgue measure on S_3. The Szegö kernel with respect to this measure has the form (see, for example, [53])

$$h(X,w) = \det{}^{-2}(\mathrm{I}^{(2)} - wX^*).$$

The domain τ_0 is classical (see [191]); therefore, Theorem 12.9 is valid for it. Let

$$D^j_{q_1,q_2}(w) = \left[\frac{(j+q_1)!\,(j-q_1)!}{(j+q_2)!\,(j-q_2)!}\right]^{1/2} \times \sum_s \binom{j+q_2}{s}\binom{j-q_2}{s-q_1-q_2} w_{11}^s w_{12}^{j+q_1-s} w_{21}^{j+q_2-s} w_{22}^{s-q_1-q_2}.$$

Applying Lemma 12.10, we then get that the characteristic formula for the functions $f \in R(D)$ is the *Vladimirov formula*

$$f(z) = \int_{S(\tau_0)} \left[\frac{2}{\det^2(\mathrm{I}^{(2)} - wX^*)} - 1\right] d\nu_x + i \operatorname{Im} f(0),$$

where $z \in \tau_0$, the measure ν is nonnegative, and

$$\int_{S(\tau_0)} (\det X)^m D^j_{q_1,q_2}(X)\, d\nu = 0,$$

$$2j = 2, 3, \ldots,\ m = -2j+1, \ldots, -1,\ -j \leqslant q_1, q_2 \leqslant j.$$

LEMMA 12.10. *The functions*

$$\sqrt{2j+1}\,(\det X)^m D^j_{q_1,q_2}, \qquad X \in S(\tau_0), \tag{12.23}$$

$$2j = 2, 3, \ldots,\ m = -2j+1, \ldots, -1,$$
$$-j \leqslant q_1, q_2 \leqslant j. \tag{12.24}$$

are representable as linear combinations of the polynomials (12.16), *and, conversely, the polynomials of the form* (12.16) *are representable as linear combinations of the functions* (12.23) *under the condition* (12.24).

PROOF. On $S(\tau_0)$ we have (see [53])

$$\overline{D^j_{q_1,q_2}(X)} = (-1)^{q_1+q_2}(\det X)^{-2j} D^j_{-q_1,-q_2}(X), \tag{12.25}$$

and the fucntions (12.23) form an orthogonal basis in $H^2_\mu(\tau_0)$ when $m = 0, 1, 2, \ldots, 2j = 0, 1, 2, \ldots$, and $-j \leqslant q_1, q_2 \leqslant j$.

From (12.25) and the orthogonality of this basis it follows that

$$\int_{S(\tau_0)} \varphi(X) u(X)\, d\mu = 0, \tag{12.26}$$

where φ is a function of the form (12.23) under the condition (12.24) and the u are functions in the basis or their conjugates. Therefore, (12.26) holds for all $u \in Ph_C(\tau_0)$. Thus, under condition (12.24) the functions (12.23) belong to $[Ph_C(\tau_0)]_\mu^\perp$ and, by Proposition 12.4, are linear combinations of the polynomials (12.16), since the functions (12.23) are polynomials in the elements of the matrices X and $\overline{X}$, due to the fact that $(\det X)^m = (\det \overline{X})^{-m}$ on $S(\tau_0)$.

Suppose now that $P(X, \overline{X})$ is a polynomial of the form (12.16). It can be expressed as a linear combination of the functions

$$(\det X)^m D^j_{q_1,q_2}(X)\,\overline{(\det X)^k D^l_{p_1,p_2}(X)}$$
$$= (-1)^{p_1-p_2}(\det X)^{m-k-2l} D^j_{q_1,q_2}(X) D^l_{-p_1,-p_2}(X).$$

The homogeneous polynomials $D^j_{q_1,q_2}(X)D^l_{-p_1,-p_2}(X)$ are linear combinations of polynomials in the basis; hence,

$$P(X, \overline{X}) = \sum_{\substack{m=-\infty \\ j,q_1,q_2}}^{\infty} a^{j,m}_{q_1,q_2}(\det X)^m D^j_{q_1,q_2}(X), \tag{12.27}$$

and only finitely many of the coefficients $a^{j,m}_{q_1,q_2}$ are nonzero. If $m \geqslant 0$ or $m \leqslant -2j$, then the function (12.23) is the trace on $S(\tau_0)$ of a function in $Ph_C(\tau_0)$ (for $m \leqslant -2j$ this follows from (12.25)); therefore, from (12.25), (12.27), and the fact that $P(X, \overline{X}) \in [Ph_C(\tau_0)]_\mu^\perp$ it follows that for $m \geqslant 0$ or $n \leqslant -2j$

$$a^{j,m}_{q_1,q_2} = \frac{1}{2j+1}\int_{S(\tau_0)} P(X, \overline{X})(\det \overline{X})^m D^j_{q_1,q_2}(\overline{X})\, d\mu = 0.$$

This means that in (12.27) the coefficients $a^{j,m}_{q_1,q_2}$ are nonzero only for indices satisfying condition (12.24). □

CHAPTER III

The Theory of Residues

§13. Statement of the problem

One of the main problems in the theory of multidimensional residues is the problem of studying and computing integrals (periods) of closed differential forms on a complex analytic manifold that have singularities on analytic subsets of this manifold when the integrals are over cycles disjoint from the singular sets.

An important case of such integrals involves integrals of holomorphic functions of n complex variables over n-dimensional cycles (closed n-dimensional surfaces) located in the domain where these functions are holomorphic. Here an n-fold integral of a holomorphic function f of the complex variables $z = (z_1, \ldots, z_n)$ can be regarded as an integral of the closed holomorphic form $f(z)\,dz = f(z_1, \ldots, z_n)\,dz_1 \wedge \cdots \wedge dz_n$.

There is a more abstract approach to the theory of residues which studies the duality of homomorphisms of the homology and cohomology groups of a complex analytic manifold, and of the groups of the analytic subsets of it and their complements (see [316], [243], and other references).

Stokes' formula and the Cauchy-Poincaré integral theorem enable us to replace the integral of a closed form over a cycle by the integral of a simpler form cohomologous to the given one over a simpler cycle homologous to the original one. Directly from Corollary 0.5 we have

PROPOSITION 13.1. *Let ω be a closed form of degree q on a manifold X, and suppose that it has singularities on a set T and that $\{\gamma_j\}$ is a basis for the q-dimensional homology of the manifold $X \setminus T$. Then for any cycle $\gamma \in Z_q(X \setminus T)$*

$$\int_\gamma \omega = \sum_j k_j \int_{\gamma_j} \omega, \tag{13.1}$$

where the k_j are the expansion coefficients of the cycle γ in the basis

$$\{\gamma_j\} : \gamma \approx \sum k_j \gamma_j .$$

Thus, the problem of computing the integral of a closed form over a cycle reduces to:

1) the study of the corresponding homology group of the complement $X \setminus T$ of the singular set T (the determination of its dimension and the construction of a homology basis);

2) the computation of the integrals of the form over the basis cycles; and

3) the determination of the expansion coefficients of the cycle of integration in the basis.

The solution of problems 1) and 3) in the multidimensional case (unlike in the case of a single complex variable, where they can be solved trivially by intuitive geometric arguments) involves significant topological difficulties. In some cases the solution of these problems is essentially facilitated by using the Alexander-Pontryagin duality (see §§14 and 18) or the de Rham duality (§§15 and 19). Problem 2) cannot always be solved completely; sometimes it is possible only to reduce the dimension of the integral (see §§16 and 18). It is often possible to replace the cycle of integration directly (without determining a homology basis) by another cycle over which the integral can be computed or at least simplified. An important means of reducing integrals to a simpler form is to lower the order of the poles of the integrand form: cohomological reduction of periods (§17).

§14. Application of Alexander-Pontryagin duality

1°. We first present some concepts and facts from combinatorial topology (see [30] and [31]). Let X be an n-dimensional orientable differentiable manifold with a fixed orientation. If the smooth nondegenerate simplexes $U_r = \{\Delta_r, \varphi\}$ and $V_q = (\Delta_q, \psi)$, $r + q = n$, in X intersect transversally at some point a that is interior for both of them, then the *intersection index* of these simplexes is defined to be

$$\chi(U_r, V_q) = \operatorname{sgn} \left. \frac{\partial(x_1, \dots, x_n)}{\partial(t_1, \dots, t_r, \tau_1, \dots, \tau_q)} \right|_a , \tag{14.1}$$

where $(x_1, \dots, x_n)$ are local coordinates in a neighborhood of a that determine the orientation of X, and $(t_1, \dots, t_r)$ and $(\tau_1, \dots, \tau_q)$ are the coordinates in Δ_r and Δ_q determining the orientation of the simplexes U_r and V_q. Transversality means that the Jacobian in (14.1) is nonzero. If U_r and V_q do not intersect, then let $\chi(U_r, V_q) = 0$. The *intersection index of two chains* $c_r = \sum_i m_i U_r^{(i)}$ and

$l_q = \Sigma_j n_j V_q^{(j)}$, $r + q = n$, whose simplexes are nondegenerate and intersect pairwise at no more than one point, and then transversally, is defined by the formula

$$\chi(c_r, l_q) = \sum_{i,j} m_i n_j \chi\big(U_r^{(i)}, V_q^{(j)}\big).$$

The definition of the intersection index leads immediately to the following properties of it:

1) $\chi(\lambda_1 c_r^{(1)} + \lambda_2 c_r^{(2)}, l_q) = \lambda_1 \chi(c_r^{(1)}, l_q) + \lambda_2 \chi(c_r^{(2)}, l_q)$;

2) $\chi(c_r, l_q) = (-1)^{rq} \chi(l_q, c_r)$;

3) $\chi(c_r, \partial l_{q+1}) = (-1)^r \chi(\partial c_r, l_{q+1})$ (this is less obvious, and has the next properties as consequences);

4) if σ_r and γ_q are cycles in X and $\gamma_q \approx 0$, then $\chi(\sigma_r, \gamma_q) = 0$; and

5) if the chains c_r' and c_r'' have a common boundary ($\partial c_r' = \partial c_r''$) and $\gamma_q \approx 0$, then $\chi(c_r', \gamma_q) = \chi(c_r'', \gamma_q)$.

The *linking coefficient* of two cycles $\sigma_{r-1} \in B_{r-1}(X)$ and $\gamma_q \in B_q(X)$ such that $r + q = n$ and $|\sigma_{r-1}| \cap |\gamma_q| = \varnothing$ is defined by

$$\mathfrak{v}(\sigma_{r-1}, \gamma_q) = \chi(c_r, \gamma_q), \tag{14.2}$$

where c_r is a chain in X that is transversal to γ_q and satisfies $\partial c_r = \sigma_{r-1}$. Property 5) of the intersection index implies that the linking coefficient is independent of the choice of the chain c_r. The following properties of the linking coefficient are consequences of (14.2) and the properties of the intersection index:

1) $\mathfrak{v}(\lambda_1 \sigma_{r-1}' + \lambda_2 \sigma_{r-1}'', \gamma_q) = \lambda_1 \mathfrak{v}(\sigma_{r-1}', \gamma_q) + \lambda_2 \mathfrak{v}(\sigma_{r-1}'', \gamma_q)$;

2) $\mathfrak{v}(\sigma_{r-1}, \gamma_q) = (-1)^r \chi(\sigma_{r-1}, b_{q+1})$, where $\partial b_{q+1} = \gamma_q$;

3) $\mathfrak{v}(\sigma_{r-1}, \gamma_q) = (-1)^{(r-1)q-1} \mathfrak{v}(\gamma_q, \sigma_{r-1})$.

THEOREM 14.1([1]) (Alexander-Pontryagin duality). *Let S^n be a manifold homeomorphic to the n-dimensional sphere and T a polyhedron in it. Then for $r + q = n$ $(r = 0, 1, \ldots, n)$ the weak homology groups $H_{r-1}(T)$ and $H_q(S^n \backslash T)$ are isomorphic. Moreover, for every $(r-1)$-dimensional homology basis $\sigma_1, \ldots, \sigma_p$ of T there exists a corresponding dual q-dimensional homology basis $\gamma_1, \ldots, \gamma_p$ of the open set $S^n \backslash T$ such that $\mathfrak{v}(\sigma_i, \gamma_j) = \delta_{ij}$, $i, j = 1, \ldots, p$.*

2°. We shall apply Theorem 14.1 and Proposition 13.1 to the following case. Let $X = \mathbf{C}^n$, and suppose that the form ω is regular in the domain $D = \mathbf{C}^n \backslash T$ (T is a singular set of ω). Since $\mathbf{C}^n$ is homeomorphic to the $2n$-dimensional sphere with a "puncture," it can be assumed that $\mathbf{C}^n$ is imbedded in the

---

([1])See [30] or [31] for a proof.

spherical space $S^{2n} = \dot{\mathbf{C}}^n = \mathbf{C}^n \cup \{\infty\}$. We regard the exterior of an arbitrary ball in $\mathbf{C}^n$ as a neighborhood of the point ∞ at infinity. If S^{2n} admits a triangulation in which the set $\dot{T} = T \cup \{\infty\}$ is a subpolyhedron of S^{2n}, then Theorem 14.1 can be applied to $\dot{T}$ and $S^{2n}\setminus\dot{T} = \mathbf{C}^n\setminus T = D$. To find the q-dimensional Betti number of D (the dimension of the group $H_q(D)$) it suffices in this case to find the $(r-1)$-dimensional Betti number p of the singular set $\dot{T}$ for $r = 2n - q$. Instead of a q-dimensional homology basis $\{\gamma_j\}_1^p$ of D it suffices to construct an $(r-1)$-dimensional homology basis $\{\sigma_j\}_1^p$ of the singular set $\dot{T}$. The expansion coefficients of the integration cycle $\gamma \in Z_q(D)$ in the basis $\{\gamma_j\}$ dual to the basis $\{\sigma_j\}$ are found as the linking coefficients:

$$\mathfrak{v}(\sigma_j, \gamma) = \mathfrak{v}\Big(\sigma_j, \sum_l k_l\gamma_l\Big) = \sum_{l=1}^{p} k_l \mathfrak{v}(\sigma_j, \gamma_l)$$

$$= \sum_{l=1}^{p} k_l \delta_{lj} = k_j.$$

To compute the integral of the form ω over the basis cycles γ_j it is also unnecessary to determine directly the basis $\{\gamma_j\}$. To do this it suffices to choose p homologically independent q-dimensional cycles $\Gamma_1, \ldots, \Gamma_p$ in D such that the integrals over them are sufficiently simple. Then the integrals over the basis cycles are found from the system of linear equations

$$\int_{\Gamma_j} \omega = \sum_{l=1}^{p} k_{jl} \int_{\gamma_l} \omega, \qquad j = 1, \ldots, p,$$

where $k_{jl} = \mathfrak{v}(\sigma_l, \Gamma_j)$. Here the inequality $\det\|k_{jl}\| \neq 0$ is a condition for the cycles $\Gamma_1, \ldots, \Gamma_p$ to be homologically independent.

If the form ω is regular in the domain $D = G\setminus T$, where G is a domain in $\mathbf{C}^n$ that is homeomorphic to a ball, then G can be complemented to form a spherical space S^{2n} by identifying its boundary ∂G with a single ideal point (∞). Instead of the group $H_{r-1}(\dot{T})$ we can consider the relative homology group $H_{r-1}(T, \partial G)$ isomorphic to it. Now let $r = q = n$ and $\omega = f(z)\,dz$, where $f \in A(D)$. The preceding arguments and Proposition 13.1 then give us

THEOREM 14.2 (on residues). *Suppose that $f \in A(D)$ and that the singular set $\dot{T} = T \cup \{\infty\}$, where $T = \mathbf{C}^n\setminus D$, is a subpolyhedron of the spherical compactification $S^{2n} = \mathbf{C}^n \cup \{\infty\}$ of $\mathbf{C}^n$. If $\{\sigma_j\}_1^p$ is an $(n-1)$-dimensional homology basis of the singular set $\dot{T}$ and $\{\gamma_j\}_1^p$ is a corresponding dual n-dimensional homology basis of D, then for any cycle $\gamma \in Z_n(D)$*

$$\int_\gamma f(z)\,dz = (2\pi i)^n \sum_{j=1}^{n} k_j R_j,$$

where $k_j = \mathfrak{v}(\sigma_j, \gamma)$, *and*

$$R_j = \frac{1}{(2\pi i)^n} \int_{\gamma_j} f(z)\,dz.$$

By analogy with the one-variable case, R_j can be called *the residue of the function* $f(z)$ *with respect to the basis cycle* γ_j. The integral $\int_\gamma f(z)\,dz$ is also called the *period* of the form $f(z)\,dz$ with respect to the cycle γ.

COROLLARY 14.3. *If* $f \in A(\mathbf{C}^n \setminus T)$ *and* $H_{n-1}(\dot{T}) \simeq 0$, *where* $\dot{T} = T \cup \{\infty\}$, *then* $\int_\gamma f(z)\,dz = 0$ *for any cycle* $\gamma \in Z_n(\mathbf{C}^2 \setminus T)$.

EXAMPLE 1. Let $f(w, z)$ be an entire function of two complex variables, and let k and m be relatively prime positive integers. Then

$$\int_\gamma \frac{f(w, z)\,dw \wedge dz}{w^k - z^m} = 0$$

for any cycle $\gamma \in Z_2(\mathbf{C}^2 \setminus T)$, where $T = \{(w, z)\colon w^k - z^m = 0\}$. Indeed, the set $\dot{T} = T \cup \{\infty\}$ is topologically equivalent to the Riemann surface of the function $w = z^{m/k}$, which is homeomorphic to the 2-dimensional sphere S^2, and $H_1(S^2) \simeq 0$.

EXAMPLE 2. If $f(z)$ is an entire function on $\mathbf{C}^n$, then

$$\int_\gamma \frac{f(z)\,dz}{\prod_{j=1}^N (\sum_{k=1}^n a_k z_k + b_j)^{m_j}} = 0$$

for any cycle $\gamma \in Z_n(\mathbf{C}^n \setminus T)$, where

$$T = \left\{ z \in \mathbf{C}^n\colon \prod_{j=1}^N \left(\sum_{k=1}^n a_k z_k + b_j \right) = 0 \right\}.$$

Indeed, since T is a family of N parallel analytic planes of complex dimension $n-1$, the set $\dot{T}$ is homeomorphic to N spheres of dimension $2n-2$ that have the single common point ∞ (a bouquet of spheres). Consequently, $H_{n-1}(\dot{T}) \simeq 0$.

EXAMPLE 3. Let $T = \{(w, z) \in \mathbf{C}^2\colon wz = 1\}$. Then $\dot{T} = T \cup \{\infty\}$ is homeomorphic to the two-dimensional sphere with two points identified (the point $z = 0$ of the Riemann sphere, where $w = 1/z = \infty$, and the point $z = \infty$ are identified with the point $\infty \in \dot{T}$). Consequently, the dimension of the groups $H_1(\dot{T}) \simeq H_2(\mathbf{C}^2 \setminus T)$ is equal to 1. As dual basis cycles we can take

$$\sigma = \{(w, z)\colon y = v = 0,\, u = 1/x,\, 0 \leqslant x < \infty\} \in Z_2(\dot{T}),$$

where $x = \operatorname{Re} z$, $y = \operatorname{Im} z$, $u = \operatorname{Re} w$, $v = \operatorname{Im} w$, and

$$\gamma = \{(w, z)\colon w = 2e^{it},\, z = 2e^{i\tau},\, 0 \leqslant t, \tau \leqslant 2\pi\} \in Z_2(\mathbf{C}^2 \setminus T).$$

Obviously, $\mathfrak{v}(\sigma, \gamma) = \chi(c, \gamma) = 1$, where $c = \{(w, z): y = v = 0,\ xu \geqslant 1,\ x \geqslant 0,\ u \geqslant 0\}$, because $\partial c = \sigma$ and $|\gamma| \cap |c|$ consists of a single point (2.2) at which

$$\operatorname{sgn} \frac{\partial(u, v, x, y)}{\partial(x, u, t, \tau)} = 1$$

(we can take $\mathfrak{v}(\sigma, \gamma) = \chi(\sigma, b)$, where $b = \{(w, z): w = re^{it}, z = 2e^{i\tau}, 0 \leqslant r \leqslant 2,\ 0 \leqslant t,\ \tau \leqslant 2\pi\}$ and $\partial b = \gamma$). Theorem 14.2 gives us that for an entire function $f(w, z)$ and a cycle $\Gamma \in Z_2(\mathbf{C}^2 \setminus T)$

$$\int_\gamma \frac{f(w, z)\, dw \wedge dz}{wz - 1} = (2\pi i)^2 mR,$$

where $m = \mathfrak{v}(\sigma, \Gamma)$, and

$$\begin{aligned} R &= \frac{1}{(2\pi i)^2} \int_\gamma \frac{f(w, z)\, dw \wedge dz}{wz - 1} \\ &= \sum_{k=0}^{\infty} \frac{1}{(2\pi i)^2} \int_\gamma \frac{f(w, z)\, dw \wedge dz}{(wz)^{k+1}} \\ &= \sum_{k=0}^{\infty} \frac{1}{(k!)^2} \frac{\partial^{2k} f(0, 0)}{\partial w^k \partial z^k} \end{aligned}$$

is the residue of the function $f(w, z)/(wz - 1)$ with respect to the cycle γ.

Application of the Alexander-Pontryagin duality is especially efficient when $n = 2$ and T is an analytic set, because in this case the study of the two-dimensional homology of a domain in four-dimensional Euclidean space reduces to the study of the one-dimensional homology group of a surface of real dimension two (see §18).

§15. Application of de Rham duality

The following result establishes a duality between the homology groups and the cohomology groups of a differentiable manifold X.

THEOREM 15.1([2]) (de Rham). *For any homomorphism $\lambda: H_p(X) \to \mathbf{C}^1$, or $\mathbf{R}^1$ there exists a unique element $\{\omega\} = h^* \in H^p(X)$ such that*

$$\lambda(h) = \int_h h^* = \int_\gamma \omega$$

for any $h = \{\gamma\} \in H_p(X)$.

Here $H_p(X)$ and $H^p(X)$ are understood to be the homology and cohomology groups with real or complex coefficients. We remark that every closed

([2])See [144] for a proof.

form $\omega \in Z^p(X)$, or rather its class $\{\omega\} \in H^p(X)$, gives a homomorphism λ: $H_p(X) \to \mathbf{C}^1$, or $\mathbf{R}^1$ defined by $\lambda(h) = \int_h \{\omega\}$.

COROLLARY 15.2. *A form $\omega \in Z^p(X)$ is exact if and only if $\int_\gamma \omega = 0$ for any cycle $\gamma \in Z_p(X)$.*

The necessity coincides with the assertion of Corollary 0.3. Let us prove the sufficiency. Suppose that $\int_\gamma \omega = \int_h \{\omega\} = 0$ for any class $\{\gamma\} = h \in H_p(X)$ for some form $\omega \in h^* \in H^p(X)$. This means that ω determines the zero homomorphism 0: $H_p(X) \to \mathbf{C}^1$. The uniqueness assertion in the de Rham theorem implies that $\{\omega\} = h^* = 0$. □

COROLLARY 15.3. *A cycle $\gamma \in Z_p(X)$ is weakly homologous to zero if and only if $\int_\gamma \omega = 0$ for any form $\omega \in Z^p(X)$.*

The necessity is Corollary 0.4. We prove the sufficiency. If $\gamma_0 \not\approx 0$, then $h_0 = \{\gamma_0\} \neq 0$. Since $H_p(X)$ is a torsion-free group in the case of real or complex coefficients, for any nonzero $h_0 \in H_p(X)$ there is a homomorphism λ: $H_p(X) \to \mathbf{C}^1$ such that $\lambda(h_0) \neq 0$. By Theorem 14.1, there exists a class $h^* = \{\omega\} \in H^p(X)$ such that $\lambda(h) = \int_h \{\omega\}$. Then $\int_{\gamma_0} \omega = \int_{h_0} \{\omega\} \neq 0$. □

The next result is useful for applications to multidimensional residues.

PROPOSITION 15.4. *If for the cycles $\gamma_j \in Z_p(X)$, $j = 1, \ldots, q$, and the forms $\omega_j \in Z^p(X)$, $j = 1, \ldots, q$, the condition $\det\|a_{ij}\| \neq 0$ holds, where $a_{ij} = \int_{\gamma_i} \omega_j$, and any form $\omega \in Z^p(X)$ has a representation $\omega \approx c_1\omega_1 + \cdots + c_p\omega_p$ for some complex numbers c_j, then the cycles $\{\gamma_j\}_1^q$ make up a p-dimensional homology basis, and the forms $\{\omega_j\}_1^q$ make up a p-dimensional cohomology basis for the manifold X.*

PROOF. The condition $\det\|a_{ij}\| \neq 0$ implies immediately that the cycles $\{\gamma_j\}$ are homologically independent and the forms $\{\omega_j\}$ are cohomologically independent. Then it follows from the hypotheses that $\{\omega_j\}$ is a p-dimensional cohomology basis. Let us show that the cycles $\{\gamma_j\}$ form a p-dimensional homology basis. Using linear transformations, we can choose the cycles $\{\gamma_j\}$ and the forms $\{\omega_j\}$ in such a way that

$$\int_{\gamma_i} \omega_j = \delta_{ij}, \qquad i, j = 1, \ldots, q. \tag{15.1}$$

We take an arbitrary cycle $\gamma \in Z_p(X)$ and consider the cycle

$$\gamma_0 = \gamma - \sum_{j=1}^{q} k_j \gamma_j, \qquad \text{where } k_j = \int_\gamma \omega_j.$$

Obviously, $\int_{\gamma_0} \omega_j = 0$, $j = 1, \ldots, q$. But since $\{\omega_j\}$ is a cohomology basis, $\int_{\gamma_0} \omega = 0$ for any form $\omega \in Z^p(X)$. By Corollary 15.3, $\gamma_0 \approx 0$. Thus, $\gamma \approx \sum_1^q k_j \gamma_j$, i.e., $\{\gamma_j\}$ is a homology basis. □

Bases $\{\gamma_j\}_1^q$ and $\{\omega_j\}_1^q$ satisfying condition (15.1) are said to be *dual in the de Rham sense*.

With the help of Corollary 0.5 we can get the following theorem, which is analogous to Theorem 14.2.

THEOREM 15.5. *If $\{\gamma_j\}_1^q$ is a p-dimensional homology basis and $\{\omega_j\}_1^q$ is a p-dimensional cohomology basis dual to it, then for any cycle $\gamma \in Z_p(X)$ and any form $\omega \in Z^p(X)$*

$$\int_\gamma \omega = \sum_{j=1}^q N_j R_j, \tag{15.2}$$

where the $N_j = \int_\gamma \omega_j$ are the expansion coefficients of the cycle γ in the basis $\{\gamma_j\}$: $\gamma \approx \sum_1^q N_j \gamma_j$, and the $R_j = \int_{\gamma_j} \omega$ are the expansion coefficients of the form ω in the basis $\{\omega_j\}$: $\omega \approx \sum R_j \omega$.

It will be useful for us to have a theorem of Serre establishing an isomorphism between the de Rham cohomology and the cohomology of holomorphic forms for a Stein manifold.([3]) We note that a domain of holomorphy in $\mathbf{C}^n$ is a particular case of a Stein manifold.

THEOREM 15.6 (Serre). *If X is a Stein manifold, then for any closed form $\omega \in Z^p(X)$ there exists a holomorphic form $\omega' \in Z^p(X)$ cohomologous to it.*

On a complex analytic manifold X of complex dimension n any holomorphic form of degree $p > n$ is identically equal to zero, so Theorems 15.1 and 15.6 give us

COROLLARY 15.7. *If X is a Stein manifold of complex dimension n, then $H^p(X) \simeq H_p(X) \simeq 0$ for $p > n$.*

EXAMPLE. We show that in the domain $D = \mathbf{C}^n \setminus \{z\colon z_1 \cdots z_n = 0\}$ the cycle $\gamma_0 = \{z\colon |z_1| = \cdots = |z_n| = 1\}$ and the form $\omega_0 = (2\pi i)^{-n(dz/z)}$ make up dual n-dimensional homology and cohomology bases. Indeed, since D is a domain of holomorphy, any form $\omega \in Z^n(D)$ is cohomologous to a holomorphic form

([3])See [189] or [201] for the definitions and properties of a Stein manifold, or a holomorphically complete complex analytic manifold.

of degree n that has the form $f(z)\,dz$, where $f \in A(D)$. The function $f(z)$ can be expanded in D in a multiple Laurent series:

$$f(z) = \sum_{\alpha} c_\alpha z^\alpha = \sum_{\alpha_1, \ldots, \alpha_n = -\infty}^{\infty} c_{\alpha_1 \cdots \alpha_n} z_1^{\alpha_1} \cdots z_n^{\alpha_n}.$$

If $\alpha \neq -I = (-1, \ldots, -1)$, for example, if $\alpha_j \neq -1$, then

$$c_\alpha z^\alpha\, dz = c_\alpha(\alpha_j + 1)^{-1} d\left(z_1^{\alpha_1} \cdots z_j^{\alpha_j + 1} \cdots z_n^{\alpha_n}\, dz_{[j]}\right) \approx 0.$$

Thus, $\omega \approx f(z)\,dz \approx c_{-I}\,dz/z$. Obviously, $\int_{\gamma_0} \omega_0 = 1$. It remains to apply Proposition 15.4.

§16. The Leray theory of residues

1°. Suppose that X is a complex analytic manifold with complex dimension n, and that $S_1, \ldots, S_m$ are complex analytic submanifolds of it with dimension $n-1$ (codimension 1) which intersect at common points *transversally*, or, in other words, are "*in general position*". This means that if U_a is a neighborhood of an arbitrary point $a \in S_{i_1} \cap \cdots \cap S_{i_k}$ and $S_j(z, a)$ are holomorphic functions determining in U_a the submanifolds S_j: $S_j \cap U_a = \{z\colon S_j(z, a) = 0\}$, then the vectors $\operatorname{grad} S_j(z, a) = (\partial S_j/\partial z_1, \ldots, \partial S_j/\partial z_n)$, $j = i_1, \ldots, i_k$, are linearly independent. Here $z = (z_1, \ldots, z_n)$ are local coordinates in U_a. We consider integrals of the form

$$\int_\gamma \varphi, \tag{16.1}$$

where $\gamma \in Z_p(X \backslash S_1 \cup \cdots \cup S_m)$ and $\varphi \in Z^p(X \backslash S_1 \cup \cdots \cup S_m)$.

In this section the main points are the concepts of the Leray coboundary, residue forms, and residue classes, which generalize the concept of the residue of a holomorphic function of one variable, along with the Leray residue theorems, which enable us to reduce the integral (16.1) to an integral of lower dimension of a residue form over a cycle lying on a singular submanifold in the case when γ is a coboundary cycle.

2°. An exact sequence and the Leray coboundary. Let us first consider the case $m = 1$, with the notation $S_1 = S$. Since S has real codimension 2, we can construct for S a neighborhood V with smooth boundary ∂V in such a way that V is a *locally trivial bundle* with base S and fiber V_a ($a \in S$) homeomorphic to a disk, while ∂V is a bundle with base S and fiber $\delta_a = \partial V_a$ homeomorphic to a circle. The last part means that $V = \bigcup_{a \in S} V_a$ and $\partial V = \bigcup_{a \in S} \delta_a$, where $V_a \cap V_b = \varnothing$ for $a \neq b$, and any point a has a neighborhood U_a such that $V \cap U_a$ is homeomorphic to $(S \cap U_a) \times V_a$ and $\partial V \cap U_a$ is homeomorphic to $(S \cap U_a) \times \delta_a$. To construct such a bundle we choose a Riemannian metric on

X and take V_a to be the union of the geodesic segments issuing from a that are orthogonal to S and of length $\rho(a)$, where $\rho(a)$ is sufficiently small, and the function $\rho(a)$ is assumed to be of class $C^{(1)}$.

We define the mappings

$$\mu\colon \overline{V} \to S, \qquad \nu\colon X\setminus S \to X\setminus V$$

as follows. If $z \in V_a$, then $\mu(z) = a$. If $z \in X\setminus V$, then $\nu(z) = z$, and if $z \in V\setminus S$, then $\nu(z)$ is the point where the geodesic passing through z and orthogonal to S intersects ∂V. Obviously, μ and ν are deformation retracts. By Corollary 0.20,

$$\nu_*\colon H_p(X\setminus S) \to H_p(X\setminus V)$$

is an isomorphism. We define a homomorphism

$$\tilde{\mu}\colon C_p(S) \to C_{p+2}(X, X\setminus V) \tag{16.2}$$

as follows. If σ is a p-dimensional simplex in S, then $\mu^{-1}(\sigma)$ is a $(p+2)$-dimensional cell in $\overline{V}$. If we assume that σ is a sufficiently small simplex, then $\mu^{-1}(\sigma)$ is homeomorphic to $V_a \times \sigma$. The orientation of the cell $\mu^{-1}(\sigma)$ is determined in the natural way by the orientations of V_a and σ. We take $\tilde{\mu}(\sigma) = \{\mu^{-1}(\sigma)\} \in C_{p+2}(X, X\setminus V)$. Obviously, $\partial \circ \tilde{\mu} = \tilde{\mu} \circ \partial$. Indeed,

$$\partial\mu^{-1}(\sigma) = \partial(V_a \times \sigma) = \partial V_a \times \sigma + V_a \times \partial\sigma \in \tilde{\mu}(\partial\sigma)$$

since $\partial V_a \times \sigma \in C_{p+1}(X\setminus V)$. Thus, (16.2) induces a homomorphism

$$\mu_*\colon H_p(S) \to H_{p+2}(X, X\setminus V). \tag{16.3}$$

This homomorphism is an isomorphism, because it has an inverse homomorphism induced by intersection of the chains in $C_{p+2}(X)$ with the submanifold S (it is assumed that the chains intersect S transversally). Here the intersection of a $(p+2)$-dimensional cycle with S is a p-dimensional cycle, and the intersection of the boundary of a chain is the boundary of the intersection of this chain. According to Theorem 0.21, we have the exact sequence

$$\xrightarrow{i_*} H_{p+2}(X) \xrightarrow{j_*} H_{p+2}(X, X\setminus V) \xrightarrow{\partial_*} H_{p+1}(X\setminus V) \xrightarrow{i_*} H_{p+1}(X) \to \cdots. \tag{16.4}$$

If in (16.4) we replace the groups $H_{p+1}(X\setminus V)$ and $H_{p+2}(X, X\setminus V)$ by the groups $H_{p+1}(X\setminus S)$ and $H_p(S)$ isomorphic to them, then we get

THEOREM 16.1 (Leray). *The sequence of homomorphisms*

$$\to H_{p+2}(X) \xrightarrow{\overline{\omega}} H_p(S) \xrightarrow{\delta} H_{p+1}(X\setminus S) \xrightarrow{i} H_{p+1}(X) \to \cdots, \tag{16.5}$$

where $i = i_* \circ \nu_*$, $\overline{\omega} = \mu_*^{-1} \circ j_*$ *and* $\delta = \nu_*^{-1} \circ \partial_* \circ \mu_*$, *is exact.*

This sequence is called a *Leray exact homology sequence*. The homomorphisms i, $\overline{\omega}$, and δ in it can be interpreted as homomorphisms induced as follows: i is induced by the imbedding $X\setminus S\subset X$; $\overline{\omega}$ is induced by intersection of chains in X with the submanifold S; δ is induced by replacing each simplex $\sigma\in C_p(S)$ by the cell $\delta\sigma=\partial\mu(\sigma)=\bigcup_{z\in\sigma}\delta z$. Most interesting is the homomorphism δ, called the *Leray coboundary*. A cycle $\gamma\in Z_{p+1}(X\setminus S)$ is called a *coboundary cycle* if $\{\gamma\}=\delta h\in\delta H_p(S)$, i.e., if it is the boundary of a chain in X that is transversal to S and whose intersection with S is a cycle $\gamma_0\in h\in H_p(S)$.

3°. Residue forms. The Leray residue formula. A form

$$\varphi\in\Omega^p(X\setminus S_1\cup\cdots\cup S_m)$$

is said to have poles of orders $r_1,\ldots,r_m$ on $S_1,\ldots,S_m$ if for any point $a\in S_{i_1}\cap\cdots\cap S_{i_k}$, $\{i_1,\ldots,i_k\}\subset\{1,\ldots,m\}$, $a\notin S_j$, $j\neq i_1,\ldots,i_k$, the form $\varphi(z)S_{i_1}^{r_1}(z,a)\cdots S_{i_k}^{r_k}(z,a)$ extends to some neighborhood U_a of a as a $C^{(\infty)}$-form. Such a form is said to be *semimeromorphic* in $X\setminus S_1\cup\cdots\cup S_m$. Again we set $m=1$ and $S_1=S$.

THEOREM 16.2 (existence of the residue form; Leray). *Suppose that $\varphi(z)$ is a closed regular form on $X\setminus S$ that has on S a polar singularity of first order. Then in some neighborhood U_a of an arbitrary point $a\in S$ there exist regular forms $\psi(z,a)$ and $\theta(z,a)$ such that in U_a*

$$\varphi(z)=\frac{ds(z,a)}{s(z,a)}\wedge\psi(z,a)+\theta(z,a),\tag{16.6}$$

where $s(z,a)\in A(U_a)$, grad *$s(z,a)\neq 0$, and $S\cap U_a=\{z\colon s(z,a)=0\}$. Moreover, the form $\psi(z,a)|_S$ is closed and uniquely determined by φ. If φ is holomorphic in $X\setminus S$, then the form $\psi(z,a)|_S$ is holomorphic in $U_a\cap S$.*

The closed form on S defined in a neighborhood of each point $a\in S$ by the restriction $\psi(z,a)|_S$ is called *the residue form of the form* φ and denoted by $\operatorname{res}[\varphi]$.

LEMMA 16.3 (on division of forms). *A form φ that is regular on X can be represented in some neighborhood U_a of a point $a\in S$ as*

$$\varphi(z)=ds(z,a)\wedge\psi(z,a),$$

where $\psi(z,a)$ is a form that is regular in U_a, if and only if

$$ds(z,a)\wedge\varphi(z)=0$$

in U_a. Moreover, $\psi(z,a)|_S$ depends only on φ and $s(z,a)$. If φ is a holomorphic form, then the form $\psi(z,a)$ can be chosen to be holomorphic.

PROOF. We choose a local system of coordinates $z_1, \ldots, z_n$ in some neighborhood U_a in such a way that $s(z, a) = z_1$. Then the form φ can be represented in U_a as $\varphi(z) = dz_1 \wedge \psi(z, a) + \theta(z, a)$, where ψ and θ are forms that are regular in U_a.

Here it can be assumed that ψ and θ do not contain dz_1, and $dz_1 \wedge \varphi = dz_1 \wedge \theta$. Hence, $dz_1 \wedge \varphi = 0$ if and only if $\theta = 0$, i.e., $\varphi = dz_1 \wedge \psi$. If $\varphi = 0$, then $\psi = dz_1 \wedge \psi_1$ by the preceding. Then $\psi|_{z_1=0} = 0$, i.e., the form $\psi|_S$ is unique. $\square$

PROOF OF THEOREM 16.2. Since $d\varphi = 0$ and the form $s(z, a)\varphi(z)$ is regular in U_a, the form

$$d(s\varphi) = ds \wedge \varphi \tag{16.7}$$

is regular in U_a. Multiplication by ds takes this form to zero. According to Lemma 16.3, there is a form $\theta(z, a)$ which is regular in U_a and such that $ds \wedge \varphi = ds \wedge \theta$. Hence, $ds \wedge (s\varphi - s\theta) = 0$. But since the form $s\varphi - s\theta$ is regular in U_a, Lemma 16.3 gives us the existence of a form $\psi(z, a)$ that is regular in U_a and such that the equality $s\varphi - s\theta = ds \wedge \psi$, or (16.6), holds.

If the form φ is holomorphic, then the form (16.7), being holomorphic in $U_a \setminus S$ and regular in U_a, extends holomorphically to U_a. According to Lemma 16.3, the forms θ and ψ can then be chosen to be holomorphic.

The form $\psi(z, a)$ is uniquely determined by the form φ and the function $s(z, a)$. Indeed, suppose that $\varphi = 0$. Then $ds/s \wedge \psi + \theta = 0$, or $ds \wedge \psi + s\theta = 0$. Therefore, $ds \wedge \theta = 0$ and $\theta = ds \wedge \omega$, where ω is a form that is regular in U_a. Consequently, $ds \wedge (\psi + s\omega) = 0$, and $\psi + s\omega = ds \wedge \omega_1$, where the form ω_1 is regular in U_a. Thus, the restriction $\psi|_S$ is 0.

Suppose that $s_1(z, a) = 0$ is another equation for $S \cap U_a$. Then $\varphi = ds_1/s_1 \wedge \psi + \theta_1$, where $\theta_1 = \theta + d\ln(s/s_1) \wedge \psi$ is a regular form in U_a, because $s/s_1 \in A(U_a)$ and $s/s_1 \neq 0$. Thus, $\psi|_S$ does not depend on the choice of $s(z, a)$.

Finally, we prove that the form $\psi|_S$ is closed. Taking the exterior differential in (16.6), we get that $0 = -ds/s \wedge d\psi + d\theta$. In view of the proved uniqueness, this gives us that $d(\psi|_S) = d\psi|_S = 0$. $\square$

EXAMPLE 1. Suppose that $X = \mathbf{C}^n$, $S = \{z\colon z \in \mathbf{C}^n\colon s(z) = 0\}$ and $\varphi(z) = (f(z)/s(z))\, dz$, where $f, s \in A(\mathbf{C}^n)$, $\operatorname{grad} s|_S \neq 0$. Then

$$\operatorname{res}[\varphi] = (-1)^{j-1}(f/s'_{z_j})|_S\, dz_{[j]}$$

at the points where $s'_{z_j} \neq 0$.

EXAMPLE 2. Let X be an analytic manifold of complex dimension 1, φ a meromorphic form of degree 1 having only poles of first order, and $S = \{a_1, \ldots, a_N\}$ its set of poles. Then $\operatorname{res}_a[\varphi] = \{\operatorname{res}_a \varphi,\ a \in S\}$ is a set of numbers that are ordinary residues of the form φ: $\operatorname{res}_a \varphi = \operatorname{res}_a f(z)$, where $\varphi = f(z)\,dz$ is the representation of φ in terms of the local parameter z in a neighborhood of a.

The following *Leray residue formula* is valid for a closed form having a pole of first order on S and for a coboundary cycle.

THEOREM 16.4 (Leray). *Suppose that the form $\varphi \in Z^{p+1}(X \backslash S)$ has a first-order pole on S, and that $\gamma \in Z_p(S)$. Then*

$$\int_{\delta\gamma} \varphi = 2\pi i \int_{\gamma} \operatorname{res}[\varphi]. \tag{16.8}$$

PROOF. We consider a family of neighborhoods V_ε of S that depends continuously on ε $(0 < \varepsilon \leqslant 1)$ and is such that $V_1 = V$ and $V_\varepsilon \to S$ as $\varepsilon \to 0$. For this we take $V_\varepsilon = \bigcup_{a \in S} V_{\varepsilon,a}$, where $V_{\varepsilon,a}$ is the union of the geodesic segments of length $\varepsilon\rho(a)$ issuing from the point a and orthogonal to S. To each ε there corresponds a coboundary operator δ_ε which assigns to each $a \in S$ the closed curve $\delta_\varepsilon a = \partial V_{\varepsilon,a}$, which is homeomorphic to a circle. Since $\delta_\varepsilon a$ can be obtained from δ_a by a continuous deformation in $V_a \subset X \backslash S$, it follows that $\delta\gamma \sim \delta_\varepsilon\gamma$ in $X \backslash S$ for any cycle $\gamma \in Z_p(S)$. Consequently,

$$\int_{\delta\gamma} \varphi = \int_{\delta_\varepsilon\gamma} \varphi.$$

It remains to show that

$$\lim_{\varepsilon \to 0} \int_{\delta_\varepsilon\gamma} \varphi = 2\pi i \int_{\gamma} \operatorname{res}[\varphi].$$

It suffices to prove this equality for small simplexes making up γ. We may assume that the simplex σ lies in the coordinate neighborhood U_a. According to Theorem 16.2, φ can be represented in the form (16.6) in U_a. Choose a local system of coordinates $z_1, \ldots, z_n$ in U_a in such a way that $z_1 = s(z, a)$ and the length of the vector dz is expressed by the formula

$$|dz| = \left(|dz_1|^2 + \cdots + |dz_n|^2\right)^{1/2}.$$

Then $\delta_\varepsilon\sigma$ can be represented in the form

$$\delta_\varepsilon\sigma = \{z\colon {}'z = (z_2, \ldots, z_n) \in \sigma, |z_1| = \varepsilon\rho('z)\}$$

to within small quantities of second order. Since the forms ψ and θ are continuous in U_a, $\operatorname{meas}|\delta_\varepsilon\sigma| \to 0$ as $\varepsilon \to 0$, and $\operatorname{meas} \delta_\varepsilon\sigma/|z_1|$ is bounded above, it follows that

$$\lim_{\varepsilon\to 0}\int_{\delta_\varepsilon\sigma}\varphi(z) = \lim_{\varepsilon\to 0}\int_{\delta_\varepsilon\sigma}\left[\frac{dz_1}{z_1}\wedge\psi(z,a)+\theta(z,a)\right]$$
$$= \int_{\delta_\varepsilon\sigma}\frac{dz_1}{z_1}\wedge\psi(\mu_\varepsilon(z),a) = \oint\frac{dz_1}{z_1}\int_\sigma\psi(z,a)|_S = 2\pi i\int_\sigma \operatorname{res}[\varphi],$$

where $\mu_\varepsilon(z)$ is the projection of a point $z\in\delta_\varepsilon\sigma$ on $S\cap U_a=\{z\colon z_1=0\}$. □

4°. Residue classes. For arbitrary closed regular forms in $X\setminus S$ we have

THEOREM 16.5 (existence of the residue class; Leray). *For every form $\varphi\in Z^{p+1}(X\setminus S)$, $p=1,2,\ldots$, there exist forms cohomologous to it in $X\setminus S$ that have a first-order pole on S. The residue forms of such forms belong to a single cohomology class on S.*

This cohomology class is called *the residue class of the form* φ and denoted by $\operatorname{Res}[\varphi]$. Since the operator res is linear, Res gives rise to a homomorphism

$$\operatorname{Res}\colon H^{p+1}(X\setminus S)\to H^p(S). \tag{16.9}$$

REMARK. It is not possible to restrict oneself to holomorphic forms in Theorem 16.5. For a holomorphic form φ the residue class $\operatorname{Res}[\varphi]$ may not contain holomorphic forms (see [113], §59, for a counterexample). However, Theorem 16.7 is valid in the class of holomorphic forms for Stein manifolds. This follows from Serre's Theorem 15.6.

See §17 about a constructive determination of the residue class for semi-meromorphic and rational forms.

We precede the proof of Theorem 16.5 by some lemmas.

LEMMA 16.6. *For every form $\chi(z)\in Z^p(S)$ there exists a form $\omega(z)\in\Omega^{p+1}(X\setminus S)$ such that 1) $\omega(z)|_{X\setminus V}=0$, and 2) for any $a\in S$ there is a neighborhood U_a in which*

$$\omega(z)=\frac{ds(z,a)}{s(z,a)}\wedge\psi(z,a)+\theta(z,a), \tag{16.10}$$

where ψ and θ are regular forms in U_a, $\psi(z,a)|_S=\chi(z)$, and $d\psi(z,a)=0$ near S.

PROOF. The form $\chi(\mu(z))$, where $\mu\colon\overline{V}\to S$ is the retraction defined in 2°, is closed in $\overline{V}$. Let $f\colon X\to\mathbf{R}^1$ be a $C^{(\infty)}$-function that equals 0 on $X\setminus V$ and 1 in a neighborhood of S. Such a function exists as a consequence of the "partition-of-unity" theorem. Let $\psi(z)=f(z)\chi(\mu(z))$. Obviously, $d\psi=0$ in a neighborhood of S, and $\psi|_S=\chi$. We take a "partition of unity" $\{\varphi_a\}_{a\in X}$ subordinate to a covering $\{U_a\}_{a\in X}$ of X by coordinate neighborhoods in which local equations exist: $S \cap U_a = \{z\colon s(z, a) = 0\}$. Then the form

$$\omega(z) = \sum_{a\in X} \varphi_a(z)\frac{ds(z, a)}{s(z, a)} \wedge \psi(z)$$

satisfies the conditions of the lemma. Moreover, $\psi(z, a) = \psi(z)$, and

$$\theta(z, a) = \sum_{b\neq a} \varphi_b(z)d\log\frac{s(z, b)}{s(z, a)} \wedge \psi(z). \quad \square$$

According to Theorems 0.22 and 15.1, to the exact sequence (16.5) there corresponds the exact sequence of dual homomorphisms

$$\cdots \to H^{p+1}(X) \xrightarrow{i^*} H^{p+1}(X\setminus S) \xrightarrow{\delta^*} H^p(S) \xrightarrow{\bar{\omega}^*} H^{p+2}(X) \to \cdots. \tag{16.11}$$

LEMMA 16.7. *For the forms χ and ω considered in Lemma* 16.6

$$\bar{\omega}^*\{\chi\} = -\frac{1}{2\pi i}\{d\omega\}, \tag{16.12}$$

where $\{\chi\}$ and $\{d\omega\}$ are the cohomology classes of the forms χ and $d\omega$.

PROOF. We take an arbitrary homology class $\{\gamma\} \in H_{p+2}(X)$ and consider the integral

$$J = \int_\gamma d\omega = \int_{\{\gamma\}} \{d\omega\}. \tag{16.13}$$

Since $\omega = 0$ outside V, the integral (16.13) depends only on the class $h \in H_{p+2}(X, X\setminus V)$ determined by the cycle γ. But since (16.2) is an isomorphism, there is a cycle $\sigma \in Z_p(S)$ such that $\tilde{\mu}\sigma \in h$. Then $J = \int_\gamma d\omega = \int_h d\omega = \int_{\tilde{\mu}\sigma} d\omega$. As in the proof of Theorem 16.4, we consider a neighborhood V_ε of S and operators $\tilde{\mu}_\varepsilon$ and δ_ε. The form $d\omega$ is regular in X, so, by Stokes' formula

$$J = \int_{\tilde{\mu}\sigma} d\omega = \lim_{\varepsilon\to 0}\int_{\tilde{\mu}\sigma - \tilde{\mu}_\varepsilon\sigma} d\omega = \int_{\delta\sigma}\omega - \lim_{\varepsilon\to 0}\int_{\delta_\varepsilon\sigma}\omega.$$

Since $\omega|_{X-V} = 0$, it follows that $\int_{\delta\sigma}\omega = 0$. Taking (16.10) into account, we get exactly as in the proof of Theorem 16.4 that

$$J = -\lim_{\varepsilon\to 0}\int_{\delta_\varepsilon\sigma}\omega = -2\pi i\int_\sigma \chi.$$

By construction, $\{\sigma\} = \bar{\omega}\{\gamma\}$. Thus,

$$\int_{\{\gamma\}}\{d\omega\} = -2\pi i\int_{\bar{\omega}\{\gamma\}}\{\chi\} = -2\pi i\int_{\{\gamma\}}\bar{\omega}^*\{\chi\}.$$

By Corollary 15.2, the equality (16.12) holds. $\square$

PROOF OF THEOREM 16.5. Let $\varphi \in Z^{p+1}(X \setminus S)$ be an arbitrary form. It determines a homomorphism $\varphi^*\colon H_{p+1}(X \setminus S) \to \mathbf{C}^1$ given by $\varphi^*(h) = \int_h \{\varphi\} = \int_\gamma \varphi$ for each $h = \{\gamma\} \in H_{p+1}(X)$. Then the homomorphism $\delta^*\varphi^*\colon H_p(S) \to \mathbf{C}^1$ is determined by

$$\delta^*\varphi^*(h_1) = \int_{\delta h_1} \{\varphi\} = \int_{\delta\gamma_1} \varphi,$$

where $h_1 = \{\gamma_1\} \in H_p(S)$. By the de Rham Theorem 15.1, there exists a cohomology class $\{\chi\} = \delta^*\{\varphi\} \in H^p(S)$ for which

$$\delta^*\varphi^*(h_1) = \int_{h_1} \delta^*\{\varphi\} = \int_{\gamma_1} \chi.$$

For the form χ we construct a form ω according to Lemma 16.6. Since $\{\chi\} = \delta^*\{\varphi\}$, Lemma 16.7 and the exactness of the sequence (16.11) give us that

$$-\frac{1}{2\pi i}\{d\omega\} = \overline{\omega}^*\{\chi\} = \overline{\omega}^*\delta^*\{\varphi\} = 0,$$

i.e., the form $d\omega$ is cohomologous to zero in X. Consequently, there exists a form $\omega_1 \in \Omega^{p+1}(X)$ such that $d\omega = d\omega_1$. Then the form $\varphi_1 = \omega - \omega_1$ is closed in $X \setminus S$ and has a first-order pole on S, and res $\varphi_1 = \chi$. From the definition of the class $\{\chi\}$ and Theorem 16.6 it follows that $\int_{\delta\gamma_1} \varphi = \int_{\delta\gamma_1} \varphi_1$ for any cycle $\gamma_1 \in Z_p(S)$, or

$$\int_{\delta\{\gamma_1\}} (\varphi - \varphi_1) = \int_{\{\gamma_1\}} \delta^*\{\varphi - \varphi_1\} = 0.$$

By Corollary 15.2, this gives us that $\delta^*\{\varphi - \varphi_1\} = 0$. Since the sequence (16.11) is exact, $\{\varphi - \varphi_1\} \in i^*H^{p+1}(X)$, i.e., there exists a form $\varphi_2 \in Z^{p+1}(X)$ such that $\varphi - \varphi_1 \sim \varphi_2$ in $X \setminus S$. Hence, the form φ is cohomologous in $X \setminus S$ to the closed form $\varphi_3 = \varphi_1 + \varphi_2$, which has a first-order pole on S. Moreover, $\operatorname{res}[\varphi_3] = \chi$. Let φ_4 be another form that is cohomologous to φ and has a first-order pole on S. Then

$$\int_{\delta\gamma_1} \varphi = \int_{\delta\gamma_1} \varphi_3 = \int_{\delta\gamma_1} \varphi_4 = 2\pi i \int_{\gamma_1} \operatorname{res}[\varphi_3] = 2\pi i \int_{\gamma_1} \operatorname{res}[\varphi_4]$$

for any cycle $\gamma_1 \in Z_p(S)$. By Corollary 15.2, this gives us that $\operatorname{res}[\varphi_3] \sim \operatorname{res}[\varphi_4]$. □

The next result follows from Theorems 16.4 and 16.5 and the definition of a residue class.

THEOREM 16.8. *Suppose that* $\varphi \in Z^{p+1}(X\setminus S)$ *and* $h \in H_p(S)$. *Then*

$$\int_{\delta h}\{\varphi\} = 2\pi i\int_h \operatorname{Res}[\varphi]. \tag{16.14}$$

5°. Composite residues. Let us now pass to the case $m > 1$. The manifold X and the submanifolds $S_1,\dots,S_m$ of it give rise to the following consecutive pairs of complex analytic manifolds and submanifolds of them with codimension 1:

$$S_1 \cap \cdots \cap S_{m-1} \supset S_1 \cap \cdots \cap S_m;$$
$$(S_1 \cap \cdots \cap S_{m-2})\setminus S_m \supset (S_1 \cap \cdots \cap S_{m-1})\setminus S_m;$$
$$\cdots\cdots\cdots\cdots\cdots\cdots$$
$$S_1\setminus(S_3 \cup \cdots \cup S_m) \supset (S_1 \cap S_2)\setminus(S_3 \cup \cdots \cup S_m);$$
$$X\setminus(S_2 \cup \cdots \cup S_m) \supset S_1\setminus(S_2 \cup \cdots \cup S_m).$$

Applying the results of 2° to these pairs and taking account of the fact that

$$\begin{aligned}&[(S_1 \cap \cdots \cap S_{j-1})\setminus(S_{j+1} \cup \cdots \cup S_m)]\\ &\qquad\setminus[(S_1 \cap \cdots \cap S_j)\setminus(S_{j+1} \cup \cdots \cup S_m)]\\ &\qquad= (S_1 \cap \cdots \cap S_{j-1})\setminus(S_j \cup \cdots \cup S_m),\end{aligned}$$

we get the sequence of Leray coboundary homomorphisms

$$\begin{aligned}&H_p(S_1 \cap \cdots \cap S_m) \xrightarrow{\delta_m} H_{p+1}(S_1 \cap \cdots \cap S_{m-1}\setminus S_m)\\ &\qquad\xrightarrow{\delta_{m-1}} \cdots H_{p+j}((S_1 \cap \cdots \cap S_j)\setminus(S_{j+1} \cup \cdots \cup S_m))\\ &\qquad\to \cdots \xrightarrow{\delta_2} H_{p+m-1}(S_1\setminus(S_2 \cup \cdots \cup S_m))\\ &\qquad\xrightarrow{\delta_1} H_{p+m}(X\setminus S_1 \cup \cdots \cup S_m),\end{aligned}$$

whose composition determines the *composite Leray coboundary*

$$H_p(S_1 \cap \cdots \cap S_m) \xrightarrow{\delta^m} H_{p+m}(X\setminus S_1 \cup \cdots \cup S_m).$$

Since the coboundary operator δ is defined also for simplexes and chains, δ^m extends also to simplexes and chains. The composite Leray coboundary δ^m assigns to each simplex a cell $\delta^m\sigma$ which is a bundle with base σ and fiber $\delta^m a = \delta_m a \times \cdots \times \delta_1 a$, homeomorphic to the topological product of m circles and is taken with the natural orientation. If the submanifolds $S_1,\dots,S_m$ are taken in another order $S_{j_1},\dots,S_{j_m}$, then the cell $\delta^m\sigma$ changes orientation by -1 raised to a power equal to the parity of the permutation $(j_1,\dots,j_m)$.

In exactly the same way we define the *composite residue class*

$$\operatorname{Res}^m\colon H^{p+m}(X\setminus S_1\cup\cdots\cup S_m)\to H^p(S_1\cap\cdots\cap S_m)$$

to be the composition of the homomorphisms

$$H^{p+m}(X\setminus S_1\cup\cdots\cup S_m)\overset{\operatorname{Res}_1}{\to}H^{p+m-1}(S_1\setminus S_2\cup\cdots\cup S_m)$$

$$\overset{\operatorname{Res}_2}{\to}\cdots\overset{\operatorname{Res}_{m-1}}{\to}H^{p+1}(S_1\cap\cdots\cap S_{m-1}\setminus S_m)$$

$$\overset{\operatorname{Res}_m}{\to}H^p(S_1\cap\cdots\cap S_m). \tag{16.15}$$

If the form $\varphi\in Z^{p+m}(X\setminus S_1\cup\cdots\cup S_m)$ has first-order poles on $S_1,\dots,S_m$, then repeated application of Theorem 16.2 enables us to define the form

$$\operatorname{res}^m[\varphi]=\operatorname{res}_m\circ\cdots\circ\operatorname{res}_1[\varphi]\in\operatorname{Res}^m[\varphi],$$

called the *composite residue form*. The possibility of repeated application of Theorem 16.2 is ensured by

LEMMA 16.9. *If the form φ is closed on $X\setminus S_1\cup\cdots\cup S_m$ and has first-order poles on $S_1,\dots,S_m$, then the form $\operatorname{res}_1[\varphi]$ is closed on $S_1\setminus(S_2\cup\cdots\cup S_m)$ and has first-order poles on $S_1\cap S_2,\dots,S_1\cap S_m$.*

PROOF. The fact that the form $\operatorname{res}_1[\varphi]$ is closed follows from Theorem 16.2. Suppose that $a\in S_1\cap S_{i_1}\cap\cdots\cap S_{i_k}$, $\{i_1,\dots,i_k\}\subset\{2,\dots,m\}$, and $a\notin S_j$ for $j\neq i_1,\dots,i_k,1$. We choose local coordinates $z_1,\dots,z_n$ in some neighborhood U_a of a in such a way that $S_j\cap U_a=\{z_j=0\}$, $j=i_1,\dots,i_k$. In a neighborhood of an arbitrary point $b\in S_1\setminus(S_{i_1}\cup\cdots\cup S_{i_k})$ the form φ can, by Theorem 16.2, be represented as

$$\varphi=\frac{dz_1}{z_1}\wedge\psi+\theta,$$

where ψ and θ are regular forms in $U_b\setminus(S_{i_1}\cup\cdots\cup S_{i_k})$. The forms ψ and θ can be chosen to be independent of dz_1. Then any coefficient $c_1(z)$ of ψ looks like $z_1\cdot c(z)$, where $c(z)$ is a coefficient of φ. The corresponding coefficient of $\operatorname{res}_1[\varphi]=\psi|_{S_1\setminus S_{i_1}\cup\cdots\cup S_{i_k}}$ is the restriction of $c_1(z)=z_1\cdot c(z)$ to $S_1\setminus S_{i_1}\cup\cdots\cup S_{i_k}$ and does not depend on the point b. By assumption, the form $\varphi z_1z_{i_1}\cdots z_{i_k}$ and, consequently, its coefficient $z_1z_{i_1}\cdots z_{i_k}c(z)=z_{i_1}\cdots z_{i_k}c_1(z)$ are regular in U_a. But then the restriction $z_{i_1}\cdots z_{i_k}c_1(z)|_{S_1}$ and, hence, the form $z_{i_1}\cdots z_{i_k}\operatorname{res}_1[\varphi]$ are regular at the point a. □

We remark that if

$$\varphi = \frac{ds_1 \wedge \cdots \wedge ds_m \wedge \omega}{s_1 \cdots s_m},$$

then $\operatorname{res}^m[\varphi] = \omega|_{S_1 \cap \cdots \cap S_m}$. Obviously, $\operatorname{res}_{i_1} \circ \cdots \circ \operatorname{res}_{i_m}$ differs from $\operatorname{res}_1 \circ \cdots \circ \operatorname{res}_m$ only by the factor -1 raised to a power equal to the parity of the permutation $(i_1, \ldots, i_m)$.

Applying the residue formulas (16.8) and (16.14) m times in succession, we get the *Leray composite residue formula.*

THEOREM 16.10. *If the form $\varphi \in Z^{p+m}(X \backslash S_1 \cup \cdots \cup S_m)$ has first-order poles on $S_1, \ldots, S_m$, then the formula*

$$\int_{\delta^m\{\gamma\}} \varphi = (2\pi i)^m \int_\gamma \operatorname{res}^m[\varphi] \tag{16.16}$$

holds for any cycle $\gamma \in Z_p(S_1 \cap \cdots \cap S_m)$. For an arbitrary form

$$\varphi \in Z^{p+m}(X \backslash S_1 \cup \cdots \cup S_m)$$

the following formula holds:

$$\int_{\delta^m\{\gamma\}} \varphi = (2\pi i)^m \int_\gamma \operatorname{Res}^m[\varphi]. \tag{16.17}$$

6°. The Leray residue formulas (16.8), (16.14), (16.16), and (16.17) enable us to lower the multiplicity of the integral (16.1) in the case when the cycle γ is a Leray coboundary. Therefore, the question naturally arises of conditions under which a cycle $\gamma \in Z^{p+m}(X \backslash S_1 \cup \cdots \cup S_m)$ is a coboundary cycle, i.e., $\{\gamma\} \in \delta^m H^p(S_1 \cap \cdots \cap S_m)$.

In the case $m = 1$ a simple criterion for a Leray coboundary follows directly from the Leray exact sequence.

PROPOSITION 16.11. *A cycle $\gamma \in Z_{p+1}(X \backslash S)$ belongs to a coboundary class $\{\gamma\} \in \delta H_p(S)$ if and only if it is weakly homologous to zero in X.*

Indeed, our assertion is equivalent to the equality

$$\delta H_p(S) = \ker\{i \colon H_{p+1}(X \backslash S) \to H_{p+1}(X)\},$$

which follows from the exactness of (16.5). □

COROLLARY 16.12. *If $H_{p+1}(X) \simeq 0$, then any cycle $\gamma \in Z_{p+1}(X \backslash S)$ is a Leray coboundary. If, moreover, $H_{p+2}(X) \approx 0$, then $\delta\colon H_p(S) \to H_{p+1}(X \backslash S)$ is an isomorphism.*

The last condition holds, for example, in any dimension p, when $X = \mathbf{C}^n$. If $X = CP^n$ (the complex projective space), then we have

PROPOSITION 16.13 (Griffiths). *If S is a submanifold (algebraic) of CP^n, then the homomorphism*

$$\delta\colon H_{n-1}(S) \to H_n(CP^n \backslash S)$$

is surjective for any n and injective for even n. For odd n the kernel of δ is generated by the section $CP^m \cap S$, where CP^m is a subspace of CP^n of dimension $m = (n+1)/2$.

In the case $m > 1$ we have

PROPOSITION 16.14. *If a cycle $\gamma \in Z_{p+m}(X \setminus S_1 \cup \cdots \cup S_m)$ is a composite Leray coboundary, i.e., $\{\gamma\} \in \delta^m H_p(S_1 \cap \cdots \cap S_m)$, then*

$$\gamma \approx 0 \quad \text{in } X \setminus S_1 \cup \cdots [S_j] \cdots \cup S_m, \qquad j = 1, \ldots, m. \tag{16.18}$$

PROOF. Let $\gamma = \delta^m \gamma_0$, where $\gamma_0 \in Z_p(S_1 \cap \cdots \cap S_m)$. Then $\gamma = \delta_1(\delta^{m-1}\gamma_0)$, where

$$\delta_1 \colon H_{p+m-1}(S_1 \setminus (S_2 \cup \cdots \cup S_m)) \to H_{p+m}(X \setminus (S_1 \cup \cdots \cup S_m)).$$

By Proposition 16.13, $\gamma \approx 0$ in $X \setminus (S_2 \cup \cdots \cup S_m)$. Considering $S_1, \ldots, S_m$ in another order, we get (16.18) for $j = 2, \ldots, m$. □

In the general case the conditions (16.18) are not sufficient for the condition

$$\{\gamma\} \in \delta^m H_p(S_1 \cap \cdots \cap S_m)$$

to hold.

EXAMPLE 3. We take $X = \{z \in \mathbf{C}^n \colon |z| > 1\}$ and $S_j = \{z \colon z_j = 0\}$, $j = 1, \ldots, n$. Then the cycle $\gamma = \{z \colon |z_1| = \cdots = |z_n| = 2\}$ satisfies the conditions (16.18), but $\{\gamma\} \notin \delta^n H_0(S_1 \cap \cdots \cap S_n)$, since $S_1 \cap \cdots \cap S_n = \varnothing$, while $\gamma \not\approx 0$ in $X \setminus (S_1 \cup \cdots \cup S_n)$.

However, we do have the following result, for example.

PROPOSITION 16.15 (Yuzhakov). *If X is a Stein manifold of dimension n and*

$$\gamma \in Z_{p+m}(X \setminus (S_1 \cup \cdots \cup S_m)), \qquad m + p \geqslant n,$$

then (16.18) *is equivalent to the condition*

$$\{\gamma\} \in \delta^m H_p(S_1 \cap \cdots \cap S_m).$$

Under certain assumptions on X and $S_1, \ldots, S_m$ the structure of the group

$$H_p(X \setminus (S_1 \cup \cdots \cup S_m))$$

is described by the following *decomposition theorem*.

THEOREM 16.16 (Froissart). *Let $S_0, \ldots, S_m$ and $\Sigma_1, \ldots, \Sigma_k$ be two families of $(n-1)$-dimensional complex analytic submanifolds in general position in the n-dimensional complex projective space CP^n, where S_0 is a complex hyperplane in CP^n identified with the hyperplane $CP^{n-1}_\infty = CP^n \setminus C^n$ at infinity. Let $X = \Sigma_1 \cap \cdots \cap \Sigma_k$ and $Y = \mathbf{C}^n \cap X$. Then*

$$\begin{aligned} H_p\left(X \setminus \bigcup_{j=0}^m S_j\right) &= H_p\left(Y \setminus \bigcup_{j=1}^m S_j\right) \\ &\simeq \bigoplus_{h \in \{1, \ldots, m\}} \delta^{|h|} H_{p-|h|}\left(\bigcap_{j \in h} S_j \cap Y\right) \\ &= H_p(Y) \oplus \sum_{j=1}^m \delta h_{p-1}(Y \cap S_j) \oplus \sum_{1 \leqslant i < j \leqslant m} \delta^2 H_{p-2}(Y \cap S_i \cap S_j) \oplus \cdots, \end{aligned}$$

where $|h|$ is the number of elements in a set h.

EXAMPLE 4. Let $Y = \mathbf{C}^n$, and let

$$S_j = \left\{ z \colon \varphi_j(z) = \sum_{k=1}^n a_{jk} z + b_j = 0 \right\}, \qquad j = 1, \ldots, m,$$

be analytic hyperplanes in general position in $\mathbf{C}^n$. The last condition means that if $S_{j_1} \cap \cdots \cap S_{j_k} \neq \varnothing$, then the functions $\varphi_{j_1}, \ldots, \varphi_{j_k}$ are linearly independent. Since $H_n(\mathbf{C}^n) \simeq 0$

$$H_{n-r}(S_{i_1} \cap \cdots \cap S_{i_r}) \simeq 0 \quad \text{for } r < n$$

($S_{i_1} \cap \cdots \cap S_{i_r}$ is a plane of real dimension $2(n-r)$), and $H_0(S_{i_1} \cap \cdots \cap S_{i_n}) \simeq Z$ if $S_{i_1} \cap \cdots \cap S_{i_n} \neq \varnothing$, it follows from Theorem 16.16 that the cycles of the form

$$\delta^n(S_{i_1} \cap \cdots \cap S_{i_n}) = \{z \in \mathbf{C}^n : |\varphi_{i_1}(z)| = \cdots = |\varphi_{i_n}(z)| = \varepsilon\} = \gamma_{i_1 \cdots i_n}$$

make up a basis of the group $H_n(\mathbf{C}^n \backslash S_1 \cup \cdots \cup S_m)$, where ε is sufficiently small, and $\{i_1, \ldots, i_n\}$ is an arbitrary subset of $\{1, \ldots, m\}$ such that the functions $\varphi_{i_1}, \ldots, \varphi_{i_n}$ are linearly independent. The forms

$$\omega_{i_1 \cdots i_n} = \frac{1}{(2\pi i)^n} \frac{d\varphi_{i_1}(z)}{\varphi_{i_1}(z)} \wedge \cdots \wedge \frac{d\varphi_{i_n}(z)}{\varphi_{i_n}(z)}$$

make up a cohomology basis dual to the above basis in the de Rham sense.

It is of interest to generalize Theorem 16.16 to the case when $S_1, \ldots, S_m$ are arbitrary analytic sets in Y. One such generalization is given by the next theorem (see also §18) for the particular case when Y is a domain in $\mathbf{C}^2$. Let $S = \bigcup_1^m S_j$ be an analytic set in $Y \subset \mathbf{C}^2$, let the S_j be irreducible components of S, let $\tilde{S}_j$ be the Riemann surface determined by the component S_j (the Riemann surface of an implicit analytic function whose graph is S_j), and let Q be the set of points at which S is locally reducible. For each point $\mathcal{P} \in Q$ we choose a local system of coordinates in such a way that in this system the boundary $\partial\Delta_{\mathcal{P}}$ of a sufficiently small polycylinder $\Delta_{\mathcal{P}}$ with center at $\mathcal{P}$ intersects each local component of S in a smooth curve that is homeomorphic to a circle. Let $A_{\mathcal{P}}$ denote the set $\partial\Delta_{\mathcal{P}} \cap S$ with one connected component removed, and let $A = \bigcup_{\mathcal{P} \in Q} A_{\mathcal{P}}$.

THEOREM 16.17 (Pedan). *If the homomorphisms* $\bar{\omega}$: $H_{p+1}(Y \backslash Q) \to H_{p-1}(S \backslash Q)$ *are zero* ($\bar{\omega} H_{p+1}(Y \backslash Q) = 0$), *then*

$$H_2(Y \backslash S) = H_2(Y) \oplus \sum_{j=1}^m \delta H_1(\tilde{S}_j) \oplus \delta H_1(A),$$

$$H_1(Y \backslash S) = H_1(Y) \oplus \sum_{j=1}^m \delta H_0(S_j).$$

Here $\bar{\omega}$ *and* δ *are the homomorphisms defined in* 2°.

§17. Cohomological reduction of certain semimeromorphic and rational forms

1°. Theorem 16.5 only proves the existence of the residue class for a form $\varphi \in Z^{p+1}(X \backslash S)$, and does not give constructive algorithms for finding it. The situation with the composite residue class is analogous. Theorem 16.2, Example 1 in §16, and Lemma 16.9 enable us to find the residue class in the case of semimeromorphic forms having first-order poles. Therefore, the problem of determining the residue class of a form $\varphi \in Z^{p+1}(X \backslash S)$ (respectively, the multiple residue class of a form $\varphi \in Z^{p+m}(X \backslash (S_1 \cup \cdots \cup S_m))$) reduces to the problem of determining a form cohomologous to it which has first-order poles on S (respectively, on $S_1, \ldots, S_m$), i.e., to the problem of cohomological reduction of the form φ. It suffices to solve this problem in a neighborhood of S (respectively, $S_1 \cap \cdots \cap S_m$). In a number of cases the problem of cohomological reduction can be solved constructively if φ has poles on S (respectively, $S_1, \ldots, S_m$), i.e., is a semimeromorphic form.

2°. If the submanifold S is given by a global equation, then the following result enables us to realize a lowering of the order of poles.

LEMMA 17.1. *Let $S = \{z: z \in V, s(z) = 0\}$, where V is a neighborhood of S, and $s \in A(V)$,* grad $s \neq 0$. *Then every form $\varphi \in Z^{p+1}(X \setminus S)$ having a pole of order k on S can be represented as*

$$\varphi = \frac{ds}{s^k} \wedge \psi + \frac{\theta}{s^{k-1}}, \tag{17.1}$$

where ψ and θ are regular forms in V.

PROOF. By the assumption, the form $s^k\varphi$ is regular in V. But, since φ is a closed form, $d(s^k\varphi) = ks^{k-1}\,ds \wedge \varphi$. Multiplication of the last form by ds carries it to zero. Consequently, by Lemma 16.3, for any $a \in S$ there is a form θ_a that is regular in some neighborhood U_a of a and such that $s^{k-1}\,ds \wedge \varphi = ds \wedge \theta_a$. The form $s^k\varphi - s\theta_a$ is regular in U_a, and $ds \wedge (s^k\varphi - s\theta_a) = 0$. By Lemma 16.3, there exists a form ψ_a that is regular in U_a and such that $s^k\varphi - s\theta_a = ds \wedge \psi_a$, or

$$\varphi = \frac{ds}{s^k} \wedge \psi_a + \frac{\theta_a}{s^{k-1}}. \tag{17.2}$$

In a neighborhood $U_a \subset V \setminus S$ of a point $a \in V \setminus S$ we represent φ in the form (17.2), where $\psi_a = 0$ and $\theta_a = \varphi s^{k-1}$. We take a partition of unity $\{\chi_a\}_{a \in V}$ subordinate to the covering $\{U_a\}_{a \in V}$ and construct forms $\psi = \Sigma_{a \in V} \chi_a \psi_a$ and $\theta = \Sigma_{a \in V} \chi_a \theta_a$ that are regular in V. For these forms (17.1) follows from (17.2). □

The form (17.1) can be transformed to the expression

$$\varphi = d\left(-\frac{\psi}{(k-1)s^{k-1}}\right) + \frac{\theta_1}{s^{k-1}},$$

where $\theta_1 = d\psi/(k-1) + \theta$. Thus, in $V \setminus S$ the form φ is cohomologous to the form θ_1/s^{k-1}, which has a pole of order $k-1$ on S. By induction φ can be cohomologically reduced in $V \setminus S$ to a form having a first-order pole on S.

3°. The *Gel'fand-Shilov construction* enables us to find the composite residue class of certain semimeromorphic closed forms.

THEOREM 17.2 (Gel'fand, Shilov and Leray). *Let* $S_j = \{z \in X\colon s_j(z) = 0\}$, $j = 1,\ldots,m$, *be submanifolds in general position and given by global equations, and* $\omega(z)$ *a regular form on* X *satisfying the condition*

$$ds_1(z) \wedge \cdots \wedge ds_m(z) \wedge d\omega(z) = 0 \tag{17.3}$$

in a neighborhood of $S = S_1 \cap \cdots \cap S_m$. *Then there exist forms* $\omega_1,\ldots,\omega_m, \omega_{11},\ldots,\omega_{1m}, \omega_{21},\ldots,\omega_{j\cdots k}$ $(j, k = 1,\ldots,m)$ *that are regular in a neighborhood of* S *and such that*

$$\begin{aligned} d\omega &= ds_1 \wedge \omega_1 + \cdots + ds_m \wedge \omega_m; \\ d\omega_1 &= ds_1 \wedge \omega_{11} + \cdots + ds_m \wedge \omega_{1m}; \\ &\cdots\cdots\cdots\cdots \\ d\omega_{p\cdots q} &= ds_1 \wedge \omega_{p\cdots q,1} + \cdots + ds_m \wedge \omega_{p\cdots q,m}. \end{aligned} \tag{17.4}$$

Moreover,

$$\frac{1}{r_1!\cdots r_m!}\omega_{j\cdots k}|_S \in \operatorname{Res}\left[\frac{ds_1 \wedge \cdots \wedge ds_m \wedge \omega}{s_1^{r_1+1}\cdots s_m^{r_m+1}}\right], \tag{17.5}$$

where r_i *is the number of indices* i $(i = 1,\ldots,m)$ *contained among* $j,\ldots,k$.

LEMMA 17.3. *If* φ *is a regular form on* X *and satisfies the condition* $ds_1 \wedge \cdots \wedge ds_m \wedge \varphi = 0$ *in a neighborhood of* $S = S_1 \cap \cdots \cap S_m$, *then in this neighborhood there exist regular forms such that*

$$\varphi(z) = ds_1(z) \wedge \psi_1(z) + \cdots + ds_m(z) \wedge \psi_m(z). \tag{17.6}$$

PROOF. In a neighborhood U_a of an arbitrary point $a \in S$ we choose local coordinates $z_1,\ldots,z_n$ in such a way that $S_j \cap U_a = \{z\colon z_j = 0\}$, $j = 1,\ldots,m$. Then in U_a the form φ can be represented as

$$\varphi(z) = \psi_0(a, z) + dz_1 \wedge \psi_1(a, z) + \cdots + dz_m \wedge \psi_m(a, z), \tag{17.7}$$

where ψ_0 is a form that does not contain $dz_1,\ldots,dz_m$. By the assumption, $dz_1 \wedge \cdots \wedge dz_m \wedge \varphi = dz_1 \wedge \cdots \wedge dz_m \wedge \psi_0 = 0$, and so $\psi_0 = 0$. Using a partition of unity subordinate to the covering $\{U_a\}_{a\in S}$, we pass from the local representation (17.7), where $\psi_0 = 0$, to the global representation (17.6) in the neighborhood $V = \bigcup_{a\in S} U_a$ of the submanifold S. □

PROOF OF THEOREM 17.2. Note that condition (17.3) gives us immediately that the form $ds_1 \wedge \cdots \wedge ds_m \wedge \omega/s_1^{r_1+1}\cdots s_m^{r_m+1}$ is closed. From Lemma 17.3 and condition (17.3) it follows that there are forms $\omega_1,\ldots,\omega_m$ such that the first of the equalities (17.4) holds. Differentiating it, we get that

$$ds_1 \wedge d\omega_1 + \cdots + ds_m \wedge d\omega_m = 0.$$

Hence, $ds_1 \wedge \cdots \wedge ds_m \wedge d\omega_j = 0$. According to Lemma 17.3, there exist forms $\omega_{j1}, \ldots, \omega_{jm}$ such that $d\omega_j = ds_1 \wedge \omega_{j1} + \cdots + ds_m \wedge \omega_{jm}$, and so on. Thus, (17.4) is obtained successively.

The relation (17.5) will be proved by induction on the number of indices $j \cdots k$, which is equal to $r_1 + \cdots + r_m$. For $r_1 = \cdots = r_m = 0$ we have (see §16, 5°)

$$\omega|_S = \operatorname{res}^m\left[\frac{ds_1 \wedge \cdots \wedge ds_m \wedge \omega}{s_1 \cdots s_m}\right] \in \operatorname{Res}^m\left[\frac{ds_1 \wedge \cdots \wedge ds_m \wedge \omega}{s_1 \cdots s_m}\right].$$

Suppose that (17.5) holds for $r_1, \ldots, r_m$; we prove it for $r_1, \ldots, r_j + 1, \ldots, r_m$. Observing that

$$\frac{ds_1 \wedge \cdots \wedge ds_m \wedge \omega}{s_1^{r_1+1} \cdots s_j^{r_j+2} \cdots s_m^{r_m+1}} = \frac{(-1)^j}{r_j + 1} d\left(\frac{ds_1 \wedge \cdots [j] \cdots \wedge ds_m \wedge \omega}{s_1^{r_1+1} \cdots s_j^{r_j+1} \cdots s_m^{r_m+1}}\right)$$
$$+ \frac{(-1)^{m-j}}{r_j + 1} \frac{ds_1 \wedge \cdots [j] \cdots \wedge ds_m \wedge d\omega}{s_1^{r_1+1} \cdots s_j^{r_j+1} \cdots s_m^{r_m+1}}$$
$$\approx \frac{ds_1 \wedge \cdots \wedge ds_m \wedge \omega_j}{(r_j + 1) s_1^{r_1+1} \cdots s_j^{r_j+1} \cdots s_m^{r_m+1}},$$

we get by the induction hypothesis that

$$\operatorname{Res}\left[\frac{ds_1 \wedge \cdots \wedge ds_m \wedge \omega}{s_1^{r_1+1} \cdots s_j^{r_j+2} \cdots s_m^{r_m+1}}\right] = \frac{1}{r_j + 1} \operatorname{Res}\left[\frac{ds_1 \wedge \cdots \wedge ds_m \wedge \omega_j}{s_1^{r_1+1} \cdots s_j^{r_j+1} \cdots s_m^{r_m+1}}\right]$$
$$= \frac{1}{r_1! \cdots (r_j + 1)! \cdots r_m!} \omega_{jp \cdots q}|_S,$$

where the indices $1, \ldots, j, \ldots, m$ are encountered $r_1, \ldots, r_j + 1, \ldots, r_m$ times, respectively, among the indices $p \cdots q$.(⁴) □

EXAMPLE 1. If $X = \mathbf{C}^n$, $m = n$, $S_j = \{z\colon z_j = 0\}$, and $\omega(z) = f(z) \in A(\mathbf{C}^n)$, then

$$\omega_{p \cdots q} = \frac{\partial^{r_1 + \cdots + r_n} f(z)}{\partial z_1^{r_1} \cdots \partial z_m^{r_m}}.$$

The devices given above for lowering the order of the poles of a closed semimeromorphic form are not completely constructive, since the determination of the representations (17.1) and (17.6) is based on a partition of unity.

(⁴)The partial derivative apparatus worked out by Leray and Norguet for exterior differential forms provides an equivalent method; see [113] (§§7, 46 and 47) and [316].

4°. For rational forms in $\mathbf{C}^n$ the problem of cohomological reduction by an analogous scheme can be solved with the use of the Hilbert Nullstellensatz.

THEOREM 17.4 (Leĭnartas and Yuzhakov). *Let $Q_1(z),\dots,Q_m(z)$ be irreducible polynomials in $\mathbf{C}^n$, and let $S_j = \{z \in \mathbf{C}^n\colon\ Q_j(z) = 0\}$, $j = 1,\dots,m$, be manifolds in general position. Then the form*

$$\omega = \frac{P(z)\,dz}{Q_1^{r_1}\cdots Q_m^{r_m}}, \tag{17.8}$$

where P is a polynomial (entire function) and $dz = dz_1 \wedge \cdots \wedge dz_n$, is cohomologous in $\mathbf{C}^n\setminus(S_1 \cup \cdots \cup S_m)$ to a form

$$\omega^* = \sum_{\mathcal{J}} \frac{P_{\mathcal{J}}\,dz}{Q_{j_1}\cdots Q_{j_k}}, \tag{17.9}$$

where $\mathcal{J} = \{j_1,\dots,j_k\} \subset \{1,\dots,m\}$, $k \leqslant n$, and the $P_{\mathcal{J}}$ are polynomials (entire functions).

PROOF. A. Suppose that $m \leqslant n$. Since $S_1,\dots,S_m$ are in general position, the matrix $\|\partial Q_j/\partial z_k\|_{j=1,\dots,m;\,k=1,\dots,n}$ has rank m at points in $S_1 \cap \cdots \cap S_m$. Consequently, the polynomials $Q_1,\dots,Q_m$ and $\partial(Q_1,\dots,Q_m)/\partial(z_{j_1},\dots,z_{j_m})$, $1 \leqslant j_1 < \cdots < j_m \leqslant n$, do not have common zeros in $\mathbf{C}^n$. By the Hilbert Nullstellensatz (see [48], Chapter XI, §79), there exist polynomials H_j and $H_{\mathcal{J}}$, $\mathcal{J} = \{j_1,\dots,j_m\}$, such that

$$1 \equiv \sum_{j=1}^{n} H_j Q_j + \sum_{\mathcal{J}} H_{\mathcal{J}} \frac{\partial(Q_1,\dots,Q_m)}{\partial(z_{j_1},\dots,z_{j_m})}. \tag{17.10}$$

Substituting (17.10) into (17.8), we get

$$\begin{aligned}\omega = &\sum_{j=1}^{m} \frac{P_j\,dz}{Q_1^{r_1}\cdots Q_j^{r_j-1}\cdots Q_m^{r_m}}\\ &+ \sum_{\mathcal{J}} \frac{(-1)^{\varepsilon[\mathcal{J}]} P_{\mathcal{J}}\,dQ_1 \wedge \cdots \wedge dQ_m \wedge dz[\mathcal{J}]}{Q_1^{r_1}\cdots Q_m^{r_m}},\end{aligned} \tag{17.11}$$

where $P_j = H_j \cdot P$, $P_{\mathcal{J}} = H_{\mathcal{J}} \cdot P$, $dz[\mathcal{J}] = dz_1 \wedge \cdots [j_1,\dots,j_m] \cdots \wedge dz_n$, and $\varepsilon[\mathcal{J}]$ is the parity of the permutation $(j_1,\dots,j_m, 1,\dots[j_1,\dots,j_m]\dots,n)$ of

$(1,\dots,n)$. If at least one exponent r_j is greater than 1, then it can be shown just as in 3° that

$$\frac{P_{\mathfrak{J}}\,dQ_1\wedge\cdots\wedge dQ_m\wedge dz[\mathfrak{J}]}{Q_1^{r_1}\cdots Q_m^{r_m}} \approx \frac{(-1)^{j-1}\,dP_{\mathfrak{J}}\wedge dQ_1\wedge\cdots[j]\cdots\wedge dQ_m\wedge dz[\mathfrak{J}]}{(r_j-1)Q_1^{r_1}\cdots Q_j^{r_j-1}\cdots Q_m^{r_m}}. \tag{17.12}$$

If we successively lower the degrees of the denominators in this way, we arrive at a form like (17.9) in finitely many steps.

B. If $m>n$, then the assumption of general position implies that $V_1\cap\cdots\cap V_m=\varnothing$, i.e., the polynomials $Q_1^{r_1},\dots,Q_m^{r_m}$ do not have common zeros in $\mathbf{C}^n$. By the Hilbert theorem, there then exist polynomials $H_1(z),\dots,H_m(z)$ such that

$$1\equiv H_1(z)Q_1^{r_1}(z)+\cdots+H_m(z)Q_m^{r_m}(z).$$

Substituting this into (17.8), we represent (17.8) as

$$\omega=\sum_{j=1}^{m}\frac{P_j\,dz}{Q_1^{r_1}\cdots[j]\cdots Q_m^{r_m}},$$

where $P_j=H_j\cdot P$. Applying this device $m-n$ times, we come to case A. □

REMARK. Since the Hilbert theorem has a constructive proof, the polynomials H_j, $H_{\mathfrak{J}}$, and, consequently, the form (17.9) can be determined effectively. In essence, part B presents a device for expanding rational functions in partial fractions. In the general case this problem is solved by

THEOREM 17.5 (Leĭnartas). *Every rational function of n variables can be represented as a sum of fractions whose denominators contain at most n irreducible factors each.*

EXAMPLE 2. In $\mathbf{C}^2$ consider the form

$$\omega=dw\wedge dz/w^3(wz-1)(w^2+z^2-1)^2.$$

Taking into account that $1\equiv wz-1(wz-1)$, we apply the arguments in B successively. This gives us

$$\omega=\frac{z^3\,dw\wedge dz}{(wz-1)(w^2+z^2-1)}-\frac{dw\wedge dz}{w^3(w^2+z^2-1)^2}-\frac{z\,dw\wedge dz}{w^2(w^2+z^2-1)^2}-\frac{z^2\,dw\wedge dz}{w(w^2+z^2-1)}.$$

Using the identities

$$1 \equiv w \cdot w - 1 \cdot (w^2 + z^2 - 1) + \frac{z}{2} \cdot 2z,$$

$$1 \equiv -\tfrac{4}{3}(1 + wz)(wz - 1) + \tfrac{1}{3}(w^2 + z^2 + 1)(w^2 + z^2 - 1) + \tfrac{1}{6}(w^2 - z^2)(2z^2 - 2w^2),$$

where

$$2z = \frac{\partial(w, w^2 + z^2 - 1)}{\partial(w, z)}, \qquad 2z^2 - 2w^2 = \frac{\partial(wz - 1, w^2 + z^2 - 1)}{\partial(w, z)},$$

along with (17.12), we find that

$$\omega \approx \frac{\partial w \wedge dz}{4w(w^2 + z^2 - 1)} + \frac{(5z^3 - 6z^5)\, dw \wedge dz}{6(wz - 1)(w^2 + z^2 - 1)}.$$

Theorem 17.4 can be generalized to meromorphic forms in domains of holomorphy.

THEOREM 17.6. *Suppose that P and $Q_1, \ldots, Q_n$ are holomorphic functions in the domain of holomorphy $D \subset \mathbf{C}^n$, and that $S_j = \{z \in D\colon Q_j(z) = 0\}$, $j = 1, \ldots, m$, are manifolds in general position. Then a form like* (17.8) *is cohomologous in $D \setminus (S_1 \cup \cdots \cup S_m)$ to a form like* (17.9) *with $P_{\mathfrak{J}} \in A(D)$.*

The proof is similar to that of Theorem 17.4, except that instead of the Hilbert Nullstellensatz we use the following assertion (see [201], p. 296): if the functions $Q_1, \ldots, Q_m$ are holomorphic in a domain of holomorphy $D \subset \mathbf{C}^n$ and do not have common zeros in D, then there exist functions $P_1, \ldots, P_m$ holomorphic in D such that $1 \equiv P_1 Q_1 + \cdots + P_m Q_m$. □

In complex projective space CP^n we have

THEOREM 17.7 (Griffiths). *Let $Q(z)$ be an irreducible polynomial in $z = (z_1, \ldots, z_n)$, and let $S = \{z \in CP^n\colon Q(z) = 0\}$ be a manifold. Then it is possible to cohomologically reduce any form ω rational in $CP^n \setminus S$ to a rational form like $\omega_1 = P\, dz/Q^k$, where $k \leqslant n - E(n/q)$, $q = \deg Q$, $E(a)$ being the integer part of a.*

§18. Residues of rational functions of two variables

1°. Let us apply the method described in §14 to integrals of the form

$$\iint_\gamma \frac{P(w, z)}{Q(w, z)}\, dw \wedge dz, \tag{18.1}$$

where P and Q are polynomials in two variables (it can be assumed that P is an entire function), $\gamma \in Z_2(\mathbf{C}^n \setminus T)$, and $T = \{(w, z) \in \mathbf{C}^2\colon Q(w, z) = 0\}$. We find the dimension of the isomorphic homology groups $H_2(\mathbf{C}^2\setminus T) \simeq H_1(\dot{T})$, where $\dot{T} = T \cup \{\infty\}$ is the closure of the singular set T in the space $\mathbf{C}^2$, completed to form the sphere $S^4 = \dot{\mathbf{C}}_2 = \mathbf{C}^2 \cup \{\infty\}$ by adding a single point ∞ at infinity (see §14), and we determine dual (in the Alexander-Pontryagin sense) bases for these groups. We also compute the residues (periods) of the form $P\,dw \wedge dz/Q$ with respect to the basis cycles.

2°. The genus of a Riemann surface. Let us first recall some facts about Riemann surfaces of algebraic functions (see [198] and [66]). Suppose that $Q(w, z) = a_0(z)w^n + a_1(z)w^{n-1} + \cdots + a_n(z)$ is an irreducible polynomial of degree n with respect to the variable w. Its discriminant is $D(z) = R(Q, Q'_w) \not\equiv 0$. The equation

$$Q(w, z) = 0 \tag{18.2}$$

determines in the domain $G = \{z \in \mathbf{C}^1\colon\ D(z) \neq 0\}$ an n-valued analytic function $w = \varphi(z)$, called an *algebraic function*. At the points of the set $\mathfrak{S} = \{z\colon\ D(z) = 0\} \cup \{\infty\} = \{z_1, \ldots, z_N, \infty\} \in \overline{\mathbf{C}}$ (where $\overline{\mathbf{C}}$ is the Riemann sphere) this function can only have poles or algebraic branch points.

We take an arbitrary point $z_0 \in G$ and make cuts on $\overline{\mathbf{C}}$ along segments joining Z_0 to the points of $\mathfrak{S}$ (it can be assumed that no two points of $\mathfrak{S}\setminus\{\infty\}$ lie on a single ray emanating from z_0). This gives a simply connected domain D. Adjoining to it the sides of the cuts, we get a simply connected closed $2(N+1)$-gon $\overline{D}$. The algebraic function $w = \varphi(z)$ splits into n branches $\varphi_1(z), \ldots, \varphi_n(z)$, each holomorphic and single-valued in D and continuously extendible to $\overline{D}$ if its values are taken in $\overline{\mathbf{C}}$. We assign to each branch φ_j its own copy $\overline{D}_j$ $(j = 1, \ldots, n)$ of the polygon $\overline{D}$ and identify pairwise the sides (edges of the cuts) of these copies on which the values of the corresponding branches coincide. This gives an n-sheeted Riemann surface S that is a 2-dimensional compact connected orientable manifold on which $w = \varphi(z)$ is single-valued.

A point of the Riemann surface S lying over a point $a \in \mathfrak{S}$ is called a *branch point of order* k if vertices of k distinct polygons $\overline{D}_j$ are identified there. In a neighborhood of such a point the algebraic function $w = \varphi(z)$ can be represented as a series

$$w = \sum_{m=m_0}^{\infty} c_m (z-a)^{m/k},$$

where m_0 is an integer, positive or negative. This expansion can be found with the help of a Newton diagram (see [198]).

We determine the genus of S. The polygons $\overline{D}_j$ $(j = 1, \ldots, n)$ form a cell complex of the manifold S. There are $\alpha_2 = n$ two-dimensional cells, $\alpha_1 = n(N+1)$ one-dimensional cells, and $\alpha_0 = r + n$ zero-dimensional cells, where

r is the number of points of S lying over points of $\mathfrak{S}$ (exactly n points of S correspond to z_0). Since the zero-dimensional and two-dimensional Betti numbers of S satisfy $p_0 = p_2 = 1$, the one-dimensional Betti number of S, which is twice its genus, can be found from the Euler-Poincaré formula (0.9):

$$\begin{aligned} p_1 = 2\rho &= p_0 + p_2 - \alpha_0 + \alpha_1 - \alpha_2 \\ &= 2 - r + (N+1)n - 2n. \end{aligned} \tag{18.3}$$

Taking into account that the sum of the orders of the branch points of S lying over a single point of $\mathfrak{S}$ is equal to n, we get that

$$n(N+1) - r = \sum_{j=1}^{r} (k_j - 1), \tag{18.4}$$

where the k_j are the orders of the corresponding branch points. From (18.3) and (18.4) we get

THEOREM 18.1 (Riemann). *The genus of the Riemann surface of an algebraic function is given by the formula*

$$\rho = \frac{\omega}{2} - n + 1, \tag{18.5}$$

where n is the number of sheets, and $\omega = \sum_{j=1}^{r}(k_j - 1)$ is the ramification index of the Riemann surface.

3°. The dimension of the two-dimensional homology group of the complement of an algebraic curve in $\mathbf{C}^2$. We return to the integral (18.1). The denominator Q can be represented in the form $Q = Q_1^{r_1} \cdots Q_m^{r_m}$, where the $Q_1, \ldots, Q_m$ are irreducible polynomials.

THEOREM 18.2 (Yuzhakov). *The dimension of the groups $H_1(\dot{T})$ and $H_2(\mathbf{C}^2 \setminus T)$ is given by*

$$p = \sum_{j=1}^{m} 2\rho_j + \sum_{i=1}^{s} (q_i - 1) + l - m, \tag{18.6}$$

where ρ_j is the genus of the Riemann surface of the algebraic function determined by the equation $Q_j(w, z) = 0$, $j = 1, \ldots, m$, q_i is the number of irreducible components of the analytic set T at the self-intersection point A_i of it, s is the number of self-intersection points of T, and l is the number of elements of T at infinity (the number of connected components of the set $T \cup \{|z|^2 + |w|^2 > R\}$ for R sufficiently large).

PROOF. The set $\dot{T} = \{z\colon Q(z) = 0\} \cup \{\infty\}$ is homeomorphic to the topological sum S of the Riemann surfaces S_j of the algebraic functions $w = \varphi_j(z)$

determined by the equations([5]) $Q_j(w, z) = 0, j = 1, \ldots, m$, with certain points identified. Namely: 1) the points of the Riemann surfaces S_j that have a common projection on the z-plane and at which the functions $w = \varphi_j(z)$ take a single value $\neq \infty$ are identified with a single self-intersection point (A_i) of $\dot{T}$, with q_i points of S identified at the point $A_i \in \dot{T}$; and 2) the points of the surfaces S_j $(j = 1, \ldots, m)$ lying over the point $z = \infty$ of the Riemann sphere, along with the poles of the functions $\varphi_j(z)$ (l points in all) are identified with the single point $\infty \in \dot{T}$. In an arbitrary triangulation of the polyhedra S and $\dot{T}$ that includes the identified points as vertices, the numbers α_2 and α_1 of two-dimensional and one-dimensional simplexes do not change in passing from S to $\dot{T}$, and the number α_0 of vertices decreases by $\sum_{i=1}^{s}(q_i - 1) + l - 1$. The Betti numbers of S are $p_0 = m$ (the number of connected components of S), $p_1 = \sum_{j=1}^{m} 2\rho_j$, and $p_2 = m$. For $\dot{T}$ we have $p_0 = 1$, because all the components $T_j = \{(w, z)\colon\ Q_j(w, z) = 0\} \cup \{\infty\}$, $j = 1, \ldots, m$, have the common point $\infty \in \dot{\mathbf{C}}^2$, and $p_2 = m$. From this we get formula (18.6) for the one-dimensional Betti number of $\dot{T}$ from the Euler-Poincaré formula (0.9). According to Theorem 14.1, the two-dimensional Betti number of the domain $\mathbf{C}^2 \setminus T = \dot{\mathbf{C}}^2 \setminus \dot{T}$ is also equal to (18.6). □

Formula (18.6) can be given another form convenient for computations. Let $Q_*(w, z) = Q_1(w, z) \cdots Q_m(w, z)$. Then $T = \{Q(w, z) = 0\} = \{Q_*(w, z) = 0\}$. Let n be the degree of the polynomial Q_* in the variable w; the roots of the discriminant $D_* = R(Q_*, \partial Q_*/\partial w)$ of Q_* are denoted by $z_1, \ldots, z_N$; r_j is the number of distinct finite roots of the equation $Q_*(w, z_j) = 0, j = 1, \ldots, N$.

PROPOSITION 18.3. *The dimension of the homology groups $H_1(\dot{T}) \simeq H_2(\mathbf{C}^2 \setminus T)$ is determined by the formula*

$$p = (N - 1)n - \sum_{j=1}^{N} r_j + m. \tag{18.7}$$

PROOF. Let us represent the ρ_j $(j = 1, \ldots, m)$ in (18.6) by (18.5). This gives

$$p = \sum_r (k_r - 1) - 2\sum_{j=1}^{m} n_j + 2m + \sum_{i=1}^{s}(q_i - 1) + l - m,$$

or

$$p = \sum_r (k_r - 1) + \sum_{i=1}^{s}(q_i - 1) + l - 2n + m, \tag{18.8}$$

([5])It can be assumed that $Q_j(w, z) \not\equiv z$. Otherwise, we make a linear change of variables.

where n_j is the degree of the polynomial $Q_j(w, z)$ in the variable w, $\Sigma_1^m n_j = n$, the k_r are the orders of the branch points of the Riemann surfaces S_j ($j = 1,\ldots,m$), and the summation in Σ_r is over all branch points. The branch points of the surfaces S_j, the self-intersection points of the algebraic curve T, and the poles of the functions $w = \varphi_j(z)$ can lie only over those points of the z-plane at which $D_*(z) = 0$. If q ($q \geqslant 1$) points of the set $S = \bigcup_1^m S_j$ with orders $k_{r_1},\ldots,k_{r_q}$ are identified at the point (z_j, w_{ji}), then w_{ji} is a root of the equation $Q_*(w, z_j) = 0$ with multiplicity $\beta_{ji} = \Sigma_{t=1}^q k_{r_t}$. To it corresponds the term

$$\sum_{t=1}^{q} (k_{r_t} - 1) + (q - 1) = \beta_{ji} - 1$$

in the sum (18.8). To the point at infinity $(\infty) \in \dot{T} \subset \dot{\mathbf{C}}^2$ there corresponds the term

$$\sum_{t=1}^{l} (k_{s_t} - 1) + l = \sum_{t=1}^{l} k_{s_t} = \sum_{j=1}^{N} \delta_j + n$$

in the sum (18.8), where the k_{q_t} are the orders of branching of the poles and of the points of the surfaces S_j ($j = 1,\ldots,m$) lying over the point $z = \infty \in \overline{\mathbf{C}}$, δ_j is the multiplicity of the infinite root of the equation $Q_*(w, z_j) = 0$, i.e., the difference between n and the degree of $Q_*(w, z_j)$, which difference is equal to the sum of the orders of branching of the poles lying over z_j, and n is the sum of the orders of the branch points lying over the point $z = \infty$, with the first-order branch points included. Thus,

$$p = \sum_{j=1}^{N} \sum_{i=1}^{r_j} (\beta_{ji} - 1) + \sum_{j=1}^{N} \delta_j - n + m.$$

Taking into account that $\Sigma_{i=1}^{r_j} \beta_{ji} + \delta_j = n$, we get (18.7). □

4°. Dual homology bases. We construct a two-dimensional homology basis for a domain $\mathbf{C}^2 \setminus T$ that is a domain of regularity of the form $P\,dw \wedge dz/Q$, along with a one-dimensional homology basis dual to it in the Alexander-Pontryagin sense for the singular set $\dot{T}$.

1. On each irreducible component $T_j = \{Q_j(w, z) = 0\}$ of the algebraic curve T we construct $2p_j$ one-dimensional cycles σ_{js}^1, $s = 1,\ldots,2p_j$, $j = 1,\ldots,m$, corresponding to the "canonical" cuts of the two-dimensional manifold S_j (see Proposition 0.18 about the parallels and meridians of the handles). Here it can be assumed that the cycles $\sigma_{j,2k-1}^1$ and $\sigma_{j,2k}^1$ have a single common point and do not intersect the remaining cycles σ_{js}^1, and also that they do not pass through self-intersection points of $\dot{T}$. We take $\gamma_{js}^1 = \delta_e \omega_{js}^1$, where ω_{js}^1 is obtained from σ_{jr}^1

by a slight shift, $r = s - (-1)^s$, δ_ε is the Leray coboundary operator (see §16), and $\varepsilon > 0$ is sufficiently small. This gives $\Sigma_1^m 2\rho_j$ cycles $\sigma_{js}^1 \in Z_1(\dot{T})$, and accordingly the same number of cycles $\gamma_{js}^1 \in Z_2(\mathbf{C}^2 \setminus T)$, $s = 1, \ldots, 2\rho_j$, $j = 1, \ldots, m$; such cycles are called *canonical*.

2. Suppose that q_j irreducible elements (germs) S_{ji} $(i = 1, \ldots, q_j)$ of the algebraic curve T intersect at the point $A_j \in T$. In $\dot{T}$ we construct a closed oriented curve (one-dimensional cycle) σ_{ji}^2 that goes out from A_j along the element S_{jq_j} and goes into A_j along the element S_{ji} $(i = 1, \ldots, q_j - 1)$. If the elements S_{ji} and S_{jq_j} belong to different irreducible components T_s, $s = 1, \ldots, m$, of T, then the passage of the curve σ_{ji}^2 from one component to another takes place through the point ∞. Since the canonical cuts σ_{js}^1 and ω_{js}^1 do not decompose the surface $\dot{T}_j$ and do not pass through self-intersection points nor the point ∞, it can be assumed that σ_{ji}^2 and ω_{jr}^1 do not have common points. We construct the two-dimensional cycles $\gamma_{ji}^2 = \delta_\varepsilon \omega_{ji}^2 \in Z_2(\mathbf{C}^2 \setminus T)$, where ω_{ji}^2 is a simple closed curve on S_{ji} that encircles the point A_j (note that $S_{ji} \setminus \{A_j\}$ is homeomorphic to a punctured disk). Obviously, the cycles σ_{ji}^2 and ω_{sr}^2 do not have common points if $(j, i) \neq (s, r)$. The cycles σ_{ji}^2 and ω_{ji}^2 can be constructed in such a way that they intersect in a single point.

Thus, we get $\Sigma_{j=1}^k (q_j - 1)$ cycles $\gamma_{ji}^2 \in Z_2(\mathbf{C}^2 \setminus T)$ and accordingly the same number of cycles $\sigma_{ji}^2 \in Z_1(\dot{T})$; these are called *intersection cycles*.

3. Finally, let $\tilde{S}_{j1}, \ldots, \tilde{S}_{jl_j}$ be the elements at infinity of the irreducible component T_j (the connected components of the set $T_j \cap \{|z|^2 + |w|^2 > R^2\}$, where R is a sufficiently large positive number) of the algebraic curve T, with $\Sigma_1^m l_j = l$. On $\dot{T}_j = T_j \cup \{\infty\}$ we construct a closed oriented curve σ_{js}^3 going out from the point ∞ along the element $\tilde{S}_{jl_j}$ and returning to the point ∞ along the element $\tilde{S}_{js}$, $s = 1, \ldots, l_j - 1$, $j = 1, \ldots, m$. The cycle σ_{js}^3 can be constructed in such a way that it does not intersect the cycles ω_{jr}^1 and does not pass through the self-intersection points A_j, i.e., does not intersect the cycles ω_{ji}^2. We take cycles $\gamma_{js}^3 = \delta_\varepsilon \omega_{js}^3 \in Z_2(\mathbf{C}^2 \setminus T)$, where ω_{js}^3 is a closed curve on $\tilde{S}_{js}$ that encircles the point ∞ (the element $\tilde{S}_{js}$ is homeomorphic to a punctured disk), $s = 1, \ldots, l_j - 1$, $j = 1, \ldots, m$. Then the cycles σ_{js}^3 and ω_{ir}^3 do not intersect if $(j, s) \neq (i, r)$. It can be assumed that σ_{js}^3 and ω_{js}^3 intersect only in a single point. We also assume that if σ_{is}^2 passes from the component T_j to the component $\dot{T}_q$, then this takes place through the elements $\tilde{S}_{jl_j}$ and $\tilde{S}_{ql_q}$. Then σ_{js}^2 does not intersect the cycles ω_{jr}^3. Thus, we get $l - m = \Sigma_{j=1}^m (l_j - 1)$ cycles $\gamma_{js}^3 \in Z_2(\mathbf{C}^2 \setminus T)$ and accordingly the same number of cycles $\sigma_{js}^3 \in Z_1(\dot{T})$; these are called *polar cycles*.

THEOREM 18.4 (Yuzhakov). *The canonical cycles γ_{ji}^1, $i = 1, \ldots, 2\rho_j$, $j = 1, \ldots, m$, the intersection cycles γ_{js}^2, $s = 1, \ldots, q_j$, $j = 1, \ldots, k$, and the polar* *cycles γ_{ji}^3, $i = 1,\ldots,l_j - 1$, $j = 1,\ldots,m$, form a basis for the two-dimensional homology of the domain $\mathbf{C}^2\setminus T$, and the cycles $\{\sigma_{ji}^1\}$, $\{\sigma_{ji}^2\}$, $\{\sigma_{ji}^3\}$ form a dual (to within a sign) basis for the one-dimensional homology of the singular set $\dot{T}$.*

PROOF. Let V_ε be the operator analogous to the Leray coboundary operator, with the circle $\delta_\varepsilon a$ of radius ε replaced by a disk of radius ε (see §16). Then $\partial V_\varepsilon \omega_{ji}^r = \delta_\varepsilon \omega_{ji}^r = \gamma_{ji}^r$. From the construction of the cycles $\{\sigma_{ji}^r\}$ and $\{\gamma_{ji}^r\}$ it follows that $|\sigma_{ji}^r| \cap |V_\varepsilon \omega_{sq}^k| = |\sigma_{ji}^r| \cap |\omega_{sq}^k|$ consists of a single point if $(j, i, r) = (s, q, k)$, and is empty if $(j, i, r) \neq (s, q, k)$. Thus,

$$\mathfrak{v}\left(\sigma_{ji}^r, \gamma_{sq}^k\right) = \chi\left(\sigma_{js}^r, V_\varepsilon \omega_{sq}^k\right) = \pm 1$$

for $(j, i, r) = (s, q, k)$, and it equals zero for $(j, i, r) \neq (s, q, k)$. But since the number of cycles $\{\gamma_{ji}^r\}$ (respectively, of cycles $\{\sigma_{ji}^r\}$) is equal to

$$\sum_{j=1}^{n} 2\rho_j + \sum_{i=1}^{s} (k_j - 1) + l - m$$

and, according to Theorem 18.2, coincides with the dimension of the groups

$$H_2(\mathbf{C}^2\setminus T) \simeq H_1(\dot{T}),$$

these cycles form bases of the homology groups

$$H_2(\mathbf{C}^2\setminus T) \quad \text{and} \quad H_1(\dot{T})$$

that are dual in the Alexander-Pontryagin sense. □

5°. Computation of the residues with respect to the basis cycles. Let us compute the integral (18.1) over the basis cycle γ_{jq}^r. It can be assumed that ω_{jq}^r does not pass through the points $(w, z) \in T$ at which $\partial Q_*/\partial w = 0$. Suppose that $\omega_{jq}^r = \{(w, z)\colon w = \psi(t),\ z = \varphi(t),\ 0 \leqslant t \leqslant 1\}$, $\varphi(0) = \varphi(1)$, $\psi(0) = \psi(1)$, and $Q_s(\varphi(t), \psi(t)) \equiv 0$ (for $r = 1$ or 3, $s = j$). Then we can take

$$\gamma_{jq}^r = \delta_\varepsilon \omega_{jq}^r = \left\{(w, z)\colon z = \varphi(t),\ w = \psi(t) + \varepsilon e^{i\tau},\ 0 < t \leqslant 1,\ 0 \leqslant \tau \leqslant 2\pi\right\},$$

where ε is a sufficiently small positive number. Since the cycle ω_{jq}^r does not pass through the singular points of the algebraic curve

$$T = \left\{(w, z)\colon Q_*(w, z) = 0\right\},$$

Theorem 16.10 can be applied to the integral (18.1) with $\gamma = \gamma_{jq}^r$. The result is

$$\iint_{\gamma_{jq}^r} \frac{P(w, z)}{Q(w, z)} dw \wedge dz = 2\pi i \int_{\omega_{jq}^r} \operatorname{Res}\left[\frac{P}{Q} dw \wedge dz\right]. \tag{18.9}$$

The residue class

$$\operatorname{Res}\left[\frac{P}{Q} dw \wedge dz\right]$$

can be found by the method described in §17, 2°. It can also be computed directly as follows.

Since $\partial Q_s/\partial w \neq 0$ on $|\omega_{jq}^r|$, we can choose the system of coordinates (ζ, ω): $\zeta = z$, $\omega = Q_s(w, z)$([6]) in a neighborhood of $|\omega_{jq}^r|$. Then

$$\operatorname{Res}\left[\frac{P}{Q} dw \wedge dz\right] = \operatorname{Res}\left[\frac{F}{\omega^r} d\zeta \wedge d\omega\right] \ni \frac{1}{(r_s - 1)!} \frac{\partial^{r_s - 1} F}{\partial \omega^{r_s - 1}}\bigg|_{\omega=0} d\zeta$$

$$= \frac{1}{(r_q - 1)!} \frac{\partial^{r_s - 1} F}{\partial Q_s^{r_s - 1}}\bigg|_{Q_s(w,z)=0} dz,$$

where $F = PQ_s^{r_s}/Q$. Since $\partial^{r_s - 1}F/\partial Q_s^{r_s - 1}$ is a rational function of w and z, this gives us the following result, according to Theorem 14.2 (on residues).

THEOREM 18.5 (Poincaré). *The integral* (18.1) *over an arbitrary cycle* $\gamma \in Z_2(\mathbf{C}^2 \setminus T)$ *can be expressed in terms of periods of abelian integrals*([7]) *on the Riemann surfaces of the algebraic functions determined by the equations* $Q_j(w, z) = 0$, $j = 1, \ldots, m$, *where the* Q_j *are the irreducible factors of the polynomial* Q.

In the case where γ_{jq}^r is an intersection cycle or a polar cycle the integral (18.9) can be completely computed. Let γ_{jq}^2 be an intersection cycle. Assume that the self-intersection point A_j coincides with the origin. Then in a neighborhood of A_j the element S_{jq} can be given by an equation

$$w = g(z) = \sum_{r=0}^{\infty} c_r z^{r/k}.$$

The cycle ω_{jq}^2 is equal to $\{(w, z)\colon w = g(z),\ z = \delta e^{it},\ 0 \leqslant t \leqslant 2\pi k_q\}$. The integral (17.9) then takes the form

$$\iint_{\gamma_{jq}^2} \frac{P}{Q} dw \wedge dz = k(2\pi i)^2 \operatorname*{res}_{z=0} [\Phi(z)],$$

where

$$\Phi(z) = \frac{1}{(r_s - 1)!} \frac{\partial^{r_s - 1}}{\partial Q_s^{r_s - 1}} F(w, z)\big|_{w=g(z)}.$$

Obviously, $\Phi(z)$ has a pole at $z = 0$.

([6])Although this substitution may not be one-to-one in the large in a neighborhood of the curve ω_{jq}^r, it is locally one-to-one. And since the local coordinate systems coincide in the intersection of two coordinate neighborhoods, it is not necessary to use a partition of unity (cf. §17, Lemma 17.1).

([7])An integral of the form $\int_\gamma R(w, z)\, dz$, where R is a rational function of w and z, and w and z are connected by the relation $Q(w, z) = 0$ for some irreducible polynomial Q, is called an *abelian integral*. The value of an abelian integral over a cycle on a Riemann surface is called its *period*.

The residues with respect to the polar cycles can be computed similarly. If the element $\tilde{S}_{jq}$ corresponds to a finite pole $z = z_0$ of the algebraic function $w = \varphi_j(z)$, then $\tilde{S}_{jq}$ is given by an equation

$$w = g(z) = \sum_{r=-n_0}^{\infty} c_r(z - z_0)^{r/k_q}.$$

If $\tilde{S}_{jq}$ corresponds to the point $z = \infty$, then its equation has the form

$$w = \sum_{r=n_0}^{-\infty} c_r z^{r/k_q}, \qquad |z| > R.$$

Thus, computation of the residue of a rational function with respect to an intersection cycle or a polar cycle reduces to iterated computation of the simple residue with respect to a pole, i.e., it reduces to differentiation.

EXAMPLE. Consider the integral

$$J = \iint_\gamma \frac{P(w, z)\, dw \wedge dz}{w(w^2 - z^3)(w^2 - z^3 - z^4)}, \tag{18.10}$$

where $\gamma \in Z_2(\mathbf{C}^2 \setminus T)$, $T = \{Q(w, z) = 0\}$, and

$$Q(w, z) = w(w^2 - z^3)(w^2 - z^3 - z^4).$$

Each irreducible factor in the denominator determines a Riemann surface of genus zero. The surface T has two self-intersection points: $(0, 0)$ and $(0, -1)$. Three irreducible elements (germs) of the algebraic curve T intersect at the point $(0, 0)$, and two intersect at $(0, -1)$. The connected components $\{w = 0\}$ and $\{w^2 - z^3 = 0\}$ each have one element at infinity, and the component $\{w^2 - z^3 - z^4 = 0\}$ has two of them. Thus, the dimension of the group $H_2(\mathbf{C}^2 \setminus T)$ is equal to four, by (18.6). There are three intersection basis cycles, of which two,

$$\gamma_1 = \{z = \varepsilon e^{it},\ \omega = \delta e^{i\tau},\ 0 \leqslant t, \tau \leqslant 2\pi\},$$

$$\gamma_2 = \{z = \varepsilon e^{it},\ w = \varepsilon^{3/2} e^{3it/2} + \delta e^{i\tau},\ 0 \leqslant t \leqslant 4\pi,\ 0 \leqslant \tau \leqslant 2\pi\},$$

correspond to the point $(0, 0)$, and one,

$$\gamma_3 = \{z = -1 + \varepsilon e^{it},\ w = \delta e^{i\tau},\ 0 \leqslant t, \tau \leqslant 2\pi\},$$

corresponds to the point $(0, -1)$, and there is one polar cycle

$$\gamma_4 = \left\{z = Re^{it},\ w = R^2 e^{2it}\sqrt{1 + R^{-1}e^{-it}} + \delta e^{i\tau},\ 0 \leqslant t, \tau \leqslant 2\pi\right\}.$$

Here ε and δ are sufficiently small numbers, $\delta \ll \varepsilon$, and $R > 1$.

We compute the residues with respect to the basis cycles:

$$R_1 = \frac{1}{(2\pi i)^2} \iint_{\gamma_1} \frac{P}{Q} dw \wedge dz = \frac{1}{2\pi i} \int_{|z|=\varepsilon} \frac{P(0, z)\, dz}{z^3(z^3 + z^4)}$$

$$= \sum_{k=0}^{5} \frac{(-1)^k}{k!} \frac{\partial^k P(0,0)}{\partial z^k};$$

$$R_2 = \frac{1}{(2\pi i)^2} \iint_{\gamma_2} \frac{P}{Q} dw \wedge dz = \frac{1}{2\pi i} \int_{\substack{z=e^{it}\\ 0\leqslant t\leqslant 4\pi}} \frac{P(z^{3/2}, z)\, dz}{z^{3/2} 2 z^{3/2} z^4}$$

$$= \frac{1}{6!} \frac{\partial^6 P(0,0)}{\partial z^6} + \frac{1}{2!3!} \frac{\partial^5 P(0,0)}{\partial w^2 \partial z^3} + \frac{1}{4!} \frac{\partial^4 P(0,0)}{\partial w^4};$$

$$R_3 = \frac{1}{(2\pi i)^2} \iint_{\gamma_3} \frac{P}{Q} dw \wedge dz = \frac{1}{2\pi i} \int_{|z+1|=\varepsilon} \frac{P(0, z)\, dz}{z^3(z^3 + z^4)} = P(0, -1);$$

$$R_4 = \frac{1}{(2\pi i)^2} \iint_{\gamma_4} \frac{P}{Q} dw \wedge dz = \frac{1}{2\pi i} \int_{|z|=R} \frac{P\left(\sqrt{z^3 + z^4}, z\right)}{2(z^3 + z^4) z^4} dz$$

$$= \operatorname*{res}_{\zeta=0} P\left(\sqrt{1+\zeta}/\zeta^2, 1/\zeta\right)\zeta^6/2(\zeta+1).$$

The last residue is easy to compute for a concrete choice of the polynomial P. By Theorem 13.2, the integral (18.1) over an arbitrary cycle γ is equal to $J = (2\pi i)^2 \sum_1^4 n_j r_j$, where the n_j are integers.

6°. In order not to distinguish between intersection cycles and polar cycles, the integral (18.1) can be considered in the complex projective space CP^2. After passage to the homogeneous coordinates (z_0, z_1, z_2): $w = z_1/z_0$, $z = z_2/z_0$ the integral (17.1) takes the form

$$\iint_\gamma \frac{P(z_0, z_1, z_2)}{Q(z_0, z_1, z_2)} (z_0\, dz_1 \wedge dz_2 - z_1\, dz_0 \wedge dz_2 + z_2\, dz_0 \wedge dz_1), \tag{17.11}$$

where $\gamma \in Z^2(CP^2 \backslash T)$, with $T = \{(z_0, z_1, z_2) \in CP^2: Q(z_0, z_1, z_2) = 0\}$.

THEOREM 18.6 (Tsikh). $H_2(CP^2 \backslash T) \simeq H_1(T)$.

THEOREM 18.7. *The one-dimensional Betti number of an algebraic curve T in CP^2, i.e., the dimension of the group $H_1(T)$, is given by*

$$p = \sum_{j=1}^{m} 2\rho_j + \sum_{i=1}^{k} (\alpha_i - 1) - m + 1,$$

where m is the number of distinct irreducible factors $Q_1, \ldots, Q_m$ of the polynomial Q, ρ_j is the genus of the algebraic curve determined by the equation $Q_j(z_0, z_1, z_2) = 0$, k is the number of self-intersection points of T, and α_i is the number of irreducible components (germs) of T in a neighborhood of the self-intersection point A_i.

A basis for the homology group $H_2(CP^2\setminus T)$ can be constructed in the same way as in 3° (see [196]). This gives $\Sigma_1^m 2\rho_j$ canonical cycles and $\Sigma_1^k(\alpha_j - 1) - m + 1$ intersection cycles.

We indicate a curious analogue of Theorems 17.2 and 17.3 for a certain particular case of rational functions of n variables. Let P and Q be polynomials in $z = (z_1, \dots, z_n)$, with

$$Q(z) = Q_2^{r_2}(z_1, z_2) \cdots Q_n^{r_n}(z_1, z_n).$$

We write $T_j = \{z \in \mathbf{C}^n\colon Q_j(z_1, z_j) = 0\}$ and $\dot{T}_j = T_j \cup \{\infty\}$, which is the closure of T_j in the spherical compactification $\dot{\mathbf{C}}^n = \mathbf{C}^n \cup \{\infty\}$ of $\mathbf{C}^n$.

THEOREM 18.8 (Tsikh).

$$H_n(\mathbf{C}^n\setminus(T_2 \cup \cdots \cup T_n)) \simeq H_1(\dot{T}_2 \cap \cdots \cap \dot{T}_n).$$

Here a basis for the group $H_n(\mathbf{C}^n\setminus(T_2 \cup \cdots \cup T_n))$ can be constructed in the same way as in 3°, i.e., one-dimensional cycles are constructed on the complex curve $\dot{T}_2 \cap \cdots \cap \dot{T}_n$, and then the composite Leray coboundary δ^{n-1} is taken. By applying the Leray composite residue formula (16.7) it can be shown that the integral $\int_\gamma (P/Q)\, dz$, where $\gamma \in Z_n(\mathbf{C}^n\setminus(T_2 \cup \cdots \cup T_n))$, can be expressed in terms of periods of abelian integrals on the algebraic curve $T_2 \cap \cdots \cap T_n$ (see [193]).

§19. Local residues of certain meromorphic and rational functions on $\mathbf{C}^n$

1°. Suppose that $\varphi_1, \dots, \varphi_N$ are functions holomorphic in a neighborhood of a point $a \in \mathbf{C}^n$, $\varphi_j(a) = 0$, and

$$\left| \frac{\partial(\varphi_{\alpha_1}, \dots, \varphi_{\alpha_n})}{\partial(z_1, \dots, z_n)} \right|_{z=a} \neq 0$$

for any $\{\alpha_1, \dots, \alpha_n\} \subset \{1, \dots, N\}$. Consider the integral

$$\int_\gamma f(z)\, dz \Big/ \prod_{j=1}^N \varphi_j^{r_j}(z), \tag{19.1}$$

where $\gamma \in Z_n(U\setminus \bigcup_1^N T_j)$, U is a sufficiently small neighborhood of a, $T_j = \{z\colon \varphi_j(z) = 0\}$, and $f \in A(U)$, along with the more general integral

$$\int_\gamma F(z)\, dz, \tag{19.2}$$

where $F \in A(U\setminus \bigcup_1^N T_j)$. We shall use the approach considered in §15 for the study of these integrals. We find dual (in the de Rham sense) bases for the n-dimensional homology and cohomology of the domain $U\setminus \bigcup_1^N T_j$, and compute the residues of the functions $f/\prod_1^N \varphi_j^{r_j}$ with respect to the basis cycles.

2°. Let us first prove some auxiliary theorems on separation of singularities. The following result is a special case of a theorem on the solvability of the first Cousin problem (see [189], Theorem 5.5.1).

THEOREM 19.1. *Let* D_1 *and* D_2 *be domains in* $\mathbf{C}^n$ *such that* $D_1 \cup D_2$ *is a domain of holomorphy. Then every function* $f \in A(D_1 \cap D_2)$ *can be represented in the form* $f = f_1 + f_2$, *where* $f_j \in A(D_j)$, $j = 1, 2$.

Using this, we prove a theorem on separation of singularities lying on analytic sets.

THEOREM 19.2 (Yuzhakov). *Let* G *be a domain of holomorphy in* $\mathbf{C}^n$, *and let* $\varphi_j \in A(G)$, $j = 1, \dots, k$. *If there exist functions* $c_j \in A(G)$, $c_j(z) \neq 0$ *in* G $(j = 1, \dots, k)$, *such that* $c_1\varphi_1 + \cdots + c_k\varphi_k \equiv 0$, *then every function*

$$f \in A\left(G \setminus \cup_1^k T_j\right), \qquad T_j = \{z \in G : \varphi_j(z) = 0\},$$

can be represented in the form $f = f_1 + \cdots + f_{k-1}$, *where*

$$f_j \in A(G \setminus (T_1 \cup \cdots [j] \cdots \cup T_k)), \quad j = 1, \dots, k-1.$$

The proof is by induction on k. For $k = 3$ Theorem 19.2 is a corollary to Theorem 19.1. Indeed, let $D_1 = G \setminus (T_2 \cup T_3)$ and $D_2 = G \setminus (T_1 \cup T_3)$. Then

$$D_1 \cap D_2 = G \setminus (T_1 \cup T_2 \cup T_3),$$

and $D_1 \cup D_2 = G \setminus (T_3 \cup (T_1 \cap T_2)) = G \setminus T_3$ is a domain of holomorphy; here

$$\begin{aligned} T_1 \cap T_2 &= \{z \in G : \varphi_1(z) = \varphi_2(z) = 0\} \subset T_3 \\ &= \{z \in G : \varphi_3(z) = -[c_1(z)\varphi_1(z) + c_2(z)\varphi_2(z)]/c_3(z) = 0\}. \end{aligned}$$

We suppose that the theorem is valid for some $k \geq 3$, and prove it for $k + 1$. If $f \in A(G \setminus \cup_1^{k+1} T_j)$, then f is also holomorphic in the domain

$$G \setminus \left(T'_k \cup \bigcup_{j=1}^{k+1} T_j \right) = G' \setminus (T_1 \cup \cdots \cup T_{k-1} \cup T'_k),$$

where

$$T'_k = \{z \in G : c_k(z)\varphi_k(z) + c_{k+1}(z)\varphi_{k+1}(z) = 0\}, \qquad G' = G \setminus (T_k \cup T_{k+1}).$$

The domain G' and the functions $\varphi_1, \dots, \varphi_{k-1}$ and $c_k\varphi_k + c_{k+1}\varphi_{k+1}$ satisfy the conditions of the theorem for k. By the induction hypothesis, f can be represented in the form $f = g_1 + \cdots + g_{k-1}$, where $g_j \in A(G_j)$,

$$G_j = G' \setminus (T_1 \cup \cdots [j] \cdots \cup T_{k-1} \cup T'_k) = D_j \setminus (T_k \cup T'_k \cup T_{k+1}),$$

with

$$D_j = G \setminus (T'_1 \cup [j] \cdots \cup T_{k-1}).$$

By what was proved for $k = 3$, g_j can be represented in the form $g_j = f_j + \psi_j$, where

$$f_j \in A(D_j \setminus T_k \cup T_{k+1}) = A(G \setminus (T_1 \cup \cdots [j] \cdots \cup T_{k+1})),$$

and $\psi_j \in A(D_j \setminus T'_k \cup T_{k+1})$, $j = 1, \ldots, k-1$. Then the function $f_k = \Sigma_{j=1}^{k-1} \varphi_j = f - \Sigma_{j=1}^{k-1} f_j$ is holomorphic in the domains $G \setminus \bigcup_{j=1}^{k+1} T_j$ and

$$\bigcap_{j=1}^{k-1} (D_j \setminus (T'_k \cup T_{k+1})) = G \setminus (T_1 \cup \cdots \cup T_{k-1} \cup T'_k \cup T_{k+1}),$$

hence, also in their union

$$G \setminus (T_1 \cup \cdots \cup T_{k-1} \cup T_{k+1} \cup (T_k \cap T'_k)) = G \setminus (T_1 \cup \cdots \cup T_{k-1} \cup T_{k+1}).$$

Here $T_k \cap T'_k = \{z \in G\colon \varphi_k(z) = c_k(z)\varphi_k(z) + c_{k+1}(z)\varphi_{k+1}(z) = 0\} \subset T_{k+1} = \{z \in G\colon \varphi_{k+1}(z) = 0\}$. □

3°. We return to the notation of 1°. Let A be the set of multi-indices $\alpha = (\alpha_1, \ldots, \alpha_n)$ with $1 \leqslant \alpha_1 < \alpha_2 < \cdots < \alpha_{n-1} \leqslant N - 1$ and $\alpha_n = N$.

THEOREM 19.3. *The dimension of the groups*

$$H_n\left(U \setminus \bigcup_1^N T_j\right), \quad \text{and} \quad H^n\left(U \setminus \bigcup_1^N T_j\right)$$

is equal to $\binom{N-1}{n-1}$. *The cycles*

$$\gamma_\alpha = \{z \in U\colon |\varphi_{\alpha_1}(z)| = \cdots = |\varphi_{\alpha_{n-1}}(z)| = \delta,$$
$$|\varphi_{\alpha_n}(z)| = \varepsilon\}, \qquad \alpha \in A,$$

where ε *and* δ/ε *are sufficiently small positive numbers, and the forms*

$$\omega_\alpha = \frac{1}{(2\pi i)^n} \frac{d\varphi_{\alpha_1} \wedge \cdots \wedge d\varphi_{\alpha_n}}{\varphi_{\alpha_1} \cdots \varphi_{\alpha_n}}, \qquad \alpha \in A,$$

make up dual (in the de Rham sense) bases for the n-dimensional homology and cohomology of the domain $U \setminus \bigcup_1^N T_j$.

LEMMA 19.4. *For* $N \geqslant n$ *every function* $f \in A(U \setminus \bigcup_1^N T_j)$ *can be represented as* $f = \Sigma_{\alpha \in A} f_\alpha$ *in some neighborhood* V *of* a *with* $V \subset U$, *where* $f_\alpha \in A(V \setminus \bigcup_1^n T_{\alpha_j})$.

PROOF. We use induction on N. For $N = n$ the assertion is trivial. We assume that the lemma is valid for $N - 1 \geqslant n$ and prove it for N. From the assumption that $(\partial \varphi_\alpha / \partial z)|_a \neq 0$ it follows for any subset $\{\alpha_1, \ldots, \alpha_n\} \subset \{1, \ldots, N\}$ that in some neighborhood V of a the function φ_N can be represented in the form $\varphi_N = c_1\varphi_1 + \cdots + c_n\varphi_n$, where $c_j \in A(V)$ and $c_j(z) \neq 0$ in V. Indeed, it suffices to make the substitution $\zeta_j = \varphi_j(z)$, $j = 1, \ldots, n$, expand

φ_N in a series with respect to $\zeta_1,\ldots,\zeta_n$, and group the terms of the series according to ζ_j ($j = 1,\ldots,n$). Since $V \setminus \bigcup_1^N T_j$ is a domain of holomorphy, f can be represented (by Theorem 19.2) as $\sum_1^n f_j$, where

$$f_j \in A(V \setminus T_1 \cup \cdots [j] \cdots \cup T_N), \qquad j = 1,\ldots,n.$$

It now remains to apply the induction hypothesis to the functions f_j. □

REMARK. If f is a meromorphic function, then the functions f_α ($\alpha \in A$) can also be chosen to be meromorphic. For this, we multiply the function $f = F/\prod_1^N \varphi_j^{r_j}$ by $\sum_1^n c_j \varphi_j$ and divide by φ_N, then cancel φ_j in each term. Repeating this finitely many times, we represent f as $\sum_1^n f_j$, where

$$f_j = F_j / \left(\varphi_1^{\beta_1} \cdots [j] \cdots \varphi_N^{\beta_N}\right).$$

PROOF OF THEOREM 19.3. We assume that the neighborhood U is sufficiently small and such that for any $\alpha = (\alpha_1,\ldots,\alpha_n)$ and any $j \neq \alpha_1,\ldots,\alpha_n$ there exist functions $c_k \in A(U)$, $c_k(z) \neq 0$ in $\overline{U}$ ($k = 1,\ldots,n$), for which $\varphi_j = \sum_{k=1}^n c_k \varphi_{\alpha_k}$. For ε and δ/ε sufficiently small we have $|\gamma_\alpha| \subset U \setminus \bigcup_1^N T_j$. Indeed, it is obvious that $|\gamma_\alpha| \cap T_j = \varnothing$ for $j = \alpha_1,\ldots,\alpha_n$. For $j \in \{1,\ldots[\alpha_1,\ldots,\alpha_n],\ldots,N\}$ we have $T_j \cap |\gamma_\alpha| = \varnothing$ if $\delta/\varepsilon < \min_{z \in \overline{U}} \sum_{k=1}^{n-1} |c_k(z)/c_n(z)|$.

By Serre's theorem, every form $\omega \in Z^n(U \setminus \bigcup_1^N T_j)$ is cohomologous to a holomorphic form, $\omega \sim f(z)\,dz$, where $f \in A(U \setminus \bigcup_1^N T_j)$. Lemma 19.4 gives us that $f\,dz = \sum_{\alpha \in A} f_\alpha\,dz$, where $f_\alpha \in A(U \setminus (T_{\alpha_1} \cup \cdots \cup T_{\alpha_n}))$. Exactly as in Example 15.8 it can be shown that $f_\alpha\,dz \sim c_\alpha \omega_\alpha$, where c_α is a constant. For this, it suffices to make the substitution

$$\zeta_j = \varphi_{\alpha_j}(z), \qquad j = 1,\ldots,n, \tag{19.3}$$

and expand f_α in a Laurent series of powers of $\zeta_1,\ldots,\zeta_n$. According to Proposition 15.4, it remains to show that

$$\int_{\gamma_\alpha} \omega_\beta = \begin{cases} 1 & \text{if } \alpha = \beta, \\ 0 & \text{if } \alpha \neq \beta. \end{cases}$$

After the substitution (19.3), we get

$$\int_{\gamma_\alpha} \omega_\alpha = \frac{1}{(2\pi i)^n} \int_{\substack{|\zeta_n| = \varepsilon \\ |\zeta_1| = \cdots = |\zeta_{n-1}| = \delta}} \frac{d\zeta}{\zeta} = 1.$$

For $\beta = (\beta_1,\ldots,\beta_n) \neq \alpha = \{\alpha_1,\ldots,\alpha_n\}$ the cycle γ_α is homologous to 0 in $U \setminus (T_{\beta_1} \cup \cdots \cup T_{\beta_n})$. Indeed, in this case there is a $j \in \{1,\ldots,n-1\}$ such that $\alpha_j \neq \beta_1,\ldots,\beta_n$. Then $\gamma_\alpha = \partial B$, where

$$B = \bigl\{z \in U\colon |\varphi_{\alpha_1}(z)| = \cdots [\alpha_j] \cdots = |\varphi_{\alpha_{n-1}}(z)| = \delta,$$
$$|\varphi_{\alpha_j}(z)| \leqslant \delta,\ |\varphi_{\alpha_n}(z)| = \varepsilon\bigr\}.$$

Moreover, $|B| \subset U \setminus T_{\beta_1} \cup \cdots \cup T_{\beta_n}$, since obviously $T_{\beta_q} \cap |B| = \varnothing$ if $\beta_q = \alpha_k$, $k \neq j$, while if $\beta_q \neq \alpha_1, \ldots, \alpha_n$, then the inequality

$$\varepsilon = |\varphi_{\alpha_n}| > \sum_{k-1}^{n-1} \left| \frac{c_k}{c_n} \right| \delta \geqslant \sum_{k-1}^{n-1} \left| \frac{c_k}{c_n} \right| |\varphi_{\alpha_k}|,$$

holds on $|B|$, i.e., $\varphi_{\beta_q} \neq 0$ and again $T_{\beta_q} \cap |B| = \varnothing$. Here $\varphi_{\beta_q} = c_1\varphi_{\alpha_1} + \cdots + c_n\varphi_{\alpha_n}$, $c_j \in A(U)$, $c_j \neq 0$, $j = 1, \ldots, n$. Then $\int_{\gamma_\alpha} \omega_\beta = 0$ by Corollary 0.4. Clearly, the number of elements in the set A is equal to the number of combinations of $N - 1$ things taken $n - 1$ at a time, i.e., $\binom{N-1}{n-1}$. □

4°. Let us compute the residues of the form $f(z)\,dz$ with respect to the basis cycles γ_α $(\alpha \in A)$. We make the change of variables (19.3) in the neighborhood U of the point a. Then

$$R_\alpha = \frac{1}{(2\pi i)^n} \int_{\gamma_\alpha} f(z)\,dz = \frac{1}{(2\pi i)^n} \int_\Gamma g(\zeta)\,d\zeta,$$

where

$$\Gamma = \{\zeta \in \mathbf{C}^n : |\zeta_1| = \cdots = |\zeta_{n-1}| = \delta, |\zeta_n| = \varepsilon\},$$

$$g(\zeta) = \left[f(z) \Big/ \frac{\partial(\varphi_\alpha)}{\partial(z)} \right] \Bigg|_{z=\varphi_\alpha^{-1}(\zeta)}, \qquad \varphi_\alpha = (\varphi_{\alpha_1}, \ldots, \varphi_{\alpha_n}).$$

Since g is holomorphic in a neighborhood of the n-circular compact set $|\Gamma|$, it admits a Laurent expansion with respect to the variables $\zeta_1, \ldots, \zeta_n$ in this neighborhood. The coefficient of $\zeta^{-1} = \zeta_1^{-1} \cdots \zeta_n^{-1}$ in this expansion is obviously equal to R_α.

In the general case it is a complicated matter to find the inverse of the holomorphic mapping (19.3). Actually, in computing the residue R_α we can confine ourselves to the linear change of variables

$$\zeta_j = \mathcal{L}_{\alpha_j}(z), \qquad j = 1, \ldots, n, \tag{19.4}$$

where $\mathcal{L}_{\alpha_j}$ is the linear part of the function φ_{α_j} at the point a. Indeed, the cycle γ_α is homologous (homotopic) in $U \setminus \bigcup_1^N T_j$ to the cycle

$$\tilde{\gamma}_\alpha = \{z \in U : |\mathcal{L}_{\alpha_1}(z)| = \cdots = |\mathcal{L}_{\alpha_{n-1}}(z)| = \delta, |\mathcal{L}_{\alpha_n}(z)| = \varepsilon\}.$$

Then

$$R_\alpha = \frac{1}{(2\pi i)^n \Delta_\alpha} \int_\Gamma f(\mathcal{L}_\alpha^{-1}(\zeta))\,d\zeta,$$

where $\Delta_\alpha = (\partial(\varphi_\alpha)/\partial(z))|_a$. To find the residue R_α it remains to expand $f(\mathcal{L}_\alpha^{-1}(\zeta))$ in a multiple Laurent series in a neighborhood of the n-circular compact set $|\Gamma|$.

In the case of the meromorphic function $f = F/\prod_1^N \varphi_j^{r_j}$ this coefficient in the Laurent series can be found as follows. After the substitution (19.4) the function $f(\mathfrak{L}_\alpha^{-1}(\zeta))$ takes the form

$$\frac{\psi(\zeta)}{\prod_{j=1}^n\left(\zeta_j + \sum_{|\beta|>1} a_{\alpha_j\beta}\zeta^\beta\right)^{r_{\alpha_j}}\prod_{j=1, j\notin\alpha}^N\left(\sum_{k=1}^n a_{jk}\xi_k + \sum_{|\beta|>1} a_{j\beta}\zeta^\beta\right)^{r_j}}. \qquad (19.5)$$

For ε, δ, and δ/ε sufficiently small on Γ the following inequalities hold:

$$\left|\sum_{|\beta|>1} a_{\alpha_j\beta}\zeta^\beta\right| < |\zeta_j|, \qquad \left|\sum_{k=1}^{n-1} a_{jk}\zeta_j + \sum_{|\beta|>1} a_{j\beta}\zeta^\beta\right| < |a_{jn}\zeta_n|.$$

Thus, on Γ the fractions

$$1/\zeta_j\left(1 + \frac{1}{\zeta_j}\sum_{|\beta|>1} a_{\alpha_j\beta}\zeta^\beta\right),$$

$$1/a_{jn}\zeta_n\left[1 + \frac{1}{a_{jn}\zeta_n}\left(\sum_{|\beta|\geq 1} a_{j\beta}\zeta^\beta + \sum_{k=1}^{n-1} a_{jk}\zeta_j\right)\right]$$

can be expanded in series of geometric progressions. Multiplying out these series and the Taylor series of ψ, we get the Laurent series of the fraction (19.5). We observe that to determine the coefficient of ζ^{-I} in the Laurent series it suffices to have only finitely many terms of the indicated series. It is sometimes convenient to represent the denominator in (19.5) in the form

$$\left(\prod_{j=1[\alpha]}^N a_{jn}^{r_j}\right)\left(\prod_{j=1}^n \zeta_j^{r_j}\right)\zeta_n^{r_1+\cdots[\alpha_1,\ldots,\alpha_n]\cdots+r_N}[1 + g(\zeta)],$$

where

$$g(\zeta) = \sum_{|\beta|>r_1+\cdots+r_N} c_\beta\zeta^\beta \Big/ \left(\prod_{j=1}^N a_{j\alpha}^{r_j}\right)\left(\prod_{j=1}^n \zeta_j^{r_j}\right)\zeta_n^{r_1+\cdots[\alpha]\cdots+r_N}.$$

For ε and δ/ε sufficiently small we have $|g(\zeta)| < 1$ on $|\Gamma|$. Expanding the fraction $1/(1-g)$ in a geometric progression, we find the Laurent series of (19.5) in a neighborhood of $|\Gamma|$.

5°. The methods used here can also be applied in studying the residues of a rational function whose denominator can be decomposed into linear factors. Consider integrals of the form

$$\int_\gamma P(z)\,dz \Big/ \prod_{j=1}^N \mathfrak{L}_j^{r_j}(z) \qquad (19.6)$$

and

$$\int_\gamma f(z)\,dz, \qquad (19.7)$$

where P is a polynomial (or an entire function),

$$\mathfrak{L}_j(z) = \sum_{k=1}^{n} a_{jk} z_k + a_{j0}, \qquad j = 1, \ldots, N, \tag{19.8}$$

are linear functions, and $f \in A(\mathbf{C}^n \setminus \bigcup_1^N E_j)$, where $E_j = \{z : \mathfrak{L}_j(z) = 0\}$.

Suppose that the rank of the matrix

$$\|a_{jk}\|_{j=1,\ldots,N;\,k=1,\ldots,n} \tag{19.9}$$

is equal to r, $r \leqslant n$. Let $\Omega = \{\omega\}$ be the family of maximal subsets $\omega \subset \{1, \ldots, N\}$ of indices such that the rank of the augmented matrix $\|a_{jk}\|_{j\in\omega;\,k=0,1,\ldots,n}$ also is equal to r, i.e., $\bigcap_{j\in\omega} E_j \neq \varnothing$.

LEMMA 19.5. *Every function $f \in A(\mathbf{C}^n \setminus \bigcup_1^N E_j)$ can be represented in the form $f = \sum_{\omega\in\Omega} f_\omega$, where $f_\omega \in A(\mathbf{C}^n \setminus \bigcup_{j\in\omega} E_j)$.*

PROOF. We use induction on N. For $N = r$ the assertion is obvious. Suppose that $N > r$, and assume that the lemma is true for $N - 1$. If the rank of the augmented matrix $\|a_{jk}\|_{j=1,\ldots,N;\,k=0,1,\ldots,n}$ is equal to r, then Ω consists of the lone subset $\omega = \{1, \ldots, N\}$. Suppose that the rank of the augmented matrix is equal to $r + 1$. Then among the functions (19.8) there exist $r + 1$ functions, say $\mathfrak{L}_1, \ldots, \mathfrak{L}_{r+1}$, for which the rank of the augmented matrix is equal to $r + 1$. From them there are numbers $c_1, \ldots, c_{r+1}$ not equal to zero such that $\sum_{j=1}^{r+1} c_j \mathfrak{L}_j \equiv 1$. Setting $G = \mathbf{C}^n \setminus (E_{r+2} \cup \cdots \cup E_N)$, $k = r + 2$, $\varphi_{r+2} \equiv 1$, and $\varphi_j = \mathfrak{L}_j$, $j = 1, \ldots, r + 1$, in Theorem 19.2, we get that each function $f \in A(\mathbf{C}^n \setminus \bigcup_1^N E_j)$ can be represented as $\sum_1^{r+1} f_j$, where $f_j \in A(\mathbf{C}^n \setminus (E_1 \cup \cdots [j] \cdots \cup E_N))$. Applying the induction hypothesis to the f_j and combining the terms corresponding to one and the same $\omega \in \Omega$, we prove the lemma for N. □

We take the system $\{\mathfrak{L}_j\}_{j\in\omega}$, $\omega \in \Omega$ of linear functions, and order the set $\omega = \{j_1, \ldots, j_q\}$, $q = q(\omega)$, in such a way that the system $\{\mathfrak{L}_{j_1}, \ldots, \mathfrak{L}_{j_r}\}$ of functions is linearly independent. Let us call this system a zeroth-step system of the family $\{\mathfrak{L}_j\}_{j\in\omega}$. Obviously, $q \geqslant r$. Linearly independent systems of r functions in $\mathfrak{L}_{j_1}, \ldots, \mathfrak{L}_{j_{r+1}}$ that contain the function $\mathfrak{L}_{j_{r+1}}$ are called first-step systems, and so on. The kth-step systems are obtained by adjoining to each $(k - 1)$st-step system the function $\mathfrak{L}_{j_{r+k}}$ and forming from them all possible systems of r linearly independent functions that contain $\mathfrak{L}_{j_{r+k}}$. The $(q - r)$th-step systems $\{\mathfrak{L}_{\alpha_1}, \ldots, \mathfrak{L}_{\alpha_r}\}$ (the last step) are called *canonical systems* of the family $\{\mathfrak{L}_j\}_{j\in\omega}$. The set of multi-indices $\alpha = (\alpha_1, \ldots, \alpha_r)$ corresponding to the canonical systems is denoted by $J(\omega)$. Note that $\alpha_r = j_q$ for any $\alpha \in J(\omega)$. The recursive method of construction of the canonical systems and Theorem 19.2 give us

LEMMA 19.6. *Every function $f_\omega \in A(\mathbf{C}^n \setminus \bigcup_{j\in\omega} E_j)$ can be represented as $\sum_{\alpha\in\omega} f_\alpha$, where $f_\alpha \in A(\mathbf{C}^n \setminus (T_{\alpha_1} \cup \cdots \cup T_{\alpha_r}))$.*

PROPOSITION 19.7. *If the rank of the matrix (19.9) is equal to $r < n$, then the integral (19.6) over any cycle $\gamma \in Z_n(\mathbf{C}^n \setminus \bigcup_1^N E_j)$ is equal to zero.*

PROOF. Since $\mathbf{C}^n \setminus \bigcup_1^N E_j$ is a domain of holomorphy, Theorem 15.6 (Serre) implies that every n-dimensional homology class can be represented by a holomorphic form $f(z)\,dz$, where $f \in A(\mathbf{C}^n \setminus \bigcup_1^N E_j)$. By Lemmas 19.5 and 19.6, this form has a representation

$$f(z)\,dz = \sum_{\omega\in\Omega} \sum_{\alpha\in J(\omega)} f_\alpha(z)\,dz,$$

where $f_\alpha \in A(\mathbf{C}^n \setminus (E_{\alpha_1} \cup \cdots \cup E_{\alpha_r}))$. We make a linear substitution such that $E_{\alpha_1}, \ldots, E_{\alpha_r}$ are the coordinate planes $\zeta_1 = 0, \ldots, \zeta_r = 0$, and then we expand f_α in a Laurent series with respect to ζ. If $r < n$, this series does not contain negative powers of $\zeta_{r+1}, \ldots, \zeta_n$. Consequently, $f_\alpha\,dz \sim 0$ (see Example 15.8). It remains to apply Corollary 0.3 of Stokes' formula. □

Suppose that the rank of the matrix (19.9) is equal to n. Then for each multi-index $\alpha = (\alpha_1, \ldots, \alpha_n) \in J(\omega)$ we construct the cycle

$$\gamma_\alpha = \left\{z : |\mathfrak{L}_{\alpha_1}(z)| = \varepsilon_1, \ldots, |\mathfrak{L}_{\alpha_n}(z)| = \varepsilon_n\right\},$$

where $\varepsilon_1 < \cdots < \varepsilon_n$ are sufficiently small and are chosen in such a way that for $z \in |\gamma_\alpha|$ and $j \in \omega$

$$\sum_{k=1}^{m-1} |b_{jk}|\varepsilon_k < |b_{jm}|\varepsilon_m,$$

$b_{j1}, \ldots, b_{jn}$ being the coefficients of the expansion of $\mathcal{L}_j$ with respect to $\mathcal{L}_{\alpha_1}, \ldots, \mathcal{L}_{\alpha_n}$: $\mathcal{L}_j = \sum_{k=1}^n b_{jk}\mathcal{L}_{\alpha_j}$, with $b_{jm} \neq 0$ and $b_{jk} = 0$ for $k > m$. This implies that $|\gamma_\alpha| \cap E_j = \varnothing$. If $j \notin \omega$, then $E_j \cap |\gamma_\alpha| = \varnothing$, because the ε_j are sufficiently small. Thus, $\gamma_\alpha \in Z_n(\mathbf{C}^n \setminus \bigcup_1^N E_j)$.

Theorem 19.8. *If the rank of the matrix* (19.9) *is $r = n$, then the cycles $\{\gamma_\alpha\}$ and the forms*

$$\varphi_\alpha = \frac{d\mathcal{L}_{\alpha_1} \wedge \cdots \wedge d\mathcal{L}_{\alpha_n}}{(2\pi i)^n \mathcal{L}_{\alpha_1} \cdots \mathcal{L}_{\alpha_n}},$$

$\alpha \in J(\omega)$, $\omega \in \Omega$, *make up dual (in the de Rham sense) bases for the n-dimensional homology and cohomology of the domain $\mathbf{C}^n \setminus \bigcup_1^N E_j$.*

Proof. Obviously, $\int_{\gamma_\alpha} \varphi_\alpha = 1$. We show that $\int_{\gamma_\alpha} \varphi_\beta = 0$ for $\beta \neq \alpha$. To do this it suffices to show that the cycle γ_α is homologous to 0 in $\mathbf{C}^n \setminus (E_{\beta_1} \cup \cdots \cup E_{\beta_n})$. If $\alpha \in J(\omega_1)$ and $\beta \in J(\omega_2)$, $\omega_1 \neq \omega_2$, then $\beta_k \notin \omega_1$ for some k. For any $\beta_j \in \omega_1$ we have a representation $\mathcal{L}_j = \sum_{k=1}^m b_{jk}\mathcal{L}_{\alpha_j}$, $b_{jm} \neq 0$, $1 \leqslant m \leqslant n$. Since it is false that $\beta_j \in \omega_1$ for all $j = 1, \ldots, n$, there is a k, $1 \leqslant k \leqslant m$, such that $m = m(\beta_j) \neq k$ for $\beta_j \in \omega_1$. Then $\gamma_\alpha = \partial B$, where

$$B = \left\{z : |\mathcal{L}_{\alpha_j}(z)| = \varepsilon_j, j = 1, \ldots [k] \ldots, n, |\mathcal{L}_{\alpha_k}(z)| \leqslant \varepsilon_k\right\}, \tag{19.10}$$

and $|B| \cap E_{\beta_j} = \varnothing$, $j = 1, \ldots, n$. Indeed,

$$|\mathcal{L}_{\beta_j}| = \left|\sum_{j=1}^m b_j\mathcal{L}_{\alpha_j}\right| \geqslant |b_m|\varepsilon_m - \sum_{i=1}^{m-1} |b_i|\varepsilon_i > 0$$

for $z \in |B|$. If $\beta_j \notin \omega_1$, then $|B| \cap E_{\beta_j} = \varnothing$, because $|B|$ lies in a sufficiently small neighborhood of the point $a = \bigcap_{i \in \omega_1} E_i \in \mathbf{C}^n$, and $\mathcal{L}_{\beta_j}(a) \neq 0$. Suppose now that $\alpha, \beta \in J(\omega)$, $\omega \in \Omega$ and $\alpha \neq \beta$. For convenience of notation we set $\omega = \{1, \ldots, q\}$, $\alpha_n = \beta_n = q$. Suppose that $\alpha_j = \beta_j$, $j = r+1, \ldots, n$ and $\alpha_r \neq \beta_r$, $r < n$. Then there is at least one index $\beta_j, j \leqslant r$, such that $\mathcal{L}_{\beta_j}$ cannot be expressed linearly in terms of $\mathcal{L}_{\alpha_1}, \ldots, \mathcal{L}_{\alpha_r}$. Otherwise the system $\mathcal{L}_{\beta_1}, \ldots, \mathcal{L}_{\beta_r}$ can be obtained from the system $\mathcal{L}_{\alpha_1}, \ldots, \mathcal{L}_{\alpha_r}$ by a nonsingular transformation, because these are linearly independent systems. Since $(\mathcal{L}_{\beta_1}, \ldots, \mathcal{L}_{\beta_r})$ and $(\mathcal{L}_{\alpha_1}, \ldots, \mathcal{L}_{\alpha_r})$ are encountered in canonical systems, they must be encountered in $(\beta_r - n)$th-step systems and $(\alpha_r - n)$th-step systems, respectively, and in systems of all subsequent steps. Suppose, for example, that $\alpha_r < \beta_r$ (the argument is simdilar for $\beta_r > \alpha_r$). Then at the $(\alpha_r - n)$th-step instead of systems containing $(\mathcal{L}_{\beta_1}, \ldots, \mathcal{L}_{\beta_r})$ there appear systems containing $(\mathcal{L}_{\beta_1}, \ldots [j] \ldots, \mathcal{L}_{\beta_r}, \mathcal{L}_{\alpha_r})$, $j = 1, \ldots, r$, i.e., in the $(\alpha_r - m)$ subsequent steps there are no systems containing $(\mathcal{L}_{\beta_1}, \ldots, \mathcal{L}_{\beta_r})$. This is a contradiction. Consequently, there is a k, $1 \leqslant k \leqslant r$, such that $m = m(j) \neq k$ in the expansions $\mathcal{L}_j = \sum_{i=1}^m b_{ji}\mathcal{L}_{\alpha_i}$, with $\beta_{jm} \neq 0$ for $j = \beta_1, \ldots, \beta_n$. And then again $\gamma_\alpha = \partial B$, where B has the form (19.10) and $|B| \cap E_{\beta_j} = \varnothing$, $j = 1, \ldots, n$.

From Lemmas 19.5 and 19.6 it follows that the form $f\,dz$ is cohomologous to a linear combination of the forms $\{\varphi_\alpha\}$ (see Example 15.8). Thus, by Proposition 15.4, the cycles $\{\gamma_\alpha\}$ and the forms $\{\varphi_\alpha\}$ make up bases that are dual in the de Rham sense. □

The integrals (19.6) and (19.7) with respect to the basis cycles (the residues) can be computed by the method indicated in 4°.

CHAPTER IV

Applications to Implicit Functions, Systems of Nonlinear Equations, Computation of the Multiplicity of a Zero, and Combinatorics

§20. Expansion of implicit functions in power series and function series

1°. The logarithmic residue for holomorphic functions of a single complex variable can be used to easily derive the Lagrange expansion (see [120], [170], and [67])

$$F(z(\zeta)) = F(z_0) + \sum_{k=1}^{\infty} \frac{\zeta^k}{k!} \frac{d^{k-1}}{dz^{k-1}} [F'(z)\Psi^k(z)]_{z=z_0}, \qquad (20.1)$$

which expresses the function $F(z)$ in terms of the variable ζ connected with z by the equation $z = z_0 + \zeta\Psi(z)$ when $F(z)$ and $\Psi(z)$ are holomorphic functions in a neighborhood of the point $z_0 \in \mathbf{C}^1$. In particular, we obtain the Bürmann-Lagrange series (see [170] and [200]), which enables us to represent one holomorphic function $\Phi(z)$ as a series of powers of another holomorphic function $w = f(z) = (z - z_0)\Psi(z)$, $\Psi(z_0) \neq 0$, in a neighborhood of a point $z_0 \in \mathbf{C}^1$:

$$\Phi(z) = \Phi(z_0) + \sum_{k=1}^{\infty} \frac{w^k}{k!} \frac{d^{k-1}}{dz^{k-1}} \left[\frac{\Phi'(z)}{\Psi^k(z)} \right] \Bigg|_{z=z_0}. \qquad (20.2)$$

For $\Phi(z) \equiv z$ the series (20.2) provides an inversion of the holomorphic function $w = f(z)$ in a neighborhood of z_0.

In this section we use the multiple logarithmic residue to derive generalizations of the Lagrange expansion (20.1) for the case of arbitrary implicit functions determined by a system of equations, along with inversion formulas for a holomorphic mapping in $\mathbf{C}^n$, and we consider the determination of a single-valued branch of an implicit function in some degenerate cases and estimate the domain of convergence and the remainder for the series obtained.

2°. Generalization of the Lagrange expansion to systems of arbitrary implicit functions. Suppose that $\Phi(w, z)$ and

$$F_j(w, z), \qquad j = 1, \ldots, n, \tag{20.3}$$

are holomorphic functions of the variables $w = (w_1, \ldots, w_m)$ and $z = (z_1, \ldots, z_n)$ in a neighborhood of the point $(0,0) \in \mathbf{C}^{m+n}$, with $F_j(0,0) = 0$, $j = 1, \ldots, n$, and

$$\left.\frac{\partial F}{dz}\right|_{(0,0)} = \left.\frac{\partial(F_1, \ldots, F_n)}{\partial(z_1, \ldots, z_n)}\right|_{(0,0)} \neq 0.$$

By a theorem on the existence of implicit functions (see [189] and [201]), the system of equations

$$F_j(w, z) = 0, \qquad j = 1, \ldots, n, \tag{20.4}$$

determines uniquely in a neighborhood of $(0,0) \in \mathbf{C}^{m+n}$ a system of functions

$$z_j = \varphi_j(w), \quad j = 1, \ldots, n \; (z = \varphi(w)), \tag{20.5}$$

that are holomorphic in a neighborhood of the point $0 \in \mathbf{C}^m$. We consider the problem of finding an explicit representation of $\Phi(w, \varphi(w))$ and, in particular (for $\Phi(w, z) \equiv z_j$) of the functions (20.5). Without loss of generality it can be assumed that

$$F_j(0,0) = 0, \qquad \frac{\partial F_j(0,0)}{\partial z_k} = \delta_{jk}, \quad j, k = 1, \ldots, n. \tag{20.6}$$

In case this does not hold we can replace (20.4) by the equivalent system

$$\left\| \frac{\partial F_j(0,0)}{\partial z_k} \right\|^{-1} \cdot F(w, z) = 0.$$

THEOREM 20.1. *Suppose that the functions* (20.3) *satisfy the conditions* (20.6). *Then there exist numbers* $\varepsilon > 0$ *and* $\delta > 0$ *such that the integral representation*

$$\Phi(w, \varphi(w)) = \frac{1}{(2\pi i)^n} \int_{\Gamma_\varepsilon} \frac{\Phi(w, z)}{F^1(w, z)} \frac{\partial F}{\partial z} dz \tag{20.7}$$

holds in the closed polydisk $\overline{V}_\delta = \{w \in \mathbf{C}^m \colon |w_j| \leq \delta, j = 1, \ldots, m\}$, *where* $F^1 = F_1 \cdots F_n$ *and* $\Gamma_\varepsilon = \{z \colon |z_1| = \cdots = |z_n| = \varepsilon\}$. *For* $\Phi(w, z) \equiv z_j$ *formula* (20.6) *gives an integral representation of the implicit functions* (20.5) *determined by the equations* (20.4).

PROOF. It follows from (20.6) that the Taylor series of the function $g_j(w, z) = F_j(w, z) - z_j, j = 1, \ldots, n$, at $(0, 0)$ contains neither a free term nor terms of first degree in the variables $z_1, \ldots, z_n$. Consequently, there exist numbers $\delta > 0$ and $\varepsilon > 0$ such that for any $w \in \overline{V}_\delta$ and $z \in \Gamma_\varepsilon$

$$|g_j(w, z)| < |z_j| = \varepsilon, \qquad j = 1, \ldots, n.$$

Since the system of functions $z = (z_1, \ldots, z_n)$ has a unique simple zero at $0 \in \mathbf{C}^n$, Theorem 4.8 tells us that for any $w \in \overline{V}_\delta$ the system (20.4) has a unique solution $z = \varphi(w) \in U_\varepsilon = \{z : |z_j| < \varepsilon, j = 1, \ldots, n\}$, and (20.7) holds by (4.8). □

THEOREM 20.2 (Yuzhakov). *Under the conditions of Theorem* 20.1 *the function* $\Phi(w, \varphi(w))$ *can be represented in some polydisk* $\overline{V}_\delta$ *by an absolutely and uniformly convergent multiple function series*

$$\Phi(w, \varphi(w)) = \sum_{\beta \geq 0} \frac{(-1)^{|\beta|}}{\beta!} D_z^\beta\left[\Phi(w, z) g^\beta(w, z) \frac{\partial F}{\partial z}\right]\bigg|_{z = h(w)}, \qquad (20.8)$$

where $h = (h_1, \ldots, h_n)$ *is an arbitrary vector-valued function that is holomorphic in* $\overline{V}_\delta$ *with* $h(0) = 0$, $g^\beta = g_1^{\beta_1} \cdots g_n^{\beta_n}$, $\beta! = \beta_1! \cdots \beta_n!$, $|\beta| = \beta_1 + \cdots + \beta_n$ *and* $g_j(w, z) = F_j(w, z) - z_j + h_j(w)$, $j = 1, \ldots, n$.

REMARK. It is frequently convenient to take $h_j(w) \equiv 0$ or $h_j(w) \equiv F_j(w, 0)$, $j = 1, \ldots, n$.

PROOF. There exist $\delta > 0$ and $\varepsilon > 0$ such that $|h_j(w)| < \varepsilon/2$ and $|g_j(w, z)| < \varepsilon/2$ for $w \in \overline{V}_\delta$ and $z \in \Gamma_\varepsilon$. Then the fraction $1/F$ can be expanded in a multiple geometric progression

$$\frac{1}{\prod_{j=1}^n F_j(w, z)} = \frac{1}{\prod_{j=1}^n (z_j - h_j)[1 + g_j/(z_j - h_j)]} = \sum_{\beta \geq 0} \frac{(-1)^{|\beta|} g^\beta(w, z)}{[z - h(w)]^{\beta + I}},$$

which converges absolutely and uniformly on the compact set $\overline{V}_\delta \times \Gamma_\varepsilon$. Substituting this in (20.7) and integrating termwise, we get the series

$$\Phi(w, \varphi(w)) = \sum_{\beta \geq 0} \frac{(-1)^{|\beta|}}{(2\pi i)^n} \int_{\Gamma_\varepsilon} \frac{\Phi(w, z) g^\beta(w, z)(\partial F/\partial z)\, dz}{[z - h(w)]^{\beta + I}}, \qquad (20.9)$$

which, according to Corollary 4.2, coincides with the series (20.8). □

PROPOSITION 20.3. *The coefficients in the Taylor expansion*

$$\Phi(w, \varphi(w)) = \sum_{\alpha \geq 0} c_\alpha w^\alpha, \qquad \alpha = (\alpha_1, \ldots, \alpha_m), \qquad (20.10)$$

are determined by the formula

$$c_\alpha = \sum_{|\beta| \leqslant 2|\alpha|} \frac{(-1)^{|\beta|}}{\alpha!\beta!} D^{\alpha,\beta}_{w,z}\left[\Phi(w,z) g^\beta(w,z) \frac{\partial F}{\partial z}\right]\bigg|_{(0,0)}, \tag{20.11}$$

where $g^\beta = g_1^{\beta_1} \cdots g_n^{\beta_n}$ *and* $g_j(w,z) = F_j(w,z) - z_j$, $j = 1, \ldots, n$.

PROOF. Let us expand $\Phi(w, \varphi(w))$ in a Taylor series (20.10) with coefficients in integral form. Since the series (20.9) converges uniformly on $\widetilde{V}_\delta$, it can be integrated termwise over $\gamma_\delta = \{w : |w_1| = \cdots = |w_m| = \delta\}$. Here we set $h(w) \equiv 0$ and get

$$\begin{aligned} c_\alpha &= \frac{1}{(2\pi i)^m} \int_{\gamma_\delta} \frac{\Phi(w, \varphi(w))\, dw}{w^{\alpha+I}} \\ &= \sum_{\beta \geqslant 0} \frac{(-1)^{|\beta|}}{(2\pi i)^{m+n}} \int_{\gamma_\delta \times \Gamma_\varepsilon} \frac{\Phi(w,z) g^\beta(w,z)(\partial F/\partial z)\, dw \wedge dz}{w^{\alpha+I} z^{\beta+I}}. \end{aligned} \tag{20.12}$$

With the help of the binomial formula the integrals in the terms of (20.12) can be represented in the form

$$\sum_{0 \leqslant \mu \leqslant \beta} \frac{\beta!}{\mu!(\beta-\mu)!} \int_{\gamma_\delta \times \Gamma_\varepsilon} \frac{\Phi \Psi^\mu \theta^{\beta-\mu} (\partial F/\partial z)\, dw \wedge dz}{w^{\alpha+I} z^{\beta+I}}, \tag{20.13}$$

where $\Psi(w) = g(w, 0)$, $\theta(w,z) = g(w,z) - \Psi(w)$, $\mu = (\mu_1, \ldots, \mu_n)$ and $\beta - \mu = (\beta_1 - \mu_1, \ldots, \beta_n - \mu_n)$. Since the order $d_0(\Psi)$ of the zero is at least 1, while $d_{(0,0)}(\theta) \geqslant 2$, and $\int_{\gamma_\delta \times \Gamma_\varepsilon} w^\alpha z^\beta\, dw \wedge dz \neq 0$ only for $\alpha_1 = \cdots = \alpha_m = \beta_1 = \cdots = \beta_n = -1$, it follows that the integrals in (20.13) are equal to zero for $|\mu| > |\alpha|$ as well as for $|\mu| + 2|\beta - \mu| = 2|\beta| - |\mu| > |\alpha| + |\beta|$ and, consequently, for $|\beta| > 2|\alpha|$. Taking this and Corollary 4.2 into account, we get (20.11) from (20.12). □

For $n = 1$ the formulas (20.8) and (20.11) can be somewhat simplified.

PROPOSITION 20.4. *Suppose that* $F(w,z)$, $\Phi(w,z)$, *and* $h(w)$ *are holomorphic functions of the variables* $w = (w_1, \ldots, w_m) \in \mathbf{C}^m$ *and* $z \in \mathbf{C}^1$ *in a neighborhood of the point* $(0,0) \in \mathbf{C}^{m+1}$, *with* $F(0,0) = 0$, $F'_z(0,0) = 1$, *and* $h(0) = 0$. *Then the function* $\Phi(w, \varphi(w))$, *where* $z = \varphi(w)$ *is the implicit function determined by the equation* $F(w,z) = 0$, *can be represented by the function series*

$$\begin{aligned} \Phi(w, \varphi(w)) &= \Phi(w, h(w)) \\ &\quad + \sum_{k=1}^\infty \frac{(-1)^k}{k!} \frac{\partial^{k-1}}{\partial z^{k-1}} \left[\Phi'_z(w,z) g^k(w,z)\right]\Big|_{z=h(w)}, \end{aligned} \tag{20.14}$$

where $g(w, z) = F(w, z) - z + h(w)$, *as well as by the multiple power series* $\Phi(w, \varphi(w)) = \Sigma_{\alpha \geq 0} c_\alpha w^\alpha$ *whose coefficients are given by*

$$c_\alpha = \frac{1}{\alpha!} D_w^\alpha \Phi(0,0) + \sum_{k=1}^{2|\alpha|} \frac{(-1)^k}{k!\alpha!} D_{w,z}^{\alpha,k} \left(\Phi_z' g^k\right)\Big|_{(0,0)}, \tag{20.15}$$

where $g(w, z) = F(w, z) - z$.

PROOF. For $n = 1$ formula (20.8) can be represented in the form

$$\begin{aligned}
\Phi(w, \varphi(w)) &= \sum_{k=0}^{\infty} \frac{(-1)^k}{k!} \frac{\partial^k}{\partial z^k} \left[\Phi(w, z) g^k(w, z) \left(1 + \frac{\partial g}{\partial z}\right)\right]\Big|_{z=h(w)} \\
&= \Phi(w, h(w)) + \sum_{k=1}^{\infty} \frac{(-1)^k}{k!} \frac{d^{k-1}}{dz^{k-1}} \left[\frac{d}{dz} (\Phi g^k)\right]\Big|_{z=h(w)} \\
&\quad - \sum_{k=1}^{\infty} \frac{(-1)^k}{k!} \frac{\partial^{k-1}}{\partial z^{k-1}} \left(\Phi \frac{\partial g^k}{\partial z}\right)\Big|_{z=h(w)},
\end{aligned}$$

or in the form (20.14). Formula (20.15) can be obtained from (20.14) in the same way as in Theorem 20.3. □

Observe that the Lagrange series (20.1) is obtained from (20.14) when $m = 1$, $h(w) = 0$, $g(w, z) = -w\Psi(z)$, and $w = \zeta$, if $z - z_0$ is taken in place of z.

3°. Local inversion of a holomorphic mapping in $\mathbf{C}^n$. The following is a particular case of the problem just considered. Suppose that in a neighborhood of $0 \in \mathbf{C}^n$ we are given a holomorphic mapping

$$w_j = f_j(z), \qquad j = 1, \ldots, n \; (w = f(z)), \tag{20.16}$$

satisfying the condition

$$f_j(0) = 0, \qquad \frac{\partial f_j(0)}{\partial z_k} = \delta_{jk}, \quad j, k = 1, \ldots, n, \tag{20.17}$$

and a holomorphic function $\Phi(z)$. It is required to express $\Phi(z)$ in terms of the variables $w = (w_1, \ldots, w_n)$ and, in particular, to find the mapping

$$z_j = \varphi_j(w), \qquad j = 1, \ldots, n \; (z = \varphi(w)), \tag{20.18}$$

inverse to (20.16) in a neighborhood of $0 \in \mathbf{C}^n$.

THEOREM 20.5. *Under the above assumptions, the expansion*

$$\Phi(\varphi(w)) = \sum_{\beta \geq 0} \frac{(-1)^{|\beta|}}{\beta!} D^\beta \left[\Phi(z) g^\beta(z) \frac{\partial f}{\partial z}\right]\Big|_{z=w} \tag{20.19}$$

is valid in a neighborhood of $0 \in \mathbf{C}^n$, *where* $g^\beta = g_1^{\beta_1} \cdots g_n^{\beta_n}$, *and* $g_j(z) = f_j(z) - z_j$, $j = 1,\ldots,n$. *For* $\Phi(z) \equiv z_j$, $j = 1,\ldots,n$, *the series* (20.19) *gives an inversion formula for the holomorphic mapping* (20.16).

PROOF. In Theorem 20.2 we set $m = n$, $F_j(w, z) = f_j(z) - w_j$, and $h_j(w) = w_j$, and get (20.19). □

PROPOSITION 20.6. *The expansion coefficients of the function* $\Phi(z)$ *in a series*

$$\Phi(z) = \sum_{\alpha \geq 0} c_\alpha f^\alpha(z) \Big(\Phi(\varphi(w)) = \sum_{\alpha \geq 0} c_\alpha w^\alpha \Big) \tag{20.20}$$

with respect to the functions (20.16) *can be expressed by the formula*

$$c_\alpha = \sum_{|\beta| \leq |\alpha|} \frac{(-1)^{|\beta|}}{\alpha!\beta!} D^{\alpha+\beta}\left[\Phi(z) g^\beta(z) \frac{\partial f}{\partial z}\right]\bigg|_{z=0}. \tag{20.21}$$

PROOF. Applying the Taylor expansion to (20.19) and taking into account that $d_0(g_j) \geq 2$, we get (20.20) and (20.21). □

The series (20.19)–(20.21) can be regarded as multidimensional analogues of the Bürmann-Lagrange series (20.2). For the special case of the functions (20.16) we get a sharper analogue of the series (20.2).

THEOREM 20.7 (Stieltjes, Poincaré and Good). *If the components of the mapping* (20.16) *have the form*

$$w_j = f_j(z) = z_j \Psi_j(z), \qquad \Psi_j(0) \neq 0, \quad j = 1,\ldots,n,$$

then the following expansion is valid:

$$\Phi(\varphi(w)) = \sum_{\alpha \geq 0} \frac{w^\alpha}{\alpha!} D^\alpha \left[\frac{\Phi(z)}{\Psi^{\alpha+I}(z)} \det \left\| \delta_{jk} + \frac{z_j}{\Psi_j(z)} \frac{\partial \Psi_j}{\partial z_k} \right\| \right]\Bigg|_{z=0}. \tag{20.19'}$$

PROOF. Let $m = n$, $F_j(w, z) = f_j(z) - w_j$, and $\Phi(w, z) = \Phi(z)$ in (20.7). Expanding the fraction

$$\prod_{j=1}^n \frac{1}{[f_j(z) - w_j]} = \prod_{j=1}^n \frac{1}{f_j(1 - w_j/f_j)}$$

in a multiple geometric progression, we get

$$\Phi(\varphi(w)) = \frac{1}{(2\pi i)^n} \int_{\Gamma_\varepsilon} \frac{\Phi(z)(\partial f/\partial z)\, dz}{[f(z) - w]^I} = \sum_{\alpha \geq 0} \frac{w^\alpha}{(2\pi i)^n} \int_{\Gamma_\varepsilon} \frac{\Phi(z)(\partial f/\partial z)\, dz}{z^{\alpha+I}\Psi^{\alpha+I}(z)}.$$

Since

$$\frac{\partial f}{\partial z} = \det \left\| \delta_{jk}\Psi_j(z) + z_j \frac{\partial \Psi_j}{\partial z_k} \right\|,$$

Corollary 4.2 tells us that the series obtained can be written in the form (20.19′). □

We indicate a case of inversion of the holomorphic mapping (20.16) when $(\partial f/\partial z)|_0 = 0$.

PROPOSITION 20.8. *Suppose that the functions of the system* (20.16) *have the form*

$$w_j = z_j^{k_j}\Psi_j(z), \qquad k_j \geq 1, \qquad \Psi_j(0) = 1, \quad j = 1,\dots,n.$$

Then in some neighborhood of the point $0 \in \mathbf{C}^n$ *the inverse of the mapping* (20.16) *is* $(k_1 \cdots k_n)$*-valued, has branchings of orders* k_j *on the coordinate planes* $w_j = 0$, *and can be represented by a series in fractional powers of* w:

$$z_j = w_j^{1/k_j} + \sum_{|\alpha|\geq 2} \frac{w^{\alpha/k}}{\alpha!} D^\alpha\left[\frac{z_j}{\Psi^{(\alpha+I)/k}(z)} \det\left\|\frac{\partial\left(z_j\Psi_j^{1/k_j}\right)}{\partial z_k}\right\|\right]\Bigg|_{z=0},$$

where $\alpha/k = (\alpha_1/k_1,\dots,\alpha_n/k_n)$.

4°. Expression of the Taylor coefficients of implicit functions in terms of the Taylor coefficients of the functions appearing in the equations. Inversion of a system of power series. Suppose that the functions (20.3) are given as power series

$$F_j(w, z) = z_j + \sum_{\mu,\nu}{}' a_{j\mu\nu} w^\mu z^\nu, \qquad j = 1,\dots,n. \tag{20.22}$$

Here and below we assume that the multi-indices $\mu = (\mu_1,\dots,\mu_m)$ and $\nu = (\nu_1,\dots,\nu_n)$ in $\Sigma'_{\mu,\nu}$, $\Pi'_{\mu,\nu}$ and $\{\ \}'_{\mu,\nu}$ run through all possible tuples of nonnegative integers for which $|\mu| + |\nu| \geq 2$ or $|\mu| = 1$, $|\nu| = 0$. Let

$$\Phi(w, z) = \sum_{(\alpha,\beta)\geq(0,0)} b_{\alpha\beta} w^\alpha z^\beta. \tag{20.23}$$

THEOREM 20.9 (Yuzhakov and Bolotov). *The coefficients in the Taylor series* (20.10) *of the function* $\Phi(w, \varphi(w))$, *where* $z = \varphi(w)$ *is the implicit vector-valued function determined by* (20.4), *can be expressed in terms of the coefficients of* (20.22) *and* (20.23) *by the formula*

$$c_\alpha = \sum_{\eta,\tau,S} (-1)^{|\rho(S)|} b_{\eta\tau}\Delta(S) \prod_{j=1}^n \left[\rho(S_j) - 1\right]! \prod_{\mu,\nu}{}' \frac{1}{s_{j\mu\nu}!}(a_{j\mu\nu})^{s_{j\mu\nu}}, \tag{20.24}$$

where $S = (S_1,\dots,S_n)$, *and* $S_j = \{s_{j\mu\nu}\}'_{\mu,\nu}$ *are finitely nonzero sequences of nonnegative integers*, $\rho(S) = (\rho(S_1),\dots,\rho(S_n))$, $\rho(S_j) = \Sigma'_{\mu,\nu} s_{j\mu\nu}$, $\Delta(S) = \det\|\delta_{kj}\rho(S_j) - \sigma_k(S_j)\|_{j,k=1,\dots,n}$ *with* $\sigma_k(S_j) = \Sigma'_{\mu,\nu} \nu_k s_{j\mu\nu}$ *and* δ_{kj} *the Kronecker symbol, and the summation in* $\Sigma_{\eta,\tau,S}$ *is over all possible multi-indices* $\eta = (\eta_1,\dots,\eta_m)$, $\tau = (\tau_1,\dots,\tau_n)$ *and systems* $S = (S_1,\dots,S_n)$ *such that*

$$\begin{aligned} \eta_k + \sum_{j=1}^{n} \lambda_k(S_j) &= \alpha_k, \qquad k = 1,\dots,m, \\ \tau_k + \sum_{j=1}^{n} \sigma_k(S_j) &= \rho(S_k), \qquad k = 1,\dots,n, \end{aligned} \tag{20.25}$$

with $\lambda_k(S_j) = \Sigma'_{\mu,\nu} \mu_k s_{j\mu\nu}$. *If* S *contains an* S_j *for which* $\rho(S_j) = 0$, *then it is necessary to replace* $[\rho(S_j) - 1]!$ *by* 1 *in the corresponding term of* (20.24), *and* $\delta_{kj}\rho(S_j) - \sigma_k(S_j)$ *by* δ_{kj} *in* $\Delta(S)$.

PROOF. From (20.7) and the first part of (20.12) we get an integral formula for the coefficients in the Taylor series for $\Phi(w, \varphi(w))$:

$$c_\alpha = \frac{1}{(2\pi i)^{m+n}} \int_{\gamma_\delta \times \Gamma_\varepsilon} \frac{\Phi(w, z)(\partial F/\partial z)\, dw \wedge dz}{F^{\mathrm{I}}(w, z) w^{\alpha + \mathrm{I}}}. \tag{20.26}$$

The function

$$\frac{\Phi(w, z)}{F^{\mathrm{I}}(w, z)} \frac{\partial F}{\partial z} = \Phi(w, z) \frac{\partial \ln F(w, z)}{\partial z} \tag{20.27}$$

is holomorphic on the $(n + m)$-circular compact set $\overline{V}_\delta \times \Gamma_\varepsilon$ and can be expanded in a multiple Laurent series on it. It follows from (20.26) that the coefficient c_α is equal to the coefficient of $w^\alpha z^{-\mathrm{I}}$ in this series. The expansion of the function (20.27) in a Laurent series can be found directly. Taking into account that the inequalities $|g_j(w, z)| < |z_j| = \varepsilon$, $j = 1,\dots,n$, hold for $w \in \overline{V}_\delta$ and $z \in \Gamma_\varepsilon$, where $g_j(w, z) = F_j(w, z) - z_j = \Sigma'_{\mu,\nu} a_{j\mu\nu} w^\mu z^\nu$, we get the expansions

$$\begin{aligned} \ln F_j(w, z) &= \ln z_j + \ln[1 + g_j(w, z)/z_j] \\ &= \ln z_j + \sum_{p=1}^{\infty} (-1)^{p-1} g_j^p(w, z)/p z_j^p \\ &= \ln z_j + \sum_{p=1}^{\infty} (-1)^{p-1}(p-1)! \sum_{\rho(S_j)=p} \left[\prod_{\mu,\nu}{}' \frac{1}{s_{j\mu\nu}!} \cdot (a_{j\mu\nu})^{s_{j\mu\nu}} \right] \\ &\qquad \times w^{\lambda(S_j)} z^{\sigma(S_j) - \rho(S_j) e_j} \\ &= \ln z_j + \sum_{\rho(S_j) \geqslant 1} (-1)^{\rho(S_j)-1} [\rho(S_j) - 1]! \left[\prod_{\mu,\nu}{}' \frac{1}{s_{j\mu\nu}} (a_{j\mu\nu})^{s_{j\mu\nu}} \right] \\ &\qquad \times w^{\lambda(S_j)} z^{\sigma(S_j) - \rho(S_j) e_j}, \qquad j = 1,\dots,n, \end{aligned}$$

where $e_j = (\delta_{j1}, \ldots, \delta_{jn})$, $\lambda(S_j) = (\lambda_1(S_j), \ldots, \lambda_n(S_j)) = \Sigma'_{\mu,\nu} \mu s_{j\mu\nu}$, and $\sigma(S_j) = (\sigma_1(S_j), \ldots, \sigma_n(S_j)) = \Sigma'_{\mu,\nu} \nu s_{j\mu\nu}$. Hence,

$$\frac{\partial \ln F_j(w, z)}{\partial z_k} = \frac{\delta_{jk}}{z_j} + \sum_{\rho(S_j) \geq 1} (-1)^{\rho(S_j)} [\rho(S_j) - 1]! \times \left[\prod_{\mu,\nu}' \frac{1}{s_{j\mu\nu}!} (a_{j\mu\nu})^{s_{j\mu\nu}} \right] [\delta_{kj} \rho(S_j) - \sigma(S_j)] \times w^{\lambda(S_j)} z^{\sigma(S_j) - \rho(S_j) e_j - e_k}, \quad j, k = 1, \ldots, n. \tag{20.28}$$

The term δ_{jk}/z_j in (20.28) can be regarded as the term on the sum $\Sigma_{\rho(S_j) \geq 0} \cdots$ when $\rho(S_j) = 0$, with account taken of the conventions adopted in the formulation of the theorem. Substituting (20.28) in the Jacobian $\partial(\ln F)/\partial z$ and taking the summation signs and common factors in the rows and columns out from under the determinant sign, we get

$$\frac{\partial \ln F(w, z)}{\partial z} = (-1)^{|\rho(S)|} \Delta(S) \left\{ \prod_{j=1}^{n} [\rho(S_j) - 1]! \times \prod_{\mu,\nu}' \frac{1}{s_{j\mu\nu}} (a_{j\mu\nu})^{s_{j\mu\nu}} \right\} w^{\lambda(S)} z^{\sigma(S) - \rho(S) - I}, \tag{20.29}$$

where $\lambda(S) = \lambda(S_1) + \cdots + \lambda(S_n)$ and $\sigma(S) = \sigma(S_1) + \cdots + \sigma(S_n)$. Multiplying out the series (20.29) and (20.23), we find the Laurent series of the function (20.27). Singling out the coefficient of $w^\alpha z^{-I}$ in it, we get (20.24). □

REMARK. The coefficients of the Taylor series $z_q = \varphi_q(w) = \Sigma_{\alpha \geq 0} c_{q\alpha} w^\alpha$, $q = 1, \ldots, n$, of the implicit functions (20.5) can be expressed in terms of the coefficients of (20.22) by the formula

$$c_{q\alpha} = \sum_{S} (-1)^{|\rho(S)|} \Delta(S) \prod_{j=1}^{n} [\rho(S_j) - 1]! \prod_{\mu,\nu}' \frac{1}{s_{j\mu\nu}} (a_{j\mu\nu})^{s_{j\mu\nu}},$$

where the summation runs over all possible n-tuples $S = (S_1, \ldots, S_n)$ such that

$$\lambda_k(S_1) + \cdots + \lambda_k(S_n) = \alpha_k, \qquad k = 1, \ldots, m;$$
$$\delta_{qk} + \sigma_k(S_1) + \cdots + \sigma_k(S_n) = \rho(S_k), \qquad k = 1, \ldots, n.$$

Theorem 20.9 gives us the following result for $m = n$, $F_j(w, z) = f_j(z) - w_j$, $j = 1, \ldots, n$, and $\Phi(w, z) \equiv \Phi(z)$.

THEOREM 20.10 (Cayley, Sylvester and Sack). *Suppose that the functions* (20.16) *and* $\Phi(z)$ *are given by the power series*

$$w_j = f_j(z) = z_j + \sum_{|\alpha|\geqslant 2} a_{j\alpha} z^\alpha, \qquad j = 1,\dots,n, \tag{20.30}$$

$$\Phi(z) = \sum_{\alpha\geqslant 0} b_\alpha z^\alpha. \tag{20.31}$$

Then the coefficients of the Taylor series (20.20) *of the function* $\Phi(\varphi(w))$, *where* $z = \varphi(w)$ *is the inverse of the system* (20.30), *can be expressed in terms of the coefficients of the power series* (20.31) *by the formula*

$$c_\alpha = \sum_{\tau, S} (-1)^{|\rho(S)|} b_\tau \Delta(S) \prod_{j=1}^{n} \frac{[\rho(S_j) + \alpha_j - 1]!}{\alpha_j!} \prod_{|\nu|\geqslant 2} \frac{1}{s_{j\nu}!} (a_{j\nu})^{s_{j\nu}}, \tag{20.32}$$

where $S = (S_1,\dots,S_n)$, $S_j = \{s_{j\nu}\}_{|\nu|\geqslant 2}$, $\rho(S) = (\rho(S_1),\dots,\rho(S_n))$, $\rho(S_j) = \sum_{|\nu|\geqslant 2} s_{j\nu}$, $\Delta(S) = \det\|\delta_{kj}[\alpha_j + \rho(S_j)] + \sigma_k(S_j)\|$ *with* $\sigma_k(S_j) = \sum_{|\nu|\geqslant 2} \nu_k s_{j\nu}$, *and the summation in* $\sum_{\tau,S}$ *is over all possible multi-indices* $\tau = (\tau_1,\dots,\tau_n)$, $|\tau|\geqslant 0$, *and n-tuples* S, $\rho(S)\geqslant 0$, *such that*

$$\tau_k + \sum_{j=1}^{n} \sigma_k(S_j) = \alpha_k + \rho(S_k), \qquad k = 1,\dots,n.$$

If $\rho(S_j) + \alpha_j = 0$, *then* $[\rho(S_j) + \alpha_j - 1]!$ *should be replaced by* 1, *and* $\delta_{kj}[\alpha_j + \rho(S_j)] - \sigma_k(S_j)$ *by* δ_{kj}.

If $\Phi(w, z) = z_q$, $q = 1,\dots,n$, then (20.20) and (20.32) provide formulas for inverting the systems of power series (20.30). In particular, a formula for inverting a power series is obtained for $n = 1$.

PROPOSITION 20.11 (Sylvester). *Consider a power series* $w = \sum_1^\infty a_k z^k$, $a_1 \neq 0$. *Then the coefficients of the inverse series* $z = \sum_1^\infty c_n w^n$ *can be expressed by the formula*

$$c_n = \frac{1}{n a_1^n} \sum_{k_2,\dots,k_n} \frac{(n + k_2 + \cdots + k_n - 1)!}{k_2! \cdots k_n!} \left(-\frac{a_2}{a_1}\right)^{k_2} \cdots \left(-\frac{a_n}{a_1}\right)^{k_n},$$

where the summation is over all possible n-tuples of numbers $k_2,\dots,k_n \geqslant 0$ *such that* $k_2 + 2k_3 + \cdots + (n-1)k_n = n - 1$.

5°. Determination of a single-valued branch of a system of implicit functions for certain degenerate cases. Let us consider a case in which we determine a single-valued holomorphic branch of an implicit vector-valued function for the system (20.4) in a neighborhood of the point $(0,0) \in \mathbf{C}^{m+n}$ when $(\partial F/\partial z)|_{(0,0)} = 0$. Suppose that in a neighborhood of $(0,0)$ the functions of the system (20.3) can be represented in the form

$$F_j(w, z) = z_j P_j(w, z) + Q_j(w, z), \qquad j = 1,\ldots,n, \tag{20.33}$$

where $z_j P_j(w, z)$ is the homogeneous polynomial of least degree in the expansion of F_j in a series of homogeneous polynomials in w and z, $P_j(w,0) \not\equiv 0$. Let $w^{\lambda(j)} = w_1^{\lambda_{j1}} \cdots w_m^{\lambda_{jm}}$ be the highest term in the polynomial $P_j(w,0)$ in the lexicographical ordering of its terms (for any term $a_{j\alpha}w^\alpha$ of $P_j(w,0)$, $\alpha \neq \lambda(j)$, the conditions $\alpha_1 = \lambda_{j1},\ldots,\alpha_{k-1} = \lambda_{j,k-1}$, $\alpha_k < \lambda_{jk}$ hold for some k, $1 \leqslant k \leqslant m$). We set the coefficient of $w^{\lambda(j)}$ equal to 1. Let

$$g_j(w, z) = Fj(w, z) - z_j w^{\lambda(j)}, \quad j = 1,\ldots,n, \qquad \lambda = (\lambda_{(1)},\ldots,\lambda_{(n)});$$

$$\lambda \cdot (\beta + I) = (\lambda_{11}(\beta_1 + 1) + \cdots + \lambda_{n1}(\beta_n + 1),\ldots,\lambda_{1m}(\beta_1 + m)$$
$$+ \cdots + \lambda_{nm}(\beta_n + 1)),\ldots,\lambda_{1m}(\beta_1 + m) + \cdots$$

$$\gamma_\delta = \{w \colon |w_1| = \cdots = |w_n| = \delta\}, \quad \Gamma_\varepsilon = \{z \colon |z_1| = \cdots = |z_n| = \varepsilon\},$$

where δ and ε are sufficiently small positive numbers.

Under our assumptions we have

THEOREM 20.12 (Yuzhakov). *There exists a system of holomorphic functions*

$$z_j = \varphi_j(w) = \varphi_j(w_1,\ldots,w_m), \qquad j = 1,\ldots,n, \tag{20.34}$$

satisfying the system (20.4) *in a neighborhood of the point* $(0,0) \in \mathbf{C}^{m+n}$ *along with the conditions*

$$\varphi_j(0) = 0, \qquad \frac{\partial \varphi_j(0)}{\partial w_k} = 0, \qquad j = 1,\ldots,n, k = 1,\ldots,m, \tag{20.35}$$

if and only if

$$\sum_{\beta \geqslant 0} (-1)^{|\beta|} \int_{\gamma_\delta \times \Gamma_\varepsilon} \frac{z_j g^\beta(w, z)(\partial F/\partial z)\, dw \wedge dz}{z^{\beta+I} w^{\lambda(\beta+I)+\alpha+I}} = 0 \tag{20.36}$$

for $j = 1,\ldots,n$ *and all the integer multi-indices* $\alpha = (\alpha_1,\ldots,\alpha_m)$, $-\infty < \alpha_q < \infty$, *such that* $|\alpha| \geqslant 2$ *and* $\alpha_q < 0$ *for at least one* $q \in \{1,\ldots,m\}$. *Under the conditions* (27.36) *the expansion coefficients in the power series*

$$\Phi(w, \varphi(w)) = \sum_{\alpha \geqslant 0} c_\alpha w^\alpha, \tag{20.37}$$

where $\Phi(w, z)$ *is a holomorphic function in a neighborhood of* $(0,0)$ *and* $\varphi = (\varphi_1, \ldots, \varphi_n)$ *is the vector-valued function* (20.34), *are determined by the formula*

$$c_\alpha = \sum_{\substack{\beta \geq 0 \\ |\beta| < N(\alpha)}} \frac{(-1)^{|\beta|}}{\beta![\lambda(\beta + I) + \alpha]!} D_{z,w}^{\beta, \lambda(\beta+I)+\alpha}\left[\Phi g^\beta \frac{\partial F}{\partial z}\right]\Bigg|_{(0,0)}. \quad (20.38)$$

Equalities (20.37) and (20.38) give the Taylor series for the functions (20.34) when $\Phi(w, z) \equiv z_j$.

COROLLARY 20.13. *For* $m = 1$ *the conditions* (20.33) *always imply the existence of a single-valued branch* (20.34) *satisfying* (20.4) *and* (20.35).

For $m > 1$ the conditions (20.33) are not enough to ensure the existence of the single-valued branch of (20.34). For example, the left-hand side of the equation $z(w_1 + w_2) + w_1^4 = 0$ satisfies (20.33), but in a neighborhood of $(0,0,0) \in \mathbf{C}^3$ this equation determines only the meromorphic function $z = w_1^4/(w_1 + w_2)$.

PROOF OF THEOREM 20.12. From our assumptions it follows that there are numbers $\delta_1 > \delta_2 > \cdots > \delta_m > \varepsilon > 0$ such that the inequalities

$$|g_j(w, z)| < |w^{\lambda(j)} z_j|, \qquad j = 1, \ldots, n, \quad (20.39)$$

hold for $z \in \Gamma_\varepsilon$ and $w \in \gamma = \{w : |w_1| = \delta_1, \ldots, |w_m| = \delta_m\}$. For example, it suffices to take $\delta_1 = \delta$, $\delta_2 = \delta^2, \ldots, \delta_m = \delta^m$ and $\varepsilon = \delta^{m+1}$, where $\delta > 0$ is sufficiently small. Since $\gamma \times \Gamma_\varepsilon$ is compact, (20.39) holds also when $z \in \Gamma_\varepsilon$ and $w \in V(\gamma)$, where $V(\gamma)$ is some m-circular neighborhood of γ. By Theorem 4.8, for any $w \in V(\gamma)$ there is a unique point $z = \varphi(w) \in U_\varepsilon = \{z : \|z\| < \varepsilon\}$ satisfying the system $z_j - g_j(w, z)/w^{\lambda(j)} = 0$, $j = 1, \ldots, n$, i.e., the system (20.4). By (4.8),

$$z_j = \varphi_j(w) = \frac{1}{(2\pi i)^n} \int_{\Gamma_\varepsilon} \frac{z_j(\partial F/\partial z)\, dz}{F^1(w, z)}, \qquad j = 1, \ldots, n, \quad (20.40)$$

and

$$\Phi(w, \varphi(w)) = \frac{1}{(2\pi i)^n} \int_{\Gamma_\varepsilon} \frac{\Phi(w, z)}{F^1(w, z)} \frac{\partial F}{\partial z} dz. \quad (20.41)$$

The fact that the integrand forms in (20.40) and (20.41) are holomorphic implies that the functions $\varphi_j(w)$ and $\Phi(w, \varphi(w))$ are holomorphic in $V(\gamma)$. Consequently, the functions (20.40) can be expanded in $V(\gamma)$ in multiple Laurent series

$$z_j = \varphi_j(w) = \sum c_{j\alpha} w^\alpha, \qquad \alpha = (\alpha_1, \ldots, \alpha_m), -\infty < \alpha_q < \infty, \quad (20.42)$$

where

$$c_{j\alpha} = \frac{1}{(2\pi i)^m}\int_\gamma \frac{\varphi_j(w)\,dw}{w^{\alpha+I}} = \frac{1}{(2\pi i)^{n+m}}\int_{\gamma\times\Gamma_\varepsilon} \frac{z_j(\partial F/\partial z)\,dw\wedge dz}{F^I(w,z)w^{\alpha+I}}$$
$$= \sum_{\beta\geq 0}\frac{(-1)^{|\beta|}}{(2\pi i)^{n+m}}\int_{\gamma\times\Gamma_\varepsilon}\frac{z_j g^\beta(w,z)(\partial F/\partial z)\,dw\wedge dz}{z^{\beta+I}w^{\lambda(\beta+I)+\alpha+I}}. \tag{20.43}$$

Here we have used (20.39) to expand the fraction

$$\frac{1}{F^I(w,z)} = \frac{1}{\prod_{q=1}^n z_q w^{\lambda(q)}\left[1+g_q(w,z)/z_q w^{\lambda(q)}\right]}$$

in a multiple geometric progression.

Let us show that $c_{j\alpha}=0$ for $|\alpha|\leq 2$. To do this we expand the functions $g_j^{\beta_j}=[z_jQ_j-\theta_j]^{\beta_j}$ with $Q_j=P_j-w^{\lambda(j)}$ by the binomial formula. We obtain

$$\int_{\gamma\times\Gamma_\varepsilon}\frac{z_j g^\beta(\partial F/\partial z)\,dw\wedge dz}{z^{\beta+I}w^{\lambda(\beta+I)+\alpha+I}}$$
$$= \sum_{0\leq\mu\leq\beta}\frac{\beta!}{\mu!\,(\beta-\mu)!}\int_{\gamma\times\Gamma_\varepsilon}\frac{z_jQ^\mu\theta^{\beta-\mu}(\partial F/\partial z)\,dw\wedge dz}{z^{\beta-\mu+I}w^{\lambda(\beta-I)+\alpha+I}}. \tag{20.44}$$

Since $d_{(0,0)}(\theta_j)\geq|\lambda_{(j)}|+2$, $d_{(0,0)}(\partial F/\partial z)=\lambda(\mathrm{I})=|\lambda_{(1)}|+\cdots+|\lambda_{(n)}|$, and $d_{(0,0)}(Q_j)=\lambda_{(j)}$, the order $d_{(0,0)}(z_jQ^\mu\theta^{\beta-\mu}(\partial F/\partial z))$ of the numerator in the integral after the summation sign in (20.44) is at least $1+\lambda(\mu)+|\lambda(\beta-\mu)|+2|\beta-\mu|+\lambda(\mathrm{I})=1+\lambda(\beta+\mathrm{I})+2|\beta-\mu|$. Consequently, this integral is equal to zero when

$$1+\lambda(\beta+\mathrm{I})-2|\beta-\mu|>|\beta-\mu|+|\lambda(\beta+\mathrm{I})+\alpha|,$$

i.e., when $1+|\beta-\mu|>|\alpha|$, and also when $|\beta-\mu|<1$ with only the integration with respect to z taken into account. Thus, it is equal to zero for $|\alpha|<2$. From this it follows that under the conditions (20.36) the Laurent series (20.42) do not contain terms with negative powers, i.e., the functions (20.40) extend analytically to $\overline{V}=\{w:|w_1|\leq\delta_1,\ldots,|w_m|\leq\delta_m\}$, and, by the uniqueness theorem for holomorphic functions, these functions satisfy (20.4) in $\overline{V}\times U_\varepsilon$. Since $c_{j\alpha}=0$ for $|\alpha|<2$, these functions also satisfy (20.35).

We prove the necessity of the conditions (20.36). If there exist functions (20.34) that are holomorphic in $\overline{V}$ and satisfy (20.4) and (20.35), then they coincide in $V(\gamma)$ with (20.40) because the solution of (20.4) is unique in $V(\gamma)\times U_\varepsilon$. Consequently, the Laurent series (20.42) do not contain terms with negative powers. The conditions (20.36) follow from this and (20.43).

If the conditions (20.36) hold, then the function (20.41) is also holomorphic in $\overline{V}$. Expanding the fraction $1/F^{\mathrm{I}}$ in (20.41) in a multiple geometric progression and then expanding (20.41) in a Taylor series, we get (20.37) with the coefficients

$$c_\alpha = \sum_{\beta \geq 0} \frac{(-1)^{|\beta|}}{(2\pi i)^{n+m}} \int_{\gamma \times \Gamma_\varepsilon} \frac{\Phi(w, z) g^\beta(w, z)(\partial F/\partial z)\, dw \wedge dz}{z^{\beta + \mathrm{I}} w^{\lambda(\beta+\mathrm{I})+\alpha+\mathrm{I}}}, \tag{20.45}$$

i.e., (20.38). It can be shown that the terms of (20.45) vanish for $|\beta| > N(\alpha)$, where $N(\alpha)$ is an integer depending on α. This gives us that in the integration cycle $\gamma \times \Gamma_\varepsilon$ the numbers $\delta_1, \ldots, \delta_m$, and ε can be chosen arbitrarily, provided only that they are small enough. □

6°. Estimation of the domain of convergence and the remainder of the series (20.8) and (20.19). We need the following generalization of the multi-dimensional analogue of the Schwarz lemma.

THEOREM 20.14. *Let $B_r^{(1)} = \{w \in \mathbf{C}^m : \|w\|_1 < r\}$ and $B_R^{(2)} = \{z \in \mathbf{C}^n : \|z\|_2 < R\}$ be balls in some norms $\|\cdot\|_1$ and $\|\cdot\|_2$, and suppose that $f\colon B_r^{(1)} \to B_R^{(2)}$ is a holomorphic mapping such that $f(0) = 0$ and the order of the zero at 0 is $d_0(f) = \min\{d_0(f_1), \ldots, d_0(f_n)\} = k \geq 1$. Then $\|f(w)\|_2 \leq R\|w\|_1^k/r^k$ for any $w \in B_r^{(1)}$.*

The proof is carried out just as was done for the case $k = 1$ in the book [201] (p. 56). It is only necessary to take into account that the vector-valued function $f(\zeta w)/\zeta^k = f(\zeta w_1, \ldots, \zeta w_n)/\zeta^k$ is holomorphic in the disk $|\zeta| < 1$ for any $w \in \partial B_r^{(1)}$. □

We estimate the domain of convergence of (20.8) first under certain additional assumptions.

THEOREM 20.15 (Yuzhakov). *Suppose that the functions* (20.3) *and $\Phi(w, z)$ are holomorphic in the closed polydisk $\overline{U}_{r,R} = \{(w, z) : \|w\| \leq r, \|z\| \leq R\}$ and satisfy the conditions* (20.6) *and*

$$\frac{\partial F_j(0,0)}{\partial w_k} = 0, \qquad j = 1, \ldots, n,\ k = 1, \ldots, m. \tag{20.46}$$

Then for $h \equiv 0$ the series (20.8) *converges absolutely and uniformly on compact subsets of the polydisk $V_\rho = \{w : \|w\| < \rho\}$ of radius $\rho = \min\{rR/A, r\}$, where $A = \max_{\overline{U}_{r,R}} \|g(w, z)\|$.*

PROOF. It follows from (20.6) and (20.46) that $d_{(0,0)}(g) \geq 2$. In $\mathbf{C}^{m+n}$ we introduce the norm $\|(w, z)\|_1 = \max\{\|w\|/r, \|z\|/R\}$. Then, by Theorem 20.14, for any $(w, z) \in \overline{U}_{r,R} = \{(w, z): \|(w, z)\|_1 \leqslant 1\}$ the inequality $\|g(w, z)\| \leqslant A\|(w, z)\|_1^2$ holds. This implies the inequalities

$$|g_j(w, z)| \leqslant A\sigma^2 = \delta\varepsilon A/rR \leqslant \delta\varepsilon/\rho < \varepsilon = |z_j|,$$
$$j = 1, \ldots, n,$$

for $w \in \overline{V}_\delta$ and $z \in \Gamma_\varepsilon$, where $\delta = r\sigma$, $\varepsilon = R\sigma$ and $\sigma < \min\{r/A, 1\}$. Consequently, for $w \in \overline{V}_\delta$ and $h \equiv 0$ the series (20.9), which coincides with (20.8), is majorized by the multiple geometric progression with ratio δ/ρ. Thus, for $h \equiv 0$ the series (20.8) converges uniformly in $\overline{V}_\delta \subset V_\rho$. □

The general case considered in Theorem 20.2 can be reduced to the case when condition (20.46) holds by means of the substitution

$$\tilde{z}_j = z_j - \sum_{k=1}^{m} \frac{\partial F_j(0,0)}{\partial w_k} w_k, \qquad j = 1, \ldots, n.$$

However, it is also useful in the general case to have estimates of the radius of the polycylinder of convergence of the series (20.8).

THEOREM 20.16. *Suppose that the functions* (20.3) *and* Φ *are holomorphic in* $\overline{U}_r = \{(w, z): \|(w, z)\| \leqslant r\}$ *and satisfy the conditions* (20.6). *Let*

$$\theta(w, z) = F(w, z) - z - F(w, 0), \quad K = \max_{\|w\| \leqslant r} \|F(w, 0)\|,$$
$$A = \max_{\|(w, z)\| \leqslant r} \|\theta(w, z)\|.$$

Then the series (20.8) *with* $h \equiv 0$ *converges absolutely and uniformly on compact subsets of the polydisk* U_ρ *of radius*

$$\rho = \begin{cases} r^3/4AK & \text{if } K \geqslant r/2,\ A \geqslant r/2, \\ r(r - A)/K & \text{if } A \leqslant r/2,\ A + K \geqslant r, \\ r(r - K)/A & \text{if } K \leqslant r/2,\ A + K \geqslant r, \\ r & \text{if } A + K \leqslant r. \end{cases} \tag{20.47}$$

PROOF. The conditions (20.6) imply that $d_0(F(w, 0)) \geqslant 1$, while $d_{(0,0)}(\theta) \geqslant 2$. Then, by Theorem 20.14, $\|F(w, 0)\| \leqslant K\|w\|/r$ and $\|\theta(w, z)\| \leqslant A\|(w, z)\|^2/r^2$ for any $(w, z) \in \overline{U}_r$. If $\varepsilon < r^2 t/A$, $\delta < r\varepsilon(1 - t)/K$, $0 < \delta \leqslant \varepsilon \leqslant r$, and $0 < t < 1$, then for $w \in \overline{V}_\delta$ and $z \in \Gamma_\varepsilon$

$$\|g(w, z)\| \leqslant \|F(w, 0)\| + \|\theta(w, z)\| \leqslant K\delta/r + A\varepsilon^2/r^2 < \varepsilon = |z_j|.$$

From this inequality it follows that (20.8) and (20.9) converge absolutely and uniformly in $\overline{U}_\delta$. Thus,

$$\rho = \sup\{\delta\} = \sup_{0 < t < 1} \{r^3 t(1 - t)/KA, r^2 t/A, r^2(1 - t)/K, r\}.$$

Investigating this expression for the maximum with respect to t, we get (20.47). □

Estimates for the series (20.19) are obtained similarly.

THEOREM 20.17. *If the functions* (20.16) *and* $\Phi(z)$ *are holomorphic in the closed polydisk* $\overline{U}_r = \{z: \|z\| \leqslant r\}$, *then the series* (20.19) *converges absolutely and uniformly on compact subsets of the polydisk* $V_\rho = \{w: \|w\| < \rho\}$, *where* $\rho = \min\{r^2/4A, r/2\}$ *and* $A = \max_{\|z\| \leqslant r} \|g(z)\|$.

It is possible to get a sharper estimate: $\rho = r^2/4A$ if $A > r/2$, and $\rho = r - A$ if $A \leqslant r/2$. Obviously, the respective series (20.10) and (20.20) also converge in the indicated polydisks.

We mention one more simple particular case.

PROPOSITION 20.18. *If in Theorem* 20.2 *the functions* (20.3) *and* Φ *are holomorphic in* $\overline{U}_{r,R}$ *and*

$$g_j(w, z) = F_j(w, z) - z_j = w_j \Psi_j(w, z), \qquad j = 1, \dots, n,$$

then the series (20.8) *converges in the polydisk* V_ρ *of radius*

$$\rho = \min\{R/A, r\}, \quad \textit{where } A = \max_{U_{r,R}} \|\Psi(w, z)\|.$$

We estimate the remainders for the series (20.8) and (20.19). Let $\Psi_\beta(w)$ be the general term, $S_k(w) = \sum_{|\beta| \leqslant k} \Psi_\beta(w)$ a partial sum, and

$$r_k(w) = \Psi(w) - S_k(w) = \sum_{|\beta| \leqslant k} \Psi_\beta(w)$$

the remainder of (20.8) with $h \equiv 0$ or of (20.19).

THEOREM 20.19. *Suppose that the functions* (20.3) *and* $\Phi(w, z)$ *are holomorphic in* $\overline{U}_{\delta,\varepsilon}$ *and that* $|g_j(w, z)| < |z_j| = \varepsilon$ *for any* $w \in \overline{V}_\delta$ *and* $z \in \Gamma_\varepsilon$. *Then the remainder of the series* (20.8) *with* $h \equiv 0$ *satisfies the following estimate for any* $w \in \overline{V}_\delta$:

$$|r_k(w)| \leqslant M \cdot (\sigma/\varepsilon)^{k+1} \sum_{p=1}^{n} \binom{k+n-p}{k} (1 - \sigma/\varepsilon)^{-p}, \tag{20.48}$$

where

$$\sigma = \max_{\overline{U}_{\delta,\varepsilon}} \|g(w, z)\|, \qquad M = \max_{\overline{U}_{\delta,\varepsilon}} \left| \Phi(w, z) \frac{\partial F}{\partial z} \right|.$$

PROOF. The integral representation of the terms of (20.8) (see (20.9)) gives us the estimate $|\Psi_\beta(w)| \leqslant M(\sigma/\varepsilon)^{|\beta|}$. Then

$$\begin{aligned} |r_k(w)| &\leqslant \sum_{|\beta| \geqslant k+1} |\Psi_\beta(w)| \leqslant M \sum_{|\beta| \geqslant k+1} (\sigma/\varepsilon)^{|\beta|} \\ &= M \sum_{q=k+1}^{\infty} \binom{q+n-1}{q} (\sigma/\varepsilon)^q. \end{aligned}$$

This implies (20.48), because for $0 < a < 1$ and $\varepsilon < a$ the remainder of the Taylor series of $(1-a)^{-n}$ is

$$\begin{aligned} \sum_{q=k+1}^{\infty} \binom{q+n-1}{q} a^q &= \sum_{q=k+1}^{\infty} \frac{1}{2\pi i} \int_{|\zeta|=\varepsilon} \frac{a^q \, d\zeta}{(1-\zeta)^n \zeta^{q+1}} \\ &= -\frac{1}{2\pi i} \int_{|\zeta|=\varepsilon} \frac{a^{k+1} \, d\zeta}{(1-\zeta)^{n+1} \zeta^k} = a^{k+1} \sum_{p=1}^{n} \binom{k+n-p}{k} (1-a)^{-p}. \quad \square \end{aligned}$$

REMARK. As is clear from the proof of Theorems 20.15 and 20.16, for any $w \in V_\rho$ there exist ε and δ satisfying the conditions of Theorem 20.19. By solving the elementary extremum problem we can choose ε and δ such that σ/ε is minimal. For example, under the conditions of Theorem 20.15 it is necessary to take $\delta = \|w\|$ and $\varepsilon = \|w\| R/r$.

The next result is proved similarly.

THEOREM 20.20. *Under the conditions of Theorem* 20.17,

$$|r_k(w)| \leqslant M(\sigma/\delta)^{k+1} \sum_{p=1}^{n} \binom{k+n-p}{k} (1-\sigma/\delta)^{-p}$$

for any $w \in \overline{V}_\delta$ *and* $\delta < \rho$, *where*

$$M = \max_{\|z\| \leqslant 2\delta} |\Phi(z) \partial\Psi/\partial z|, \qquad \sigma = \max_{\|z\| \leqslant 2\delta} \|g(z)\|.$$

7°. EXAMPLES. 1. We find the implicit functions $z_j = \varphi_j(w_1, w_2)$, $j = 1, 2$, determined by the system

$$z_1 - w_1 e^{z_2 w_2} = 0, \qquad z_2 - w_2 e^{z_1 w_1} = 0.$$

By (20.8) with $h \equiv 0$,

$$z_j = \sum_{m,n=0}^{\infty} \frac{w_1^m w_2^n}{m!\,n!} \frac{\partial^{m+n}}{\partial z_1^m \partial z_2^n} \left[z_j e^{m w_2 z_2 + n w_1 z_1} \left(1 - w_1^2 w_2^2 e^{w_1 z_1 + w_2 z_2}\right) \right]_{z=0}.$$

From this we get

$$z_1 = w_1 + \sum_{m,n=1}^{\infty} \frac{1}{m!n!} n^{m-1} m^n w_1^{2m-1} w_2^{2n},$$
$$z_2 = w_2 + \sum_{m,n=1}^{\infty} \frac{1}{m!n!} m^{n-1} n^m w_1^{2m} w_2^{2n-1}. \tag{20.49}$$

Using Proposition 20.18, we find the radius ρ of the bidisk of convergence of these series. Here $\Psi_1 = e^{w_2 z_2}$, $\Psi_2 = e^{w_1 z_1}$, and $A = \max_{\overline{U}_{r,R}} \|\Psi\| = e^{rR}$. We obtain $\rho = \min\{r, e^{-rR}\}$. This quantity has a minimum equal to $\rho = 1/\sqrt{e} \approx 0.607$ at $r = 1/\sqrt{e}$, $R = \sqrt{e}$. A direct study of the series (20.49) shows that this estimate is sharp. We find an estimate of the remainder for the series (20.49) by Theorem 20.19. For $\delta = \|w\|$, $\varepsilon = 1/\delta$, and $M = \max|z_j| = \varepsilon = 1/\delta$ (the terms of the series corresponding to the second term of the Jacobian $1 - w_1^2 w_2^2 \exp(w_1 z_1 + w_2 z_2)$ cancel) we have the estimate:

$$|r_k(w)| \leqslant \frac{1}{\delta}(\delta^2 e)^{k+1}\left[\frac{k+1}{1+\delta^2\varepsilon} + \frac{1}{(1-\delta^2 e)^2}\right].$$

In particular, $|r_k(w)| < 0.004$ for $\delta = 0.2$ and $k = 3$.

2. We find the inverse of the holomorphic mapping

$$w_1 = z_1 - z_2^p, \qquad w_2 = z_2 - z_1^q, \qquad p > 1, q > 1.$$

By (20.20) and (20.21),

$$z_j = \sum_{m,n=0}^{\infty} c_{jmn} w_1^m w_2^n, \qquad j = 1, 2,$$

where

$$c_{jmn} = \sum_{m+n \geqslant k+r \geqslant 0} \frac{1}{k!r!m!n!} \frac{\partial^{m+n+k+r}}{\partial z_1^{m+k} \partial z_2^{n+r}} \left[z_j z_2^{pk} z_1^{qr} (1 - pq z_1^{q-1} z_2^{p-1})\right]_{(0,0)}.$$

The computations yield

$$z_1 = \sum_{\substack{qr+1 \geqslant k \geqslant 0 \\ pk \geqslant r \geqslant 0}} \frac{(pqrk - 1)(qr + pk - r - k + 1)!}{k!r!(qr - k + 1)!(pk - r)!} w_1^{qr-k+1} w_2^{pk-r}.$$

An analogous series is obtained for z_2.

3. Let us determine the holomorphic branches of the curve (folium of Descartes)

$$w^3 - 3wz + z^3 = 0$$

in a neighborhood of $(0,0)$. According to Theorem 20.12, a regular branch tangent to the complex curve $z=0$ is determined by the series $z=\sum_1^\infty c_n w^n$, where

$$c_n=\sum_{k\geq 0}\frac{1}{(2\pi i)^2}\iint_{\substack{|w|=\delta\\ |z|=\varepsilon}}\frac{z(z^3+w^3)^k(w-z^2)\,dw\wedge dz}{3^k z^{k+1}w^{k+n+2}},$$

i.e.,

$$z=\sum_{r=0}^{\infty}\frac{(3r)!w^{3r+2}}{3^{3r+1}r!(2r+1)!}=\frac{w^2}{3}+\frac{w^5}{3^4}+\frac{w^8}{3^6}+\frac{4w^{11}}{3^9}+\cdots.$$

A second branch tangent to $w=0$ is determined similarly.

8°. Analogues of the Lagrange expansion for the coefficients of a Weierstrass pseudopolynomial. Let $f(z)$ be a holomorphic function of the variables $z=(z_1,\dots,z_n)$ in a neighborhood of the point $0\in\mathbf{C}^n$, with $f(0)=0$, $\partial^k f(0)/\partial z_n^k=0$, $k=1,\dots,m-1$, and $\partial^m f(0)/\partial z_n^m\neq 0$. According to the Weierstrass preparation theorem (see [201]), f can be represented in a neighborhood of $0\in\mathbf{C}^n$ in the form

$$f(z)=\left[z_n^m+a_1('z)z_n^{m-1}+\cdots+a_m('z)\right]\Psi(z), \tag{20.50}$$

where $a_j('z)$ and $\Psi(z)$ are holomorphic functions, $'z=(z_1,\dots,z_{n-1})$, $a_j('0)=0$, $j=1,\dots,m$, and $\Psi(0)\neq 0$. The polynomial in z_n in (20.50) is called the *Weierstrass pseudopolynomial of f at* 0. It is well known (see [201], p. 113) that there exist numbers $\varepsilon>0$ and $\delta>0$ such that for $'z\in {}'U_\delta=\{'z\colon \|'z\|<\delta\}$ the roots of the pseudopolynomial with respect to the variable z_n lie in the disk $|z_n|<\varepsilon$, and the sum of the kth powers of these roots can be expressed by the formula

$$S_k('z)=\sum_{j=1}^{m}z_{n_j}^k('z)=\frac{1}{2\pi i}\int_\gamma\frac{z_n^k f'_{z_n}(z)\,dz}{f(z)}, \tag{20.51}$$

where $\gamma=\{z_n\colon |z_n|=\varepsilon\}$. We find explicit expressions for the functions (20.51). For convenience we assume that $\partial^m f(0)/\partial z_n^m=m!$, i.e., that the coefficient of z_n^m in the Taylor expansion of the function at the point 0 is equal to 1.

THEOREM 20.21 (Yuzhakov). *Under the above assumptions the functions* (20.51) *can be expressed by the formula*

$$S_k('z)=k\sum_{r=0}^{\infty}\frac{(-1)^r}{(mr-k)!r}\left.\frac{\partial^{mr-k}[g^r(z)]}{\partial z_n^{mr-k}}\right|_{z_n=0}, \tag{20.52}$$

where $g(z)=f(z)-z_n^m$.

The coefficients of the Weierstrass pseudopolynomial can be expressed in terms of the power sums $S_k('z)$ by the Newton formula (see §21):

$$a_k('z)=-\tfrac{1}{k}\left[S_k('z)+a_1('z)S_{k-1}('z)+\cdots+a_{k-1}('z)S_1('z)\right]. \tag{20.53}$$

PROOF. It follows from the hypotheses that $\partial^k g(0)/\partial z_n^k = 0$ for $k = 1,\dots,m$. Starting from this, we can show that ε and δ can be chosen in such a way that $|g(z)|<|z_n|^m = \varepsilon^m$ for $'z \in 'U_\delta$ and $z_n \in \gamma$. Consequently, $1/f$ can be expanded in a multiple geometric progression

$$\frac{1}{f(z)} = \frac{1}{z_n^m[1 - g(z)/z_n^m]} = \sum_{r=0}^{\infty} \frac{(-1)^r g^r(z)}{z_n^{m(r+1)}} \tag{20.54}$$

which converges uniformly on γ for any $'z \in 'U_\delta$. Substituting (20.54) into (20.51), we get

$$\begin{aligned} S_k('z) &= \sum_{r=0}^{\infty} \frac{(-1)^r}{2\pi i} \int_\gamma \frac{g^r(z)\left[mz_n^{m-1} + g'_{z_n}(z)\right] dz_n}{z_n^{m(r+1)-k}} \\ &= \sum_{r=1}^{\infty} \frac{(-1)^r m}{2\pi i} \int_\gamma \frac{g^r(z)\, dz_n}{z_n^{mr-k+1}} + \sum_{r=1}^{\infty} \frac{(-1)^{r-1}}{r 2\pi i} \int_\gamma \frac{[g^r(z)]'_{z_n}\, dz_n}{z_n^{mr-k}}. \end{aligned}$$

By the formula for integration by parts (Proposition 0.8),

$$\int_\gamma \frac{[g^r(z)]'_{z_n}\, dz_n}{r z_n^{mr-k}} = \int_\gamma \frac{[mr-k] g^r(z)\, dz_n}{r z_n^{mr-k+1}}.$$

Thus,

$$S_k('z) = \sum_{r=1}^{\infty} \frac{(-1)^r k}{r \cdot 2\pi i} \int_\gamma \frac{g^r(z)\, dz}{z_n^{mr-k+1}},$$

which is equivalent to (20.52). □

EXAMPLE. Let us find the Weierstrass pseudopolynomial of the function $f(w, z) = w^5 + w^3 - z^3 + z^5$ at the point $(0,0) \in \mathbf{C}^2$ with respect to the variable w. Here $m = 3$ and $g(w, z) = w^5 - z^3 + z^5$. By (20.52),

$$S_k(z) = k \sum_{r=1}^{\infty} \frac{(-1)^r}{r(3r-k)!} \frac{\partial^{3r-k}}{\partial w^{3r-k}} (w^5 - z^3 + z^5)^2 \big|_{w=0}, \qquad k = 1,2,3.$$

From this we get

$$\begin{aligned} S_1(z) &= \sum_{q=0}^{\infty} \frac{(-1)^{q+1}(5q+1)!}{(3q+1)!(2q+1)!} (z^3 - z^5)^{2q+1}, \\ S_2(z) &= 2 \sum_{q=0}^{\infty} \frac{(-1)^q (5q+3)!}{(3q+2)!(2q+2)!} (z^3 - z^5)^{2q+2}, \\ S_3(z) &= 3 \sum_{q=0}^{\infty} \frac{(-1)^q (5q)!}{(3q)!(2q+1)!} (z^3 - z^5)^{2q+1}. \end{aligned}$$

Then the coefficients of the Weierstrass pseudopolynomial

$$w^3 + a_1(z) w^2 + a_2(z) w + a_3(z)$$

of f at $(0,0)$ are found from (20.53):

$$\begin{aligned} a_1(z) &= (z^3 - z^5) - 5(z^3 - z^5)^3 + 66(z^3 - z^5)^5 - \cdots, \\ a_2(z) &= -(z^3 - z^5)^2 + 9(z^3 - z^5)^4 - 136(z^3 - z^5)^6 + \cdots, \\ a_3(z) &= -(z^3 - z^5) + 2(z^3 - z^5)^3 - 23(z^3 - z^5)^5 + \cdots. \end{aligned}$$

§21. Application of the multidimensional logarithmic residue to systems of nonlinear equations

1°. Let D be a bounded domain in $\mathbf{C}^n$ with piecewise smooth boundary ∂D. We consider a system of equations

$$\begin{cases} f_1(z_1,\dots,z_n) = 0 \\ \vdots \\ f_n(z_1,\dots,z_n) = 0, \end{cases} \tag{21.1}$$

where the functions $f_i \in A(\overline{D})$ do not have any common zeros on ∂D. In this case (see §2) the system (21.1) has at most finitely many roots (common zeros) in D. Denote the number of these roots by N. The formulas in §§3 and 4 enable us to compute sums of the form

$$\sum_{l=1}^{N} \varphi(z^{(l)}) \tag{21.2}$$

with the help of the integral over the boundary ∂D or over the skeleton, where φ is any function in $A_C(D)$, and the $z^{(l)}$ ($l = 1,\dots,N$) are the roots lying in D, with each term $\varphi(z^{(l)})$ taken as many times as the multiplicity([1]) of the root $z^{(l)}$.

If we set $\varphi = 1$, then the corresponding integrals will be equal to the number of roots (counting multiplicity) in D, i.e., to N. This is written briefly as follows:

$$\varphi = 1 \Rightarrow \int \cdots = N. \tag{21.3}$$

Further,

$$\begin{aligned} \varphi = z_1 &\Rightarrow \int \cdots = \sum_{l=1}^{N} z_1^{(l)}, \\ \varphi = z_1^2 &\Rightarrow \int \cdots = \sum_{l=1}^{N} \left(z_1^{(l)}\right)^2, \\ &\cdots\cdots\cdots\cdots\cdots\cdots \\ \varphi = z_1^N &\Rightarrow \int \cdots = \sum_{l=1}^{N} \left(z_1^{(l)}\right)^N. \end{aligned} \tag{21.4}$$

([1]) In sums of the form (21.2) each root is counted according to its multiplicity everywhere in this section.

In these formulas we have for definiteness written the power sums of the first coordinates of the roots. The power sums of the other coordinates are written similarly. Thus, computation of the integrals (21.3) and (21.4) enables us to replace the system (21.1) (which is in general transcendental) by an algebraic system of the simple form (we again write the system only for the unknown first coordinates of the roots):

$$\sum_{l=1}^{N} \left(z_1^{(l)}\right)^j = S_j, \qquad j = 1,\ldots,N.$$

It is possible to proceed in various ways from here. We discuss some variants.

METHOD I. The elementary symmetric polynomials which are (to within a sign) the coefficients of a polynomial $\Omega(z_1) = z_1^N + b_1 z_1^{N-1} + \cdots + b_{N-1}z_1 + b_N$, having the roots $z_1^{(1)},\ldots,z_1^{(N)}$ are expressed in terms of the power sums S_j. To find the coefficients b_j it is possible to use either the familiar Waring formula (see, for example, [128], Chapter III, §5.3) or the *Newton recursion formula*

$$jb_j = -S_j - S_{j-1}b_1 - \cdots - S_1 b_{j-1}, \qquad j = 1,\ldots,N. \tag{21.5}$$

Thus, in finding the polynomial $\Omega(z_1)$ we eliminate all the unknowns except one, with no superfluous roots added and no roots missing. The problem is reduced to the thoroughly studied problem of finding the roots of a polynomial in a single complex variable. The main difficulty in this approach is the necessity of computing the multidimensional integrals in (21.4) with sufficient accuracy.

METHOD II. The integral (21.3) is computed, then the domain D is divided into two domains, $D = D_1 \cup D_2$, and the integral (21.3) is sought. For D_1 we get the number N_1 of roots in D_1 (and, consequently, the number $N_2 = N - N_1$ in D_2); then we divide each of these domains into two parts, and so on, until a domain is divided into subdomains each containing one or two roots of the system. For such subdomains we find the first (or the first two) of the integrals in (21.4) and get at once the root (or a quadratic equation for two roots).

This method is especially convenient for the case where D is a rectangular parallelepiped and the integrals in (21.3) are taken over the whole of ∂D. The integral of interest to us splits into a finite number of integrals over faces: ordinary $(2n - 1)$-dimensional integrals. In computing them we can calculate beforehand the integrals over parts of the faces, and then it will be easier later to perform the calculations for smaller parallelepipeds. If during the division of the parallelepiped into parts one of the corresponding "partition walls" by chance hits a root of the system, then the integral becomes improper. This can be taken into account in making up a computer program. (If the integral (21.3) for a subdomain is computed to be larger than N, then the "partition wall" should be moved slightly.)

If the number of roots does not decrease during the implementation of a fairly large number of divisions of the domain, this indicates the existence of a multiple root (or of several close simple roots).

A disadvantage of this method is that it is not known beforehand how any divisions of the domain must be performed; on the other hand, the integral (21.3), computed for each division, is an integer and, consequently, need be computed only to within $\frac{1}{3}$.

METHOD III. It can be applied when all the roots (or their first coordinates) are real and it is necessary to find a root with the largest first coordinate. By a change of variables (a translation) we can make the first coordinates of all the roots positive. It is then necessary to use an elementary inequality: if $a_1,\ldots,a_n$ are positive, then

$$1 < \sqrt[k]{\Sigma_{i=1}^{N} a_i^k} / a_{\max} \leqslant \sqrt[k]{N},$$

hence, there is a k such that

$$\sqrt[k]{\Sigma_{i=1}^{N} a_i^k}$$

is sufficiently close to $a_{\max}$. Therefore, to find the largest of the first coordinates of the roots with the necessary accuracy it suffices to compute one of the integrals of the form in (21.4) with $\varphi = z_1^k$.

In conclusion we note that the integrals in §3 contain a large amount of arbitrariness (the choice of the vector-valued function w or of a set of vector-valued functions) for $n > 1$ which is not present in the one-dimensional case. This arbitrariness can be used as follows: for the system (21.1) we should choose a suitable formula for the logarithmic residue so that the corresponding integrals can be computed exactly (see the next subsection) or approximately.

2°. Consider the *system of algebraic equations*

$$Q_j(z) + P_j(z) = 0, \qquad j = 1,\ldots,n, \tag{21.6}$$

where the $Q_j(z)$ are the homogeneous polynomials of highest degree in the variables jointly and have only one common zero: the origin. The degrees of Q_j are denoted by k_j, so that the degree of P_j is less than k_j, $j = 1,\ldots,n$. By

Theorem 2.5 (Rouché), the system (21.6) has finitely many roots. Let $m_1, \dots, m_n$ be positive integers such that the equation

$$\sum_{j=1}^{n} w_j Q_j(z) = \sum_{j=1}^{n} |z_j|^{2m_j} \tag{21.7}$$

has a solution of the form

$$w_j = \sum_{k=1}^{n} a_{jk}(z) \bar{z}_k^{m_k}, \qquad j = 1, \dots, n, \tag{21.8}$$

where the a_{jk} are polynomials in z. The indicated condition is certainly satisfied by $m_j = |k| + 1 - n$, $j = 1, \dots, n$, $|k| = k_1 + \cdots + k_n$ (see the theorem of Macaulay [303]), but sometimes the m_j can be chosen to be smaller. A solution (21.8) (which, generally speaking, is not unique) of (21.7) can be found by the method of undetermined coefficients, i.e., the $a_{jk}(z)$ are determined after solving the corresponding system of linear equations. The vector-valued function w found can now be used in formula (3.18) for $D = B_r^{(m)}(0)$, and this permits us to prove the next assertion.

THEOREM 21.1 (Aĭzenberg). *If $R(z)$ is a polynomial of degree μ, then*

$$\sum_l R(z^{(l)}) = \mathfrak{M}\left[R\Delta_1\Delta_2 \sum_{j=0}^{\mu} \frac{(-1)^j}{j!} \langle w, P \rangle^j \right], \tag{21.9}$$

where Δ_1 is the Jacobian of the system (21.6), $\Delta_2 = \det\|a_{jk}\|$, *the summation on the left-hand side of* (21.9) *is over all roots of the system* (21.6), *and $\mathfrak{M}$ is the linear functional on the polynomials in $z_1, \dots, z_n, \bar{z}_1^{m_1}, \dots, \bar{z}_n^{m_n}$ that is given by*

$$\mathfrak{M}\left[z^{\beta} \bar{z}_1^{m_1\alpha_1} \cdots \bar{z}_n^{m_n\alpha_n} \right] = \begin{cases} \alpha!, & \text{if } \beta_j = m_j\alpha_j + m_j - 1, j = 1, \dots, n, \\ 0, & \text{otherwise}. \end{cases}$$

PROOF. By (3.18), for sufficiently large r

$$\sum_l R(z^{(l)}) = \frac{(n-1)!}{(2\pi i)^n} \int_{\partial B_r^{(m)}} \frac{R \sum_{k=1}^{n} (-1)^{k-1} w_k dw_{[k]} \wedge d(Q+P)}{\langle Q + P, w \rangle^n}. \tag{21.10}$$

It is not hard to establish that

$$\sum_{k=1}^{n} (-1)^{k-1} w_k \, dw_{[k]} \wedge d(Q+P) = \Delta_1\Delta_2 \sum_{k=1}^{n} (-1)^{k-1} \bar{z}_k^{m_k} d\bar{z}_{[k]}^{(m)} \wedge dz$$

and, for large r, $r^2 > |\langle w, P\rangle|$ on $\partial B_r^{(m)}(0)$; therefore, the integral in (21.10) is equal to

$$\frac{(n-1)!}{(2\pi i)^n}\int_{\partial B_r^{(m)}} \frac{R\Delta_1\Delta_2\Sigma_{k=1}^n (-1)^{k-1}\bar z_k^{m_k}\, d\bar z_{[k]}^{(m)} \wedge dz}{(r^2+\langle w, P\rangle)^n}$$

$$= \frac{1}{(2\pi i)^n}\int_{\partial B_r^{(m)}} R\Delta_1\Delta_2 \sum_{j=0}^{\infty} (-1)^j \frac{(j+n-1)!}{j!\, r^{2(j+n)}}\langle w, P\rangle^j$$

$$\times \sum_{k=1}^{n} (-1)^{k-1}\bar z_k^{m_k}\, d\bar z_{[k]}^{(m)} \wedge dz.$$

Next, we use a formula that is a certain generalization of (5.2) and is proved similarly:

$$\frac{1}{(2\pi i)^n}\int_{\partial B_r^{(m)}} z^\beta \bar z_1^{m_1\alpha_1}\cdots \bar z_n^{m_n\alpha_n}\sum_{j=1}^{n}(-1)^{j-1}\bar z_j^{m_j}\, d\bar z_{[j]}^{(m)}\wedge dz$$

$$= \begin{cases} r^{2(n+|\alpha|)}\dfrac{\alpha!}{(|\alpha|+n-1)!} & \text{if } \beta_j = m_j(\alpha_j+1)-1, \\ 0 & \text{otherwise.}\end{cases}$$

We get

$$\sum_l R(z^{(l)}) = \sum_{j=0}^{\infty} \mathfrak{M}\left[R\Delta_1\Delta_2\frac{(-1)^j}{j!}\langle w, P\rangle^j\right]. \tag{21.11}$$

It remains to show that the terms of the series on the right-hand side of (21.11) are equal to zero for $j > \mu$. We compute the degree of the polynomial in the square brackets in (21.11) with respect to z, after representing it in the form

$$R\Delta_1\Delta_2\frac{(-1)^j}{j!}\sum_{|\alpha|=j} w^\alpha P^\alpha \frac{j!}{\alpha!}.$$

The term corresponding to the multi-index α has degree in z

$$\mu + |m| - n + \sum_{p=1}^{n} m_p\alpha_p - j.$$

According to the definition of the functional $\mathfrak{M}$, if it is nonzero, then

$$\mu + |m| - n + \sum_{p=1}^{n} m_p\alpha_p - j \geqslant \sum_{p=1}^{n} m_p\alpha_p + |m| - n,$$

i.e., $\mathfrak{M}$ is nonzero only when $j \leqslant \mu$. $\square$

COROLLARY 21.2. $\Sigma_l R(z^{(l)}) = t/s^{\mu+n}$, *where* t *is a polynomial in the coefficients of the system* (21.6) *and of the polynomial* R, *while* s *is a polynomial only in the coefficients of the polynomials* $Q_1,\ldots,Q_n$ *and does not depend on* R.

Formula (21.9) has an especially simple form in the case of the system

$$z_j^{k_j} + P_j(z) = 0, \qquad j = 1,\ldots,n, \tag{21.12}$$

where the degree of P_j is less than k_j, $j = 1,\ldots,n$. Here we can take $m_i = k_i$, $i = 1,\ldots,n$, so that $w_i = \bar{z}_i^{k_i}$, and the determinant $\Delta_2 = 1$, and we arrive at the following assertion.

THEOREM 21.3 (Aĭzenberg). *For the system* (21.12) *and a polynomial* $R(z)$ *of degree* μ

$$\sum_l R(z^{(l)}) = \mathfrak{N}\left[R\Delta_1 \frac{z_1\cdots z_n}{z_1^{k_1}\cdots z_n^{k_n}} \sum_{|\alpha|=0}^{\mu} (-1)^{|\alpha|}\left(\frac{P_1}{z_1^{k_1}}\right)^{\alpha_1}\cdots\left(\frac{P_n}{z_n^{k_n}}\right)^{\alpha_n}\right], \tag{21.13}$$

where the summation on the left-hand side is over all the roots of (21.12), *and* $\mathfrak{N}$ *is the linear functional on the polynomials in* $z_1,\ldots,z_n$ *and* $1/z_1,\ldots,1/z_n$ *that assigns to each such polynomial its free term.*

PROOF. For our case (21.9) has the form

$$\sum_l R(z^{(l)}) = \mathfrak{M}\left[R\Delta_1 \sum_{|\alpha|=0}^{\mu} (-1)^{|\alpha|} \frac{(\bar{z}_1^{k_1}P_1)^{\alpha_1}\cdots(\bar{z}_n^{k_n}P_n)^{\alpha_n}}{\alpha!}\right],$$

and it is now obvious that the expression obtained is equal to the right-hand side of (21.13). □

Theorem 21.3 (and Theorem 21.4) lead to a method of elimination (see 1°) of unknowns that seems to us simpler than the familiar classical method of elimination using the resultants of polynomials.

We remark that if the system (21.12) has real coefficients, then after elimination of the unknowns we get a polynomial $\Omega(z_1)$ that also has real coefficients. If we are only interested in the real roots of (21.12), then it remains to find the real roots of $\Omega(z_1)$ with sufficient accuracy, i.e., things reduce to a thoroughly studied classical problem.

The formula (21.13) leads to especially simple computations when the degree of $R(z)$ is small.

EXAMPLE 1. If we are interested in the number N of roots of (21.12), then in (21.13) we must take $R(z) \equiv 1$ and $\mu = 0$. In this case

$$N = \mathfrak{R}\left[\Delta_1 \frac{z_1 \cdots z_n}{z_1^{k_1} \cdots z_n^{k_n}}\right],$$

and it remains to find the coefficient of the monomial $z_1^{k_1-1} \cdots z_n^{k_n-1}$ in the Jacobian Δ_1. This coefficient is obviously equal to $k_1 \cdots k_n$, and we have obtained Bézout's theorem([2]) for the given case: $N = k_1 \cdots k_n$.

EXAMPLE 2. In the three-dimensional space $\mathbf{R}^3$ we consider the three third-order surfaces:

$$\begin{cases} x_1^3 + \sum\limits_{i+j+k\leq 2} a_{ijk} x_1^i x_2^j x_3^k = 0, \\ x_2^3 + \sum\limits_{i+j+k\leq 2} b_{ijk} x_1^i x_2^j x_3^k = 0, \\ x_3^3 + \sum\limits_{i+j+k\leq 2} c_{ijk} x_1^i x_2^j x_3^k = 0, \end{cases} \tag{21.14}$$

where i, j, k are nonnegative integers, and the coefficients $a_{ijk}, b_{ijk}, c_{ijk}$ are real numbers. We require that the surfaces (21.14) be "in general position" in the sense that they have a maximal number of points of intersection in $\mathbf{R}^3$ (counting multiplicity), i.e. (see Example 1) 27 common points. Consider any fixed point $(A, B, C) \in \mathbf{R}^3$. We pose the problem of computing the mean square distance from this point to the 27 common points of the surfaces (21.14). It suffices to find the sum of the squares of the distances from (A, B, C) to the points of intersection of (21.14). By (21.13) this is easily done, because the square of the distance is a polynomial of only second order. We get

$$\sum_{l=1}^{27} \left[\left(A - x_1^{(l)}\right)^2 + \left(B - x_2^{(l)}\right)^2 + \left(C - x_3^{(l)}\right)^2\right]$$
$$\begin{aligned} &= 9\left(a_{200}^2 + b_{020}^2 + c_{002}^2\right) - 18\left(a_{100} + b_{010} + c_{001}\right) \\ &\quad + 6\left(a_{101}c_{002} + a_{110}b_{020} + a_{200}b_{110} + b_{011}c_{002} + a_{200}c_{101} + b_{020}c_{011}\right) \\ &\quad + 12\left(a_{002}c_{101} + a_{020}b_{110} + a_{110}b_{200} + b_{002}c_{011} + b_{011}c_{020} + a_{101}c_{200}\right) \\ &\quad + 27\left(A^2 + B^2 + C^2\right) + 18\left(Aa_{200} + Bb_{020} + Cc_{002}\right). \end{aligned}$$

It is curious that the answer does not depend on 12 of the 30 coefficients of the surfaces (21.14).

([2]) See Theorem 22.5.

In what follows we generalize Theorem 21.3 in order to include the case of systems of a more general form than (21.12). Although systems of a more special form than (21.6) will be considered, the result will be more clear (it is not necessary, as under the conditions of Theorem 21.1, to find solutions of (21.7)). We first consider a system of algebraic equations of the form

$$f_i(z) = z_i^{k_i} + \sum_{j=1}^{i-1} z_j\varphi_{ij}(z) + P_i(z), \qquad i = 1,\dots,n, \tag{21.15}$$

where the $\varphi_{ij}(z)$ are homogeneous polynomials of degree $k_i - 1$, and the $P_i(z)$ are polynomials of degree at most $k_i - 1$. System (21.12) is obtained from (21.15) when all the φ_{ij} are identically 0. As before, let Δ_1 be the Jacobian of system (21.15), let $\mathfrak{R}$ be the functional that assigns to a polynomial in $z_1,\dots,z_n$ and $1/z_1,\dots,1/z_n$ its free term, and let $z^{(1)},\dots,z^{(N)}$ be the roots of (21.15),

THEOREM 21.4 (Aĭzenberg and Tsikh). *The equality*

$$\sum_{l=1}^{N} R(z^{(l)}) = \mathfrak{R}\left[\frac{R\Delta_1 z_1\cdots z_n}{z_1^{k_1}\cdots z_n^{k_n}}\sum_{\alpha}(-1)^{|\alpha|}\left(\frac{s_1}{z_1^{k_1}}\right)^{\alpha_1}\cdots\left(\frac{s_n}{z_n^{k_n}}\right)^{\alpha_n}\right] \tag{21.16}$$

holds for any polynomial $R(z)$ of degree m_i in the variable z_i $(i = 1,\dots,n)$ and of degree μ in the variables jointly, where $s_i(z) = f_i(z) - z_i^{k_i}$, $i = 1,\dots,n$, and the summation on the right-hand side of (21.16) *is taken over all n-tuples α with integer coordinates in the parallelepiped* $M = \{\alpha\colon\ 0 \leqslant \alpha_1 \leqslant \mu,\ 0 \leqslant \alpha_2 \leqslant k_1(|m|+1) - m_1 - 2,\ 0 \leqslant \alpha_j \leqslant k_1\cdots k_{j-1}(|m|+1) - k_2\cdots k_{j-1}(m_1+1) - k_3\cdots k_{j-1}(m_2+1) - \cdots - k_{j-1}(m_{j-2}+1) - (m_{j-1}+1) - 1,\ j = 3, 4,\dots,n\}$.

This can be proved in the same way as Theorem 21.1 by using the $(2n-1)$-dimensional logarithmic residue (more precisely, one of the formulas for this residue with a suitable vector-valued function w). But in the present case it is also possible to take another path, namely, to use the n-dimensional logarithmic residue and Theorem 4.8. To demonstrate the different approaches we give a proof by the second method. Let Δ_r be the distinguished boundary of the polydisk U_r, i.e., $\Delta_r = \{z\colon\ |z_i| = r_i,\ i = 1,\dots,n\}$, $r = (r_1,\dots,r_n)$, $r_i > 0$, $i = 1,\dots,n$. We first choose r so that on Δ_r

$$|z_i^{k_i}| > \left|\sum_{j=1}^{i-1} z_j\varphi_{ij}(z)\right|, \qquad i = 2,\dots,n. \tag{21.17}$$

By increasing r_n, we can make (21.17) hold for $i = n$. Or we can leave r_n alone and decrease $r_1,\dots,r_{n-1}$. We then decrease $r_1,\dots,r_{n-2}$ so that (21.17) is valid for $i = n$ and $i = n - 1$, and so on. The required inequalities (21.17) are obtained in $n-1$ steps. It is now not difficult to establish the existence of an r such that on Δ_r

$$|z_i^{k_i}|>|s_i(z)|, \qquad i=1,\dots,n; \tag{21.18}$$

it suffices to use (21.17) and replace $(r_1,\dots,r_n)$ by $(Rr_1,\dots,Rr_n)$ with sufficiently large R; here it is necessary to take into account that the degree of P_i is less than k_i, $i=1,\dots,n$.

Thus, let r be such that (21.18) holds. By Rouché's theorem (Theorem 2.5), all the roots of (21.15) lie in the polydisk U_r. We next apply Theorem 4.8:

$$\sum_{l=1}^{N} R(z^{(l)}) = \frac{1}{(2\pi i)^n}\int_{\Delta_r} R(z)\frac{df(z)}{f(z)} = \frac{1}{(2\pi i)^n}\int_{\Delta_r}\frac{R\Delta_1\,dz}{\prod_{i=1}^n(z_i^{k_i}+s_i)}$$

$$= \frac{1}{(2\pi i)^n}\int_{\Delta_r}\frac{R\Delta_1}{z_1^{k_1}\cdots z_n^{k_n}}\sum_{\alpha_1=0}^{\infty}(-1)^{\alpha_1}\left(\frac{s_1}{z_1^{k_1}}\right)^{\alpha_1}\cdots\sum_{\alpha_n=0}^{\infty}(-1)^{\alpha_n}\left(\frac{s_n}{z_n^{k_n}}\right)^{\alpha_n}dz$$

$$= \sum_{\alpha}\mathfrak{N}\left[\frac{R\Delta_1 z_1\cdots z_n}{z_1^{k_1}\cdots z_n^{k_n}}(-1)^{|\alpha|}\left(\frac{s_1}{z_1^{k_1}}\right)^{\alpha_1}\cdots\left(\frac{s_n}{z_n^{k_n}}\right)^{\alpha_n}\right]. \tag{21.19}$$

It remains to show that the terms in (21.19) are equal to zero for $\alpha\notin M$. Since Δ_1, s_1 and s_j $(j=2,\dots,n)$ are polynomials in z of degree at most $|k|-n$, k_1-1, and k_j $(j=2,\dots,n)$, respectively, for a nonzero term in (21.19) it is necessary that

$$\mu+|k|+\sum_{j=1}^n k_j\alpha_j-\alpha_1\geqslant|k|+\sum_{j=1}^n k_j\alpha_j,$$

i.e., $\alpha_1\leqslant\mu$. We next compute the maximal possible degree d_i of the polynomial $P\Delta_1 z_1\cdots z_n s_1^{\alpha_1}\cdots s_n^{\alpha_n}$ in the variables $z_i, z_{i+1},\dots,z_n$ jointly. The polynomials s_j $(j=1,\dots,i)$ have degree at most k_j-1 in these variables, while the polynomials s_j $(j=i+1,\dots,n)$ have degree at most k_j. Therefore,

$$d_i=\sum_{j=1}^n m_j+|k|-2i+2+\sum_{j=1}^n k_j\alpha_j-\alpha_1-\cdots-\alpha_i. \tag{21.20}$$

In (21.20) there is a contribution only from terms with α for which

$$d_i\geqslant\sum_{j=i}^n k_j(\alpha_j+1). \tag{21.21}$$

By (21.20) and (21.21),

$$\alpha_i\leqslant\sum_{j=i}^n m_j+\sum_{j=1}^{i-1}k_j-2i+2+\sum_{j=1}^{i-1}(k_j-1)\alpha_j. \tag{21.22}$$

Formula (21.22) enables us to successively find the limits of variation of α_i when they are known for $\alpha_1, \ldots, \alpha_{i-1}$. In particular, it follows from (21.22) that

$$\alpha_2 \leqslant k_1(|m|+1) - m_1 - 2, \qquad \alpha_3 \leqslant k_1 k_2(|m|+1) - k_2 m_1 - k_2 - m_2 - 2.$$

We now carry out the induction. Suppose that there are nonzero terms in (21.19) only for α such that

$$\begin{aligned}\alpha_j \leqslant k_1 \cdots k_{j-1}(|m|+1) - k_2 \cdots k_{j-1}(m_1+1) \\ - \cdots - k_{j-1}(m_{j-2}+1) - (m_{j-1}+1) - 1,\end{aligned} \tag{21.23}$$

where $j = 3, 4, \ldots, i-1$. It is easy to see that (21.22) and (21.23) give us the estimate also for α_i. Thus, we can confine ourselves in (21.19) to a finite number of terms, namely, those with $\alpha \in M$. □

EXAMPLE 3. In the plane let us consider two "remarkable" third-order curves: the Tschiunhausen cubic (see [150], p. 89) and the Chasles cubic (see [154], p. 229):

$$\begin{cases} 2(2p+x)^3 = 27p(x^2+y^2), \\ y = ax^3 + bx^2y + cxy^2 + dy^3. \end{cases} \tag{21.24}$$

It is easy to investigate this system when $a = b = c = 0$ and to show that it has nine distinct real roots for $p = 1$ and $\frac{3}{4} < d < 1$; therefore, we can assert that the curves in (21.24) are "in general position" (i.e., have the maximal possible number (nine) of points of intersection) if $|a|, |b|, |c| < \varepsilon$, $|p-1| < \varepsilon$ and $\frac{3}{4} - \varepsilon < d < 1 + \varepsilon$ for sufficiently small $\varepsilon > 0$. We assume that the "general position" condition is satisfied.

To each generic point (x, y) of the curves (21.24) we assign the area (algebraic) of the rectangle bounded by the coordinate axes and the perpendiculars from (x, y) to these axes, i.e., the number xy. Consider the problem of computing the sum of all these areas. This problem is easy to solve with the help of (21.16):

$$\begin{aligned}\sum_{l=1}^{9} x^{(l)}y^{(l)} = {} & -c + \frac{p}{4}(22 + 121p) \\ & - \frac{p}{4d}(54a + 75b + 76c + 1329pa + 594pb + 715pc) \\ & + \frac{p}{4d^2}(54b^2 + 76c^2 + 108bc + 264pbc + 729pb^2 + 1458pab) \\ & - \frac{p}{4d^3}(54c^3 + 162bc^2 + 1323pc^3 + 2187pbc^2 + 2187pac + 2187pb^2c) \\ & + \frac{p}{4d^4}(54c^4 - 729pc^4 + 729pbc^3) - \frac{p}{4d^5}729pc^5.\end{aligned}$$

We extend the result obtained in Theorem 21.4 to the case where the polynomials $f_i - P_i$ $(i = 1,\dots,n)$ in (21.15) are not simply homogeneous, but *weighted-homogeneous with proportional weights* and of highest weighted degree. This means the existence of positive rational numbers $a_1,\dots,a_n$ and $c_1,\dots,c_n$ such that

$$z_i^{k_i} + \sum_{j=1}^{i-1} z_j \varphi_{ij}(z)$$

consists of monomials z^β such that $\beta_1/a_1 + \cdots + \beta_n/a_n = c_i$, while $P_i(z)$ consists of monomials such that $\beta_1/a_1 + \cdots + \beta_n/a_n < c_i$, $i = 1,\dots,n$. Then there is an integer $q > 0$ such that all the numbers $l_j = q\Pi_{l\neq j} a_l$, $j = 1,\dots,n$, are integers.

THEOREM 21.5 (Aĭzenberg and Tsikh). *Suppose that the system* (21.15) *satisfies the weighted-homogeneity conditions indicated above, and that $R(z)$ is any polynomial. Let μ' be the maximal of the numbers $\beta_1 l_1 + \cdots + \beta_n l_n$, with the maximum taken over all monomials z^β in $R(z)$, and let m'_j be the maximal of the numbers $\beta_j l_j$, $j = 1,\dots,n$. Then* (21.16) *holds, where the summation on the right-hand side is over all n-tuples α with integer coordinates in the parallelepiped*

$$\begin{aligned} M' = \{\alpha\colon 0 \leqslant \alpha_1 \leqslant \mu', 0 \leqslant \alpha_2 \leqslant k_1 l_1(|m'| + 1) - m'_1 - 2,\\ 0 \leqslant \alpha_j \leqslant k_1 l_1 \cdots k_{j-1} l_{j-1}(|m'| + 1) - k_2 l_2 \cdots k_{j-1} l_{j-1}(m'_1 + 1)\\ - \cdots - k_{j-1} l_{j-1}(m'_{j-2} + 1) - (m'_{j-1} + 1) - 1, j = 3, 4,\dots,n\}. \end{aligned}$$

PROOF. It is easy to verify that the change of variables $z_j(\zeta) = \zeta_j^{l_j}$, $j = 1,\dots,n$, reduces things to the case considered in Theorem 21.4. Therefore, denoting by $\zeta^{(1)},\dots,\zeta^{(N')}$ the roots of the system

$$\tilde{f}_i(\zeta) = f_i(z(\zeta)) = 0, \qquad i = 1,\dots,n, \tag{21.25}$$

we get

$$\begin{aligned} \sum_{j=1}^{N'} R(z(\zeta^{(j)})) = \mathfrak{N}\Bigg[& \frac{R(z(\zeta))\Delta_1(\tilde{f})\zeta_1 \cdots \zeta_n}{\zeta_1^{k_1 l_1} \cdots \zeta_n^{k_n l_n}} \\ & \times \sum_\alpha (-1)^{|\alpha|} \left(\frac{s_1(z(\zeta))}{\zeta_1^{k_1 l_1}}\right)^{\alpha_1} \cdots \left(\frac{s_n(z(\zeta))}{\zeta_n^{k_n l_n}}\right)^{\alpha_n}\Bigg]. \end{aligned} \tag{21.26}$$

To each root $z^{(j)}$, $j = 1,\dots,n$, of (21.15) there correspond $l_1 \cdots l_n$ roots of (21.25) under our conditions, and the coordinates of these roots are roots of degree l_q from $z_q^{(j)}$, $q = 1,\ldots,n$. Consequently,

$$\sum_{j=1}^{N} R(z^{(j)}) = \frac{1}{l_1 \cdots l_n} \sum_{j=1}^{N'} R(z(\zeta^{(j)})). \tag{21.27}$$

Taking account of the connection between the Jacobians of the systems (21.15) and (21.25), we now get (21.16) from (21.26) and (21.27) after returning to the original variables. The limits of variation of α in this case can be established just as in the proof of Theorem 21.4. □

Theorem 21.4 can be obtained from Theorem 21.5, because in the case of the usual homogeneity $l_1 = \cdots = l_n = 1$ and $\mu' = \mu$, $M' = M$. On the other hand, the proof of Theorem 21.5 amounts to reducing matters to the case considered in Theorem 21.4 by a suitable change of variables. We regard Theorem 21.4 as the main result of these two; therefore, it is formulated separately even though it is a consequence of Theorem 21.5. Formula (21.16) can be applied to solve various problems directly (see Examples 1–3; it is also not hard to show that the number of roots is again $N = k_1 \cdots k_n$ in the case of Theorem 21.5) or as a method of elimination of unknowns. It is sometimes useful to combine it with the method to be presented in §22.

Suppose that we are given a system of the form

$$f_j(z, z_{n+1}) = 0, \qquad j = 1,\ldots,n, n+1, \tag{21.28}$$

where the functions $f_j(z, z_{n+1})$, $j = 1,\ldots,n$, have the form (21.15) for fixed z_{n+1}, while f_{n+1} is an arbitrary polynomial. Let $z^{(j)}(z_{n+1})$, $j = 1,\ldots,N$, be the roots of the system of equations $f_j(z, z_{n+1}) = 0$, $j = 1,\ldots,n$, with z_{n+1} as a parameter. The holomorphic function

$$r(z_{n+1}) = \prod_{j=1}^{N} f_{n+1}(z^{(j)}(z_{n+1}), z_{n+1})$$

is called the *resultant of the polynomial* f_{n+1} *with respect to the system* $f_j = 0$, $j = 1,\ldots,n$, and its zeros are precisely the $(n+1)$st coordinates of the roots of (21.28). We can find $r(z_{n+1})$ as follows: use (21.16) to determine the power sums

$$S_q(z_{n+1}) = \sum_{j=1}^{N} f_{n+1}^q(z^{(j)}(z_{n+1}), z_{n+1}), \qquad q = 1,\ldots,N,$$

then use Newton's formula (21.5) to compute $b_N = (-1)^N r(z_{n+1})$ (from this it is clear, in particular, that $r(z_{n+1})$ is a polynomial). The determination of $r(z_{n+1})$ means the elimination of the unknowns $z_1,\ldots,z_n$, and also enables us to find out how many (finite) roots system (21.28) has.

The method of elimination based on (21.13) or (21.16) can be realized considerably more easily than the classical method, which uses resultants of polynomials. For example, these formulas contain a determinant of order n, while if elimination is performed by the classical method, at the last step it is necessary to compute a determinant of order

$$k_1^{2^{n-3}} k_2^{2^{n-4}} \cdots k_{n-3}^2 k_{n-2}(k_{n-1} + k_n).$$

Below we give an example of our method in which the computations are carried out "by hand." We do not know how to employ the classical method in this example without using a computer, since at the last step it would be necessary to compute a determinant of at least order ten.

Of course, the use of a computer extends considerably the possibilities of the method of elimination given in this section.

EXAMPLE 4. Consider the system

$$\begin{cases} f_1 = z_1 - z_3 - z_2^3 = 0, \\ f_2 = z_2^4 - z_3 - z_1 = 0, \\ f_3 = z_1^2 - z_3 - z_2^2 = 0. \end{cases} \tag{21.29}$$

It is required to eliminate z_1 and z_2. We represent $f_1 = z_1 + P_1$ and $f_2 = z_2^4 - z_1 + P_2$, where z_1 and $z_2^4 - z_1$ are weighted-homogeneous polynomials of weight $(1, 4)$, and P_1 and P_2 have, respectively, small weighted degree with respect to the variables z_1 and z_2 jointly. The system of equations $f_1 = 0, f_2 = 0$ for fixed z_3 has 4 roots; therefore,

$$\begin{aligned} S_q(z_3) &= \sum_{j=1}^{4} f_3^q(z_1^{(j)}(z_3), z_2^{(j)}(z_3), z_3) \\ &= \mathfrak{N}\left[\sum_{\alpha_1, \alpha_2} \frac{f_3^q \Delta_1}{z_2^3} \left(\frac{z_3 + z_2^3}{z_1}\right)^{\alpha_1} \left(\frac{z_3 + z_1}{z_2^4}\right)^{\alpha_2}\right], \qquad q = 1, \ldots, 4. \end{aligned}$$

Here the functional $\mathfrak{N}$ acts on the Laurent polynomial in z_1 and z_2, while the variable z_3 is a parameter. The computations give the following results:

$$\begin{aligned} S_1(z_3) &= 4z_3^2 + 10z_3, \\ S_2(z_3) &= 4z_3^4 + 244z_3^3 + 100z_3^2, \\ S_3(z_3) &= 4z_3^6 + 1410z_3^5 + 4068z_3^4 + 922z_3^3 - 12z_3^2 - 9z_3, \\ S_4(z_3) &= 4z_3^8 + 4440z_3^7 + 51012z_3^6 + 45520z_3^5 + 7744z_3^4 + 7472z_3^3 - 480z_3^2. \end{aligned}$$

With the help of the Newton recursion formula (21.5) we then find the desired polynomial:

$$r(z_3) = z_3^8 - 6z_3^7 - 67z_3^6 - 4272z_3^5 + 293z_3^4 - 1920z_3^3 + 90z_3^2.$$

Thus, elimination of z_1 and z_2 from (21.29) gives the equation $r(z_3) = 0$.

Let us consider the problem of finding a multidimensional analogue of Waring's formula (see, for example, [128], Chapter III, §5.3) for the systems (21.15), a formula which expresses the power sums S_j at once in terms of the coefficients of the corresponding polynomial; this is more cumbersome than Newton's formula (21.5) but, on the other hand, it gives the answer at once and not in a recursive way.

The power sums for the coordinates of the roots of (21.15) are denoted as follows:

$$\sum_{j=1}^{N} \left(z_l^{(j)}\right)^m = S_l^{(m)}.$$

Further, let $\sum_\alpha^j$ stand for summation over all multi-indices α such that either $|\alpha| < k_j$, or $|\alpha| = k_j$ but for some i we have $\alpha_1 = \cdots = \alpha_{i-1} = 0$, $\alpha_i \geqslant 1$, $i = 1, \ldots, j-1$. The product $\prod_\alpha^j$ is defined similarly, as are the tuples of numbers $\{\ \}_\alpha^j$, $j = 1, \ldots, n$. In this notation (21.15) can be written in the form

$$z_j^{k_j} - \sum_\alpha{}^j a_{j\alpha} z^\alpha = 0, \qquad j = 1, \ldots, n. \tag{21.30}$$

THEOREM 21.6 (Bolotov). *The following multidimensional analogue of Waring's formula is valid for the system* (21.30):

$$S_l^{(m)} = \sum_q \prod_{j=1}^{n} \left[(\beta_j(q) - 1)! \prod_\alpha{}^j \frac{a_{j\alpha}^{\lambda_{j\alpha}}}{\lambda_{j\alpha}!} \right] D(q), \tag{21.31}$$

where $q = (q_1, \ldots, q_n)$, $q_j = \{\lambda_{j\alpha}\}_\alpha^j$, *the* $\lambda_{j\alpha}$ *are nonnegative integers*, $\beta_j(q) = \sum_\alpha^j \lambda_{j\alpha}$, *and* $D(q) = \det\|\delta_{pj} k_j \beta_j(q) - \gamma_p(q_j)\|$, $p, j = 1, \ldots, n$, *with* $\gamma_p(q_j) = \sum_\alpha^j \lambda_{j\lambda}\alpha_p$. *The summation with respect to* q *in* (21.31) *runs over all* n*-tuples* q *such that*

$$\sum_{j=1}^{n} \gamma_p(q_j) - k_p\beta_p(q) + \delta_{lp} \cdot m = 0, \qquad p = 1, \ldots, n.$$

Here δ_{lp} *is the Kronecker symbol. There are only finitely many terms in* (21.31).

We do not give the proof. It is similar to that of Theorem 20.8. The formula (21.31) (with $n = 1$) contains the familiar classical Waring formula.

3°. Consider now *an arbitrary system of algebraic equations*

$$P_j(z) = 0, \qquad j = 1, \ldots, n, \tag{21.32}$$

along with an auxiliary system of the form (21.12):

$$\lambda z_j^{k_j} + P_j(z) = 0, \qquad j = 1, \ldots, n, \tag{21.33}$$

where k_j is greater than the degree of $P_j(z)$, $j = 1, \ldots, n$. Without loss of generality it can be assumed that zero is not among the roots of (21.32) (in the contrary case a translation can be made). Suppose that this system has N

(finite) roots, while (21.33) has $N' = k_1 \cdots k_n$ roots (see Example 1). We introduce linear combinations of the coordinates of the roots of (21.33):

$$a_1 z_1^{(j)}(\lambda) + \cdots + a_n z_n^{(j)}(\lambda), \qquad j = 1,\ldots,N',\ a \in \mathbf{C}^n. \qquad (21.34)$$

As $\lambda \to 0$ the "superfluous" $N' - N$ roots tend to ∞, while N roots tend to the roots of (21.32), as is not hard to establish with the help of Rouché's theorem. Therefore, for all parameters $a \in \mathbf{C}^n$ with the exception of a set of complex dimension $n - 1$ exactly $N' - N$ of the numbers (21.34) tend to ∞ as $\lambda \to 0$, while N of them tend to the corresponding (nonzero) linear combinations of roots of (21.32). Let

$$S_a^{(m)}(\lambda) = \sum_{j=1}^{N'} \langle a, z^{(j)}(\lambda)\rangle^m, \qquad m = 1,2,\ldots,N'.$$

The sums $S_a^{(m)}(\lambda)$ can be computed by Theorem 21.3. From $S_a^{(m)}(\lambda)$ and Newton's formula (21.5) we can find the elementary symmetric polynomials in (21.34), which we denote by $\sigma_a^{(m)}(\lambda)$, $m = 1,\ldots,N'$. We next consider the ratios

$$\frac{\sigma_a^{(m)}(\lambda)}{\sigma_a^{(N')}(\lambda)}, \qquad m = N' - 1, N' - 2,\ldots,1, \frac{1}{\sigma_a^{(N')}(\lambda)}, \qquad (21.35)$$

which are elementary symmetric polynomials in $\langle a, z^{(j)}(\lambda)\rangle^{-1}$, $j = 1,\ldots,N'$. They are rational functions of the coefficients of (21.33). Letting $\lambda \to 0$, we get from (21.35) that

$$\delta_1,\ldots,\delta_N,\underbrace{0,\ldots,0}_{N'-N}, \qquad (21.36)$$

where the δ_j are the elementary symmetric polynomials in $\langle a, z^{(j)}\rangle^{-1}$, the $z^{(j)}$ being the roots of (21.32), $j = 1,\ldots,N$. Hence, we have constructed a polynomial $\Omega(z_1)$ in the single complex variable z_1 that has the roots $\langle a, z^{(j)}\rangle^{-1}$, $j = 1,\ldots,N$, thereby indicating a *method of elimination for arbitrary algebraic systems* and, in particular, an algorithm that enables us to determine the total number of (finite) roots of these systems (the index N of the number δ_N after which there are only zeros in (21.36).)([3]) We note that this method works correctly for parameters $a \in \mathbf{C}^n$ outside a set of parameters of complex dimension $n - 1$. The given method of elimination seems to us (as does the method indicated in 2° for the systems (21.15)) to be simpler than the classical method which uses resultants of polynomials.

([3]) It may happen that δ_N is equal to zero also for the excluded set of parameters $a \in \mathbf{C}^n$ of complex dimension $n - 1$.

4°. We discuss the question of *the number of real roots of a nonlinear algebraic system with real coefficients*. Suppose that the coefficients of (21.32) are real. As before, the roots of this system are denoted by $z^{(1)},\dots,z^{(N)}$. Consider the linear combinations $\langle a, z^{(j)}\rangle, j = 1,\dots,N$, of the coordinates of the roots, with the parameter a taken from the real space $\mathbf{R}^n$ this time. We construct a polynomial $\Omega_a(z_1)$ of a single variable having the roots $\langle a, z^{(j)}\rangle$, $j = 1,\dots,N$ (this is easy to do by computing the sums $S_a^{(m)} = \sum_{j=1}^N \langle a, z^{(j)}\rangle^m$; for system (21.15) $S_a^{(m)}$ can be calculated for any $a \in \mathbf{R}^n$, but for the general case of the system (21.32) it can be calculated for $a \in \mathbf{R}^n$ outside an $(n-1)$-dimensional subset of $\mathbf{R}^n$). If the root $z^{(j)}$ is real, then the linear combination $\langle a, z^{(j)}\rangle$ is also real. But if $\langle a, z^{(j)}\rangle$ is real, then either the root $z^{(j)}$ is real or the parameter a satisfies the equation $\langle a, \operatorname{Im} z^{(j)}\rangle = 0$.

An algebraic system with real coefficients has, along with each complex root $z^{(j)}$, also the root $\overline{z^{(j)}}$. Therefore, we have

PROPOSITION 21.7. *The polynomial $\Omega_a(z_1)$ has as many real roots as has the system of equations under consideration with the exception of the case where the parameter a lies on a "singular" set consisting of not more than $N/2$ hyperplanes in $\mathbf{R}^n$.*

If we find the number of real roots of $\Omega_a(z_1)$ by any classical method and it turns out that for small variations in the coefficients of $\Omega_a(z_1)$ this number does not change,([4]) then we can be certain that the given number is also the number of real roots of the system. It is often sufficient just to eliminate all the unknowns except for one from the given nonlinear system.

EXAMPLE 5. Consider the system of equations

$$\begin{cases} z_1^3 - z_2 + z_3^2 - 1 = 0, \\ z_2^3 + z_1 - z_3 + 1 = 0, \\ z_3^2 - z_1 - z_2 + 2 = 0, \end{cases} \tag{21.37}$$

which has the form (21.12). It is possible to compute $\Omega_a(z_1)$ for any a by using (21.13) (first compute the sums $S_a^{(m)}$, and then use Newton's formula (21.5)). However, to determine the number of real roots of (21.37) it suffices to eliminate the unknowns z_1 and z_3. The result is

$$\begin{aligned} P(z_2) = z_2^{18} + 6z_2^{15} - 3z_2^{13} + 21z_2^{12} - 12z_2^{10} + 44z_2^9 + 3z_2^8 \\ -30z_2^7 + 63z_2^6 + 6z_2^5 - 36z_2^4 + 53z_2^3 + 9z_2^2 - 27z_2 + 27 = 0. \end{aligned}$$

([4]) Generally speaking, the degree of $\Omega_a(z_1)$ is equal to N. If for a given a it turns out to be less than N, then we regard the missing coefficients as equal to zero, and it is also necessary to vary them when we speak of varying the coefficients.

If we consider the number of sign changes in the sequence $P(z_2)$, $P'(z_2),\ldots,P^{(18)}(z_2)$ at the points $-\infty$, -1, 1, and ∞, then we see by the Budan-Fourier theorem (see, for example, [128], Chapter III, §4.2) that $P(z_2)$ has no roots on $(-\infty,-1)$ and $(1,\infty)$, and the sign sequences contain only plus and minus signs and no zeros. It is easy to estimate the polynomial $P(z_2)$ on $[-1,1]$ and to show that it also has no roots there. Since small changes in the coefficients of P preserve the inequalities used here as well as the plus and minus signs in the sign sequences, we arrive at the conclusion that none of the 18 roots of (21.37) are real.

We remark that if we eliminated z_2 and z_3, then we would get $Q(z_1) = z_1^3 - z_1 + 1 = 0$. The polynomial $Q(z_1)$ has real roots, but this does not mean that our system also has real roots (see footnote 4). In other words, the point $(1,0,0)$ is on the "singular" set of parameters (see Proposition 21.7).

EXAMPLE 6. Consider the system (21.29) in Example 4. It is easy to eliminate z_1 and z_3, with the result $P(z_2) = z_2^2Q(z_2) = 0$, where $Q(z_2) = z_2^6 + 2z_2^5 + z_2^4 - 2z_2^2 + 2z_2 - 4$. If we vary the coefficients of $P(z_2)$ a little, then the number of real roots can change (the multiple $z_2 = 0$ can pass into a pair of complex conjugate roots); therefore, it is not possible to apply Proposition 21.7 directly here. But we can proceed differently. With the help of Theorem 22.7 it is easy to see that (21.19) has $(0,0,0)$ as a root of multiplicity two. Therefore, we look for the number of real roots of it different from zero. Then we can investigate $Q(z_2)$. If we compute the number of sign changes in the sequence $Q(z_2)$, $Q'(z_2),\ldots,Q^{(6)}(z_2)$ at the points $-\infty$ and -1, then we see by the Budan-Fourier theorem that $(-\infty,-1]$ contains a single root, and there are only plus or minus signs in the sign sequences. Further, it is easy to see that $z_2 = 1$ is a simple root and there are no other roots in $[0,\infty)$. Simple inequalities give us that there are also no roots in $(-1,0)$. Thus, (21.29) has 4 real roots. We arrive at a similar conclusion if we eliminate z_1 and z_2 (see Example 4), i.e., the points $(0,1,0)$ and $(0,0,1)$ do not lie on the "singular" set of parameters.

Let us next consider the question of *the number of roots of a nonlinear algebraic system with real coefficients in a given ellipsoid in* R^n (the discussion of a method for finding the number of roots in a domain bounded by an algebraic surface is similar).

We first compute the sums

$$S_{a,b}^{(m)} = \sum_{j=1}^{N} \left[a_1\left(z_1^{(j)} - b_1\right)^2 + \cdots + a_n\left(z_n^{(j)} - b_n\right)^2\right]^m,$$

$m = 1,\ldots,N$. Then we use Newton's formula (21.5) to find the coefficients of a polynomial $\Omega_{a,b}(z_1)$ having the roots

$$a_1\left(z_1^{(j)} - b_1\right)^2 + \cdots + a_n\left(z_n^{(j)} - b_n\right)^2, \tag{21.38}$$

$j = 1, \ldots, N$. Here the parameters (a, b) form a point in R^{2n}. If the root $z^{(j)}$ is real, then the expression (21.38) is also real. But if (21.38) is real, then either $z^{(j)}$ is real or the parameters (a, b) satisfy the equation obtained by setting the imaginary part of (21.38) equal to zero. There cannot be more than $N/2$ such equations, where N is the total number of zeros of the given system, i.e., we have

PROPOSITION 21.8. *The polynomial* $\Omega_{a,b}(z_1)$ *has as many roots on* $[0, r)$ *as the given system has in the ellipsoid* $\{x\colon x \in R^n,\ a_1(x_1 - b_1)^2 + \cdots + a_n(x_n - b_n)^2 < r\}$, *with the exception of the case when the point* (a, b) *belongs to the "singular" set, which consists of at most* $N/2$ *third-order hypersurfaces in* R^{2n}.

The remark after Proposition 21.7 about how to determine that (a, b) does not lie on the "singular" set is also true in this case. Another method consists in the following. First it is necessary to use Proposition 21.7 to find the total number N_1 of real roots of the system. Then determine the number N_2 of roots of the polynomial $\Omega_{a,b}$ on $[0, \infty)$. If $N_2 = N_1$, then the parameter point (a, b) is not on the "singular" set.

EXAMPLE 7. Consider the system

$$\begin{cases} z_1^3 - z_2^2 - 1 = 0, \\ z_2^3 - z_1^2 - z_1 = 0. \end{cases} \tag{21.39}$$

If we eliminate z_2, we get $z_1^9 - 3z_1^6 - z_1^4 + z_1^3 - z_1^2 - 1 = 0$. Just as in Examples 5 and 6, it can then be established that (21.39) has a single real root. This system is not large, so $\Omega_{a,b}$ can be computed even without assigning concrete values to a or b. For example, when $b = (0, 0)$ and $a \in R^2$,

$$\begin{aligned}
\Omega = z_1^9 &- (5a_1a_2 + 6a_2^2)z_1^7 - (7a_1^3 + 7a_1^2a_2 + 14a_1a_2^2 + 6a_2^3)z_1^6 \\
&- (6a_1^4 + 42a_1^3a_2 + 19a_1^2a_2^2 + a_1a_2^3 - 9a_2^4)z_1^5 \\
&- (7a_1^5 + 29a_1^4a_2 + 91a_1^3a_2^2 + 6a_1^2a_2^3 - 27a_1a_2^4 - 17a_2^5)z_1^4 \\
&- (7a_1^6 + 29a_1^5a_2 + 39a_1^4a_2^2 + 88a_1^3a_2^3 - 4a_1^2a_2^4 - 30a_1a_2^5 - 3a_2^6)z_1^3 \\
&- (3a_1^7 + 24a_1^6a_2 + 30a_1^5a_2^2 + 12a_1^4a_2^3 + 60a_1^3a_2^4 + 13a_1^2a_2^5 \\
&\qquad - 17a_1a_2^6 + 13a_2^7)z_1^2 \\
&- (2a_1^8 + 6a_1^7a_2 + 18a_1^6a_2^2 + 7a_1^5a_2^3 - 7a_1^4a_2^4 + 36a_1^3a_2^5 + 11a_1^2a_2^6 \\
&\qquad - 10a_1a_2^7 + 12a_2^8)z_1 \\
&- a_1^9 - 2a_1^8a_2 - 3a_1^7a_2^2 - 4a_1^6a_2^3 + 12a_1^5a_2^4 + 2a_1^4a_2^5 \\
&- 9a_1^3a_2^6 - a_1^2a_2^7 + 4a_1a_2^8 - 4a_2^9.
\end{aligned}$$

It is now possible to find the number of roots of (21.39) in the ellipse $\{x: x \in R^2, a_2x_1^2 + a_2x_2^2 < r\}$.

For example, if $a_1 = a_2 = 1$, then $\Omega = z_1^9 - 11z_1^7 - 34z_1^6 - 59z_1^5 - 39z_1^4 - 126z_1^3 - 138z_1^2 - 75z_1 - 6$. As in the preceding examples, it is not hard to establish that this polynomial has one root in $[0, \infty)$; therefore, the parameters $a_1 = a_2 = 1$ and $b_1 = b_2 = 0$ do not lie on the "singular" set. Next, it is easy to see that $\Omega(4) < 0$ and $\Omega(5) > 0$, i.e., this root lies in the interval (4, 5). Our system therefore has a root in the disk $\{x: x_1^2 + x_2^2 < 5\}$, but no roots in the disk $\{x: x_1^2 + x_2^2 < 4\}$.

If $a_1 = 1$ and $a_2 = 2$, then $\Omega = z_1^9 - 34z_1^7 - 125z_1^6 - 22z_1^5 + 119z_1^4 + 291z_1^3 - 2179z_1^2 - 3678z_1 - 1421$, and it is easy to find as in the foregoing that (21.39) has a root in the ellipse $\{x: x_1^2 + 2x_2^2 < 8\}$, but no roots in the ellipse $\{x: x_1^2 + 2x_2^2 < 7\}$.

REMARK. Thus, we have given an algorithm that enables us to answer the question of the number of roots of a system in a given ellipsoid in R^n. In particular, by varying the sizes of the axes of the ellipsoid we can speak of the number of roots in a ball or, for example, of the number of roots in a neighborhood of an interval.

§22. Computation of the multiplicity of a zero of a system of holomorphic functions from their Taylor series

1°. Let

$$f(z) = (f_1(z), \ldots, f_n(z)) \tag{22.1}$$

be a system of holomorphic functions (a holomorphic mapping) in a neighborhood U_a of a point $a \in \mathbf{C}^n$ that is an isolated zero of the system with multiplicity $\mu_a(f)$ (see §2). For a holomorphic function of a single variable the multiplicity $\mu_a(f)$ of a zero always coincides with the order $d_a(f) = \min_{c_k \neq 0}\{k\}$ of the function $f = \sum_{k \geq 0} c_k(z-a)^k$, i.e., the multiplicity of a zero depends only on which Taylor coefficients of f at a are not equal to zero. The picture is more complicated in the multidimensional case. First, the multiplicity of a zero of (22.1) is not always determined by the orders of the zero a of the functions f_j. For example, for the system f: $f_j = z_j^{k_j} + \sum_{|\alpha| \geq k_j} c_{j\alpha} z^\alpha$, $j = 1, \ldots, n$, we have

$$\mu_0(f) = k_1 \cdots k_n = d_0(f_1) \cdots d_0(f_n).$$

This is easy to see from Rouché's principle (see Theorem 2.5) and from the definition of the multiplicity of a zero. On the other hand, the orders $d_a(f_i)$, $i = 1, 2$, of the functions of the system $(z_1, z_1 + z_2^2)$ are equal to 1, but $\mu_0(f) = 2$.

In 3° we give a necessary and sufficient condition on the system (22.1) for $\mu_a(f) = d_a(f_1) \cdots d_a(f_n)$ to hold.

Moreover, knowledge of the monomials with nonzero coefficients does not always enable us to determine the multiplicity of a zero. For example, the systems $(z_1 + z_2 + z_1^2, z_1 - z_2 + z_2^2)$ and $(z_1 + z_2 + z_1^2, z_1 + z_2 + z_2^2)$ have the same collection of monomials, but the multiplicity of the zero $(0, 0)$ is different for them. However, as a rule, the multiplicity of an isolated zero is determined only by the monomials appearing in the f_i with nonzero coefficients, i.e., by the sets

$$S_i = \operatorname{supp} f_i = \{\alpha \in \mathbf{N}^n \colon c_{i\alpha} \neq 0\},$$

where $\mathbf{N} = \{0, 1, 2, \ldots\}$ and $f_i = \Sigma_\alpha c_{i\alpha}(z - a)^\alpha$. Therefore, to a given collection $S = \{S_1, \ldots, S_n\}$, $S_i \subset \mathbf{N}^n$, we can assign a number $\mu_a(S)$ equal to the multiplicity of a zero for "almost all" the systems f with $\operatorname{supp} f_i \subset S_i$, $i = 1, \ldots, n$.

In 4° we solve constructively the problem of computing $\mu_a(S)$ in terms of the collection S, and also give "nondegeneracy" conditions on the Taylor coefficients of the functions f_i for the equality $\mu_a(f) = \mu_a(S)$ to hold.

We present some simple properties of the multiplicity of an isolated zero of a system of holomorphic functions. If $f(a) \neq 0$, then $\mu_a(f)$ is set equal to zero.

PROPOSITION 22.1. 1) $\mu_a(f \circ g) = \mu_0(f) \cdot \mu_a(g)$, *where* $f \circ g$ *is the composition of the mappings.*

2) $\mu_a(f_1 \cdot g_1, f_2, \ldots, f_n) = \mu_a(f_1, f_2, \ldots, f_n) + \mu_a(g_1, f_2, \ldots, f_n)$.

3) *The multiplicity of a zero of a system of holomorphic functions coincides with the multiplicity of a zero of the system of their Weierstrass pseudopolynomials.*

PROOF. 1) Suppose that $w = f(z)$, $z = g(\zeta)$, $g(a) = 0$ and $f(0) = 0$. If $B = \{|\zeta - a| < \delta\}$, then, by the argument principle (see Theorem 2.9), $(f \circ g)(\partial B) \sim \mu_a(f \circ g)\{|w| = \delta\}$ in $\mathbf{C}^n \setminus \{0\}$. But $\underset{\sim}{g}(\partial B) \sim \mu_a(g)\partial V$ and $\underset{\sim}{f}(\partial V) \sim \mu_0(f)\{|w| = \delta\}$, where $V = \{|\tilde{z}| < \delta\}$. Therefore,

$$\underset{\sim}{(f \circ g)}(\partial B) = \underset{\sim}{f}(\underset{\sim}{g}(\partial B)) \sim \mu_a(g) \cdot \underset{\sim}{f}(\partial V) \sim \mu_0(f) \cdot \mu_a(g) \cdot \{|w| = \delta\}$$

in $\mathbf{C}^n \setminus \{0\}$, and so $\mu_a(f \circ g) = \mu_a(g) \cdot \mu_0(\tilde{f})$.

2) By Theorem 4.6,

$$\begin{aligned}\mu_a(f_1 g_1, f_2, \ldots, f_n) &= \frac{1}{(2\pi i)^n} \int_\Gamma \frac{d(f_1 g_1)}{f_1 g_1} \wedge \frac{df_2}{f_2} \wedge \cdots \wedge \frac{df_n}{f_n} \\ &= \frac{1}{(2\pi i)^n} \int_\Gamma \frac{df_1}{f_1} \wedge \cdots \wedge \frac{df_n}{f_n} + \frac{1}{(2\pi i)^n} \int_\Gamma \frac{dg_1}{g_1} \wedge \frac{df_2}{f_2} \wedge \cdots \wedge \frac{df_n}{f_n} \\ &= \mu_a(f_1, \ldots, f_n) + \mu_a(g_1, f_2, \ldots, f_n),\end{aligned}$$

where $\Gamma = \{z \in S(\varepsilon): |f_2(z)| = \cdots = |f_n(z)| = \delta\}$ and $S(\varepsilon) = \{z: |z - a| = \varepsilon\}$, $\delta \ll \varepsilon$.

3) follows from 2), since $\mu_a(f)$ is symmetric with respect to the f_j ($j = 1,\ldots,n$), and, moreover, for the Weierstrass pseudopolynomials P_j we have $f_j = P_j\psi_j$, $\psi_j(a) \neq 0$, $j = 1,\ldots,n$. □

The main apparatus in this section for computing the multiplicity of a zero of a system f amounts to the Rouché principle, along with a general theorem on the multiplicity of a zero of f which constructs from f a function of a single variable with a zero of multiplicity that coincides with $\mu_a(f)$.

2°. A general theorem on the multiplicity of a zero. We need two theorems on analytic sets: a theorem on a local description, and a theorem on the relation of the dimensions of analytic sets (see [205]).

Let K be a one-dimensional analytic set containing the point $0 \in \mathbf{C}^n$.

THEOREM 22.2 (on a local description). *There exist a basis in $\mathbf{C}^n$ and a polycylinder $U_\varepsilon = \{z: |z_j| < \varepsilon_j,\ j = 1,\ldots,n\}$ (the radii ε_j are assumed to be sufficiently small) that satisfy the following conditions*:

a) $z = 0$ *is an isolated point of the set* $K \cap \{z: z_n = 0\}$.

b) *For any* $z_n^{(0)} \in U_{\varepsilon_n}\setminus\{0\} = \{z_n \in \mathbf{C}^1: 0 < |z_n| < 1\}$ *the set* $K(z_n^{(0)}) = \{z \in K \cap U_\varepsilon: z_n = z_n^{(0)}\}$ *consists of one and the same finite number p of points.*

c) *There exist pseudopolynomials $P_j(z_j, z_n)$, $j = 1,\ldots,n-1$, distinguished with respect to the z_j, such that* $K \cap U_\varepsilon \subseteq \{z \in U_\varepsilon: P_j(z_j, z_n) = 0,\ j = 1,\ldots,n-1\}$.

The local description theorem implies that in a neighborhood of each point $z_n^{(0)} \in U_{\varepsilon_n}\setminus\{0\}$ the set $K(z_n)$ is a finite tuple

$$z^{(\nu)}(z_n) = \left(z_1^{(\nu)}(z_n),\ldots,z_{n-1}^{(\nu)}(z_n), z_n\right), \qquad \nu = 1,\ldots,p,$$

of holomorphic vector-valued functions that can be continued along any path in $U_{\varepsilon_n}\setminus\{0\}$ and can be expressed there by fractional power series of the form

$$z_i^{(\nu)}(z_n) = \sum_{k=1}^{\infty} a_{ik}^{(\nu)} z_n^{k/m}, \qquad i = 1,\ldots,n-1,\ \nu = 1,\ldots,p, \tag{22.2}$$

where $m > 0$ is an integer. A one-dimensional analytic set will be called an *analytic curve*, and the vector-valued functions $z^{(\nu)}(z_n)$ *the branches of the curve*.

THEOREM 22.3. *Suppose that A is an analytic set in an open set $U \subset \mathbf{C}^n$, h is a function holomorphic in U, and $A' = \{z \in A: h(z) = 0\}$. Then* $\dim_a A' = \dim_a A$ *or* $\dim_a A' = \dim_a A - 1$ *for each point $a \in A'$, where $\dim_a$ is the complex dimension at a point a of an analytic set.*

The concept of the resultant of a function with respect to a system plays the main role in the general multiplicity theorem. If a system $\varphi = (\varphi_1, \ldots, \varphi_n)$ of functions holomorphic in a domain $D \subset \mathbf{C}^n$ has a finite set $E = \{z^{(\nu)}\}_{\nu=1}^{s}$ of zeros there, then the *resultant of a function* ψ *with respect to the system* φ is defined to be the number

$$R(\varphi, \psi) = \prod_{\nu=1}^{s} \psi^{\mu_\nu}(z^{(\nu)}),$$

where $\mu_\nu = \mu_{z^{(\nu)}}(\varphi)$ is the multiplicity of the zero $z^{(\nu)}$ of φ.

In the case where $n = 1$, $D = C^1$, and φ_1 and ψ are polynomials, the resultant $R(\varphi_1, \psi)$ coincides with the usual Sylvester resultant of φ_1 and ψ.

Suppose that $z = 0$ is an isolated zero of (22.1). Then, by Theorem 22.3, the complex dimension at 0 of the analytic set $K = \{z \in U_0 \colon f_j(z) = 0, j = 1, \ldots, n - 1\}$ is equal to 1. According to the local description theorem, there is a basis in $\mathbf{C}^n$ such that the following condition holds in the polycylinder $U_\varepsilon = \{z\colon |z_j| < \varepsilon_j,\ j = 1, \ldots, n\}$: for any $\zeta \in U_{\varepsilon_n} \setminus \{0\} = \{z_n \colon 0 < |z_n| < \varepsilon_n\}$ the set

$$K(\zeta) = \{z \in U_\varepsilon \colon f_1(z) = \cdots = f_{n-1}(z) = 0,\ z_n = \zeta\}$$

consists of p points $z^{(\nu)}(\zeta)$, $\nu = 1, \ldots, p$. Therefore, for any $z_n \in U_{\varepsilon_n} \setminus \{0\}$ we can define the resultant of the function f_n with respect to the system $'f = (f_1, \ldots, f_{n-1})$, namely,

$$R('f, f_n) = \prod_{\nu=1}^{p} f_n^{\mu_\nu}('z^{(\nu)}(z_n), z_n),$$

where the $'z^{(\nu)}(z_n) = (z_1^{(\nu)}(z_n), \ldots, z_{n-1}^{(\nu)}(z_n))$, $\nu = 1, \ldots, p$, are isolated zeros of the system $'f$ in the domain $'U_\varepsilon = \{'z \in \mathbf{C}^{n-1} \colon |z_j| < \varepsilon_j,\ j = 1, \ldots, n-1\}$, and μ_ν is the multiplicity of the zero $'z^{(\nu)}(z_n)$ of the system $'f$.

We remark that the multiplicity μ_ν does not depend on z_n, although the zeros $'z^{(\nu)}(z_n)$ do. Indeed, by Theorem 2.4, the multiplicity of the zero $'z^{(\nu)}(z_n)$ can be expressed by the integral $\mu_\nu = \int_{S_\nu(\delta)} \omega('f, '\bar{f})$, where $S_\nu(\delta) = \{'z\colon |'z - 'z^{(\nu)}(\zeta)| = \delta\}$, δ is sufficiently small, and the Martinelli-Bochner kernel $\omega('f, '\bar{f})$ is a form continuous in the parameter z_n. Therefore, $\mu_\nu = \mu('z^{(\nu)}(z_n))$, being a continuous and integer-valued function, takes a single value for all $z^{(\nu)}(z_n)$ in a connected component of $K \setminus \{0\}$.

Further, it is easy to see that $R('f, f_n)$ is a holomorphic function in the disk $U_{\varepsilon_n} = \{z_n \colon |z_n| < \varepsilon_n\}$, since it is symmetric with respect to the branches $z^{(\nu)}(z_n)$ of the curve K.

We can now state a general theorem on the multiplicity of a zero of a system of holomorphic functions.

THEOREM 22.4 (Tsikh). *The multiplicity of the isolated zero $z = 0$ of the system* (22.1) *is equal to the multiplicity of the zero $z_n = 0$ of the resultant of f_n with respect to the system $'f$: $\mu_0(f) = \mu_0(R('f, f_n))$.*

PROOF. By Theorem 4.6,

$$\mu_0(f) = \frac{1}{(2\pi i)^n}\int_\Gamma \frac{df}{f} = \frac{1}{(2\pi i)^n}\int_\Gamma \frac{d'f}{'f} \wedge \frac{df_n}{f_n},$$

where $\Gamma = \{z \in \partial U_\varepsilon\colon |f_j(z)| = \delta, j = 1,\dots,n-1\}$, $0 < \delta < \min_{z\in\partial U_\varepsilon}\|f\|$, with $\|f\| = \max\{|f_1|,\dots,|f_n|\}$. Since we are operating under the conditions of the local description theorem, for sufficiently small δ we can write the cycle Γ in the form $\Gamma = \{z\colon |f_j(z)| = \delta,\ j = 1,\dots,n-1,\ |z_n| = \varepsilon_n\}$. With the notation $\gamma = K \cap \{z\colon |z_n| = \varepsilon_n\}$ it follows that $\Gamma \sim \Gamma' = \bigcup_{w\in\gamma}\Gamma_w$ in $U_\varepsilon \setminus \{z\colon f_1(z)\cdots f_n(z) = 0\}$, where Γ_w is the connected component of the $(n-1)$-dimensional cycle $\{z \in U_\varepsilon\colon |f_j(z)| = \delta, j = 1,\dots,n-1,\ z_n = w_n\}$ (this cycle has p components if δ is sufficiently small) that lies in a neighborhood of the point $w \in \gamma$. By Theorem 4.5,

$$\frac{1}{(2\pi i)^{n-1}}\int_{\Gamma_w}\frac{d'f}{'f} = \mu(w),$$

where $\mu(w)$ is the multiplicity of the isolated zero $'w = (w_1,\dots,w_{n-1})$ of the system $'f$ for $z_n = w_n$. Consequently,

$$\mu_0(f) = \frac{1}{(2\pi i)^n}\int_{\Gamma'}\frac{d'f}{'f}\wedge\frac{df_n}{f_n} = \frac{1}{2\pi i}\int_\gamma \mu(w)\frac{df_n}{f_n}.$$

Since on γ the p points $w = z^{(\nu)}(z_n)$, $\nu = 1,\dots,p$, correspond to any given $z_n \in \gamma_0 = \{|z_n| = \varepsilon_n\}$, and $\mu(w) = \mu(z^{(\nu)}(z_n)) = \mu_\nu$ does not depend on z_n, the last integral is equal to

$$\frac{1}{2\pi i}\int_{\gamma_0}\sum_{\nu=0}^{p}\mu_\nu\frac{df_n(z^{(\nu)}(z_n))}{f_n(z^{(\nu)}(z_n))} = \frac{1}{2\pi i}\int_{\gamma_0}\frac{d\left(\prod_{\nu=1}^{p} f_n^{\mu_\nu}(z^{(\nu)}(z_n))\right)}{\prod_{\nu=1}^{p} f_n^{\mu_\nu}(z^{(\nu)}(z_n))}$$

$$= \frac{1}{2\pi i}\int_{\gamma_0}\frac{dR('f, f_n)}{R('f, f_n)}. \quad \square$$

REMARK. By Theorem 22.4 and Proposition 22.1, the multiplicity of the zero $(0,0) \in \mathbf{C}^2$ of a system of two holomorphic functions is equal to the multiplicity of a zero of the Sylvester resultant of their Weierstrass pseudopolynomials, which can be written in the form of a determinant. Since the coefficients of a Weierstrass pseudopolynomial can be found constructively (see Theorem 27.1), Theorem 22.4 gives a practical method for computing the multiplicity of a zero of any holomorphic system in $\mathbf{C}^2$.

Theorem 22.4 and the Rouché principle enable us easily to prove the familiar theorem of Bézout on the total number of roots of a system of algebraic equations in the complex projective plane $\mathbf{C}P^n$. We use this theorem in what follows.

Suppose that we are given a system of equations

$$f_j(\zeta) = 0, \qquad j = 1,\ldots,n, \tag{22.3}$$

where the $f_j(\zeta)$ are homogeneous polynomials and $\zeta = (\zeta_0, \zeta_1,\ldots,\zeta_n)$ are homogeneous coordinates in $\mathbf{C}P^n$. Proportional solutions of (22.3) determine a single point in $\mathbf{C}P^n$, called a *root* of (22.3) in $\mathbf{C}P^n$. If $\tilde{\zeta} = (\tilde{\zeta}_0, \tilde{\zeta}_1,\ldots,\tilde{\zeta}_n)$ is an isolated root of (22.3) in $\mathbf{C}P^n$ with $\tilde{\zeta}_j \neq 0$, then the multiplicity $\mu_{\tilde{\zeta}}(f)$ of this root is defined to be the multiplicity of the zero $(\tilde{\zeta}_0/\tilde{\zeta}_j,\ldots,\tilde{\zeta}_{i-1}/\tilde{\zeta}_j, \tilde{\zeta}_{i+1}/\tilde{\zeta}_j,\ldots,\tilde{\zeta}_n/\tilde{\zeta}_j)$ of the system $f_j(\zeta_0,\ldots,\zeta_{i-1}, 1, \zeta_{i+1},\ldots,\zeta_n)$, $j = 1,\ldots,n$, in $\mathbf{C}^n$.

THEOREM 22.5 (Bézout). *If the system* (22.3) *has only isolated roots in* $\mathbf{C}P^n$, *then their number, counting multiplicity, is equal to the product of the degrees of the polynomials* $f_j(\zeta)$, $j = 1,\ldots,n$.

PROOF. Let us use induction on the number of variables. For $n = 1$ the proof is elementary. We assume the validity of the theorem for $n - 1$ and prove it for n. Choose a system of homogeneous coordinates $\zeta = (\zeta_0, \zeta_1,\ldots,\zeta_n)$ in $\mathbf{C}P^n$ such that (22.3) does not have roots on the plane $\zeta_0 = 0$. Then the number of roots of this system in $\mathbf{C}P^n$ coincides with the number of zeros of the system

$$f_j(1, z_1,\ldots,z_n), \qquad j = 1,\ldots,n, \tag{22.4}$$

in $\mathbf{C}^n$. Suppose that $f_j(1, z_1,\ldots,z_n) = \sum_{|\alpha| \leq k_j} c_{j\alpha} z^\alpha$, $j = 1,\ldots,n$, with $k_j = \deg f_j$; then the system

$$Q_j(z) = f_j(0, z_1,\ldots,z_n) = \sum_{|\alpha| = k_j} c_{j\alpha} z^\alpha, \qquad j = 1,\ldots,n, \tag{22.5}$$

does not have zeros different from $z = 0$ in $\mathbf{C}^n$, for otherwise (22.3) would have a root on the plane $\zeta_0 = 0$.

We first prove that the multiplicity of the zero $z = 0$ of (22.5) is equal to $k_1 \cdots k_n$. Choose a coordinate system in $\mathbf{C}^n$ such that $K \cap \{z: z_n = 0\} = \{0\}$, where $K = \{z: Q_j(z) = 0, 1,\ldots,n-1\} = \{z = a^{(\nu)} z_n\}_{\nu=1}^p$, and $a^{(\nu)} = (a_1^{(\nu)},\ldots,a_{n-1}^{(\nu)}, 1)$ is the set of roots of the system of equations $Q_j(z) = 0$, $j = 1,\ldots,n-1$, in $\mathbf{C}P^{n-1}$ with homogeneous coordinates $z = (z_1,\ldots,z_n)$. If $\mu_\nu = \mu_{'a^{(\nu)}}('Q)$ is the multiplicity of the zero $'a^{(\nu)} = (a_1^{(\nu)},\ldots,a_{n-1}^{(\nu)})$ of the system $'Q('z, 1) = (Q_1('z, 1),\ldots,Q_{n-1}('z, 1))$, then

$$\sum_{\nu=1}^p \mu_\nu = k_1 \cdots k_{n-1}$$

by the induction hypothesis. Since $z = 0$ is an isolated zero of (22.5), it follows that $Q_n(a^{(\nu)}z_n) = z_n^{k_n}Q_n(a^{(\nu)}) \not\equiv 0$, $\nu = 1,\ldots,p$; therefore, the multiplicity of the zero $z_n = 0$ of the resultant $R('Q, Q_n) = \prod_{\nu=1}^{p} Q_n^{\mu_\nu}(a^{(\nu)}z_n)$ is equal to $k_1 \cdots k_n$. By Theorem 22.4, $\mu_0(Q) = k_1 \cdots k_n$.

It remains to prove that (22.4) and (22.5) have the same number of zeros in $\mathbf{C}^n$, counting multiplicity. We note first that $f_j(1, z_1,\ldots,z_n) = g_j(z) + Q_j(z)$, where $g_j(z) = \sum_{|\alpha|<k_j} c_{j\alpha}z^\alpha$, $j = 1,\ldots,n$. Since $\{z\colon Q_j(z) = 0,\ j = 1,\ldots,n\} = \{0\}$, there exists for any point $w \in S(1) = \{z\colon |z| = 1\}$ an index $j \in \{1,\ldots,n\}$ for which $Q_j(w) \neq 0$. There exist a neighborhood U_w of w on $S(1)$ and a number $r_w > 0$ such that for all $z \in U_w$ and $r > r_w$

$$|Q_j(rz)| = \left|r^{k_j} \sum_{|\alpha|=k_j} c_{j\alpha}z^\alpha\right| > \left|\sum_{m=0}^{k_j-1} r^m \sum_{|\alpha|=m} c_{j\alpha}z^\alpha\right| = |g_j(rz)|.$$

Choosing from the covering $\{U_w\}_{w\in S(1)}$ of the compact set $S(1)$ a finite subcovering $U_{w_1},\ldots,U_{w_N}$ and letting $r_0 = \max\{r_{w_1},\ldots,r_{w_N}\}$, we get that on the sphere $S(r) = \{z\colon |z| < r\}$ the systems Q and g satisfy the conditions of Rouché's principle for any $r > r_0$. Consequently, (22.4) and (22.5) have the same number of zeros in $\mathbf{C}^n$ counting multiplicity, and this number is equal to the product $k_1 \cdots k_n$ of the degrees of the polynomials $f_j(\zeta)$, $j = 1,\ldots,n$. □

3°. Computation of the multiplicity of a zero from the homogeneous and weighted-homogeneous principal parts. Let $f_j(z) = \sum_{|\alpha|\geqslant k_j} c_{j\alpha}z^\alpha$, $j = 1,\ldots,n$, be the Taylor expansion of f_j in (22.1) at the point 0, where $k_j = d_0(f_j)$ is the order of the zero $z = 0$ of this function. We write $P_j(z) = \sum_{|\alpha|=k_j} c_{j\alpha}z^\alpha$, $j = 1,\ldots,n$, and call the system of homogeneous polynomials

$$P(z) = (P_1(z),\ldots,P_n(z)) \tag{22.6}$$

the *homogeneous principal part of the system* (22.1).

THEOREM 22.6 (Tsikh and Yuzhakov). *The multiplicity of the isolated zero $z = 0$ of the system* (22.1) *is equal to the product of the orders of the zero $z = 0$ of the functions in this system* ($\mu_0(f) = d_0(f_1) \cdots d_0(f_n)$) *if and only if* 0 *is an isolated zero of the system* (22.6). *Moreover, the inequality $\mu_0(f) \geqslant d_0(f_1) \cdots d_0(f_n)$ always holds.*

PROOF. Let $l(z) = c_1z_1 + \cdots + c_nz_n$ be an arbitrary linear form, and $z^{(\nu)}(z_n)$, $\nu = 1,\ldots,p$, the branches of the curve $K = \{z \in U_\varepsilon\colon f_j(z) = 0,\ j = 1,\ldots,n-1\}$, represented in the form (22.2). The resultant

$$R('f, l) = \prod_{\nu=1}^{p} l^{\mu_\nu}(z^{(\nu)}(z_n))$$

is a power series in z_n whose coefficients are polynomials in $c = (c_1,\dots,c_n)$. Consequently, the set of c such that the multiplicity $\mu_0(R('f, l)) = \mu_0('f, l)$ is minimal (denote this multiplicity by μ) is open and dense in $\mathbf{C}^n$, being the complement of the zero set of the polynomial which is the coefficient of the lowest power of z_n. In the space $\mathbf{C}^n$ of variables z we choose a coordinate system such that the equalities $\mu_0('f, z_i) = \mu$ are satisfied for the forms $l(z) = z_i$, $i = 1,\dots,n$. In this coordinate system the representation (22.2) for the analytic elements of the curve K takes the form

$$z_i^{(\nu)}(z_n) = a_i^{(\nu)} z_n + o(z_n^{1+\lambda}), \tag{22.7}$$

where[5] $a_i^{(\nu)} \neq 0$, $i = 1,\dots,n-1$, $\nu = 1,\dots,p$, $o(z_n^{1+\lambda})$ is infinitesimally small with respect to $z_n^{1+\lambda}$ as $z_n \to 0$, and $\lambda > 0$ is a constant. We prove this. Since $\mu_0('f, z_n) = \mu$ and $R('f, z_n) = z_n^{\mu_1+\dots+\mu_n}$, it follows that $\mu_1 + \dots + \mu_n = \mu$. If (22.2) is written in the form $z_i^{(\nu)}(z_n) = b_i^{(\nu)} z_n^{\varepsilon_{i\nu}} + o(z_n^{\varepsilon_{i\nu}})$, $b_j^{(\nu)} \neq 0$, then for almost all $u \in \mathbf{C}^1$ we have

$$\begin{aligned}\mu_0(R('f, z_i + uz_n)) &= \sum_{\nu:\, \varepsilon_{i\nu} \geqslant 1} \mu_\nu + \sum_{\nu:\, \varepsilon_{i\nu} < 1} \mu_\nu \varepsilon_{i\nu} \\ &= \mu = \mu_1 + \dots + \mu_p,\end{aligned}$$

which implies that $\varepsilon_{i\nu} \geqslant 1$, $i = 1,\dots,n-1$, $\nu = 1,\dots,p$. But $\mu_0(R('f, z_i)) = \sum_{\nu=1}^p \mu_\nu \varepsilon_{i\nu} = \mu$, $i = 1,\dots,n-1$; consequently, all the $\varepsilon_{i\nu}$ are equal to 1, and (22.7) is proved.

We write (22.7) in the vector form $z^{(\nu)}(z_n) = a^{(\nu)} z_n + o(z_n^{1+\lambda})$, where $a^{(\nu)} = (a_1^{(\nu)},\dots,a_{n-1}^{(\nu)}, 1)$; then

$$f_j(z^{(\nu)}(z_n)) = z_n^{k_j} P_j(a^{(\nu)}) + o(z_n^{k_j}) \equiv 0,$$

and so

$$P_j(a^{(\nu)} z_n) \equiv 0, \qquad j = 1,\dots,n-1,\ \nu = 1,\dots,p. \tag{22.8}$$

We show that if the complex dimension at 0 of the set $K_1 = \{z\colon P_j(z) = 0,\ j = 1,\dots,n-1\}$ is equal to 1, then K_1 is the union of the analytic lines

$$A = \bigcup_{\nu=1}^{p} \{z = a^{(\nu)} t,\ t \in \mathbf{C}^1\}.$$

Let

$$f_j(z) = P_j(z) + \sum_{l=1}^{\infty} P_{jl}(z), \qquad j = 1,\dots,n-1,$$

([5]) The number p actually depends on the choice of the coordinate system in $\mathbf{C}^n$, but we keep the notation in (22.2).

be the expansion of f_j in a series of homogeneous polynomials, with $\deg P_{ji} = k_j = i$. Consider the system

$$F_j('z, t, \tau) = P_j('z, t) + \sum_{i=1}^{\infty} \tau^i P_{ji}('z, t) = \tau^{-k_j} f_j(\tau('z, t)), \qquad (22.9)$$

$j = 1, \ldots, n-1$, which depends on the parameters $t \in \mathbf{C}^1$, $\tau \in [0, 1]$. It is clear that $F_j('z, t, 1) = f_j('z, t)$ and $F_j('z, t, 0) = P_j('z, t)$. Further, (22.9) has p isolated zeros $'z^{(\nu)}(t, \tau)$ for fixed t and τ, and they satisfy the condition $\tau 'z^{(\nu)}(t, \tau) = 'a^{(\nu)} t\tau + O((t, \tau)^{1+\lambda})$, from which we get $'z^{(\nu)}(t, \tau) = 'a^{(\nu)} t + o(t^{1+\lambda}\tau^{\lambda})$, $\nu = 1, \ldots, p$. Therefore, for $|t|$ sufficiently small and for any $\tau \in [0, 1]$ the zero $'z^{(\nu)}(t, \tau)$ of (22.9) lies in the ball $B_\nu(t) = \{'z\colon |'z - 'a^{(\nu)} t| < |t|^{1+\delta}\}$, where $0 < \delta < \lambda$. We have thus constructed a homotopy $'F('z, t, \tau)$ between the systems $'P(z, t)$ and $'f('z, t)$ that does not have zeros on the spheres $\partial B_\nu(t)$ for small $|t|$. Therefore, these systems have the same number (counting multiplicity) of zeros in $B(t) = \bigcup_1^p B_\nu(t)$. The system $'f('z, t)$ has $\mu = \mu_1 + \cdots + \mu_p$ zeros in $B(t)$, counting multiplicity. But we are under the conditions of the local description theorem; therefore, the system $'f('z, t)$ does not have zeros on the boundary of the polycylinder $'U_\varepsilon = \{'z\colon |z_j| < \varepsilon_j, j = 1, \ldots, n-1\}$ for any fixed t with $|t| \leqslant \varepsilon_n$. Consequently, $\mu_0('f('z, 0)) = \mu$. It can be assumed that $\{'z \in \mathbf{C}^{n-1}\colon P_j('z, 0) = 0, j = 1, \ldots, n-1\} = \{0\}$. If we now use induction on the number of variables (i.e., assume the validity of the theorem for $n-1$ variables), then we can conclude that $\mu_0('f('z, 0)) = k_1 \cdots k_{n-1}$. Hence, the system $'P('z, t)$ has $k_1 \cdots k_{n-1}$ zeros (counting multiplicity) in $B(t)$ for sufficiently small $|t|$. By Bézout's theorem, all the zeros of $'P('z, t)$ lie in $B(t)$. Since the radii of the balls $B_\nu(t)$ tend to zero as $t \to 0$ more rapidly than the distance between points lying in the hyperplane $\{z\colon z_n = t\}$ and on distinct analytic lines in K_1, it follows that $K_1 \subset A$. The reverse inclusion $A \subset K_1$ follows from (22.8).

Let $E_1 = \{z\colon P_j(z) = 0, j = 1, \ldots, n\}$. If $E_1 = \{0\}$, i.e., the system (22.6) has $z = 0$ as an isolated zero, then K_1 is one-dimensional and equal to A. Consequently, $P_n(a^{(\nu)} z_n) \not\equiv 0$, $\nu = 1, \ldots, p$. Since all the branches $z^{(\nu)}(z_n)$ of the curve K have the form (22.7), it follows that

$$f_n(z^{(\nu)}(z_n)) = z_n^{k_n} P_n(a^{(\nu)}) + o(z_n^{k_n}),\ P_n(a^{(\nu)}) \neq 0.$$

By what has been proved, $\sum_{\nu=1}^p \mu_\nu = k_1 \cdots k_{n-1}$, and so the multiplicity of the zero $z_n = 0$ of the resultant $R('f, f_n) = \prod_{\nu=1}^p f_n^{\mu_\nu}(z^{(\nu)}(z_n))$ is equal to $k_1 \cdots k_n$. Theorem 22.4 gives us the sufficiency.

We prove the necessity by induction on the number of variables. Suppose that (22.6) has a nonisolated zero; then $\dim_0 K_1 \geqslant 1$. If $\dim_0 K_1 = 1$, then

$K_1 = A$. Since $E_1 \neq \{0\}$, there exists a ν_0 $(1 \leqslant \nu_0 \leqslant p)$ such that $P_n(a^{(\nu_0)}) = 0$, from which we obtain $f_n(z^{(\nu_0)}(z_n)) = o(z_n^{k_n})$. Therefore,

$$\mu_0(R('f, f_n)) = \mu_0(f) > k_1 \cdots k_n.$$

If $\dim_0 K_1 > 1$, then $\dim_0\{z\colon P_j('z, 0) = 0,\ j = 1, \ldots, n-1\} \geqslant 1$. By the induction hypothesis, $\mu_0('f('z, 0)) > k_1 \cdots k_{n-1}$. On the other hand, $\mu_0('f('z, 0))$ is equal to the number of zeros (counting multiplicity) of the system $'f('z, z_n)$ in the polycylinder $'U_\varepsilon$, i.e., to the number $\sum_1^p \mu_\nu$; therefore, $\mu_0(R('f, f_n)) = \mu_0(f) > k_1 \cdots k_n$. $\square$

REMARK. The sufficiency of the condition in Theorem 22.6 is easy to prove by means of Rouché's principle and Bézout's theorem.

Theorem 22.6 is easily generalized to the case which uses the concept of weighted homogeneity instead of ordinary homogeneity. Let $a_1, \ldots, a_n$ be positive rational numbers. We recall that a polynomial $g(z)$ is said to be *weighted-homogeneous with respect to the weight* $(1/a_1, \ldots, 1/a_n)$ if all its monomials $z_1^{i_1} \cdots z_n^{i_n}$ have the same weighted degree $d = i_1/a_1 + \cdots + i_n/a_n$ with respect to this weight.

THEOREM 22.7. *Suppose that $f_j(z) = P_j(z) + \theta_j(z)$, where $P_j(z)$ is a weighted-homogeneous polynomial of degree d_j with respect to the weight $(1/a_1, \ldots, 1/a_n)$, and $\theta_j(z)$ is a series of monomials having weighted degree greater than d_j with respect to the same weight, $j = 1, \ldots, n$. Then*

$$\mu_0(f) = \mu_0(f_1, \ldots, f_n) \geqslant a_1 d_1 \cdots a_n d_n,$$

and equality holds if and only if $z = 0$ is an isolated zero of the system $(P_1, \ldots, P_n)$.

Theorem 22.7 can be obtained from Theorem 22.6 by the change of variables $z_j = \zeta_j^{s_j}$, $j = 1, \ldots, n$, where $s_j = q\prod_{k \neq j} a_k$, and q is chosen so that all the s_j $(j = 1, \ldots, n)$ are integers. Then $f_j(z)$ becomes $\tilde{f}_j(\zeta) = \tilde{P}_j(\zeta) + \tilde{\theta}_j(\zeta)$, where $\tilde{P}_j(\zeta)$ is a homogeneous polynomial of degree $d_j q a_1 \cdots a_n$. Therefore, $\mu_0(\tilde{f}) = q^n d_1 \cdots d_n a_1^n \cdots a_n^n$ if and only if $\zeta = 0$ is an isolated zero of the system $(\tilde{P}_1, \ldots, \tilde{P}_n)$. But the property of being an isolated zero does not change under the indicated change of variables, and, moreover, by Proposition 22.1,

$$\mu_0(f) = \frac{\mu_0(\tilde{f})}{\mu_0(z(\zeta))} = \frac{q^n(\prod_{j=1}^n d_j)(\prod_{j=1}^n a_j)^n}{q^n(\prod_{j=1}^n a_j)} = a_1 d_1 \cdot \cdots \cdot a_n d_n. \quad \square$$

4°. Computation of the multiplicity of a zero from Newton polygons. In analogy with homogeneous principal parts we describe a larger class of systems of polynomials that play the role of principal parts. Let $\mathbf{N}$ and $\mathbf{R}_+$ denote the sets of nonnegative integers and nonnegative real numbers, respectively. For a function $g = \Sigma_\alpha c_\alpha z^\alpha$ and a compact subset $M \subset \mathbf{R}^n_+$ we let $\operatorname{supp} g = \{\alpha \in \mathbf{N}^n, c_\alpha \neq 0\}$ and $g_M = \Sigma_{\alpha \in M \cap \mathbf{N}^n} c_\alpha z^\alpha$.

The system (22.1) is said to be *nondegenerate on* M if the polynomials $f_{1M}, \ldots, f_{nM}$ do not vanish on the set $(\mathbf{C}\backslash 0)^n$. For any polyhedron $\Delta \subset \mathbf{R}^n_+$ the n-dimensional volume of Δ is denoted by $v(\Delta)$.

The *Newton polyhedron* $\Gamma_+(f_1, \ldots, f_n)$ *of the system* (22.1) *at zero* is defined to be the convex hull in $\mathbf{R}^n_+$ of the set

$$\left(\cup\,(\alpha + \mathbf{R}^n_+)\colon \alpha \in (\operatorname{supp} f_1 \cup \cdots \cup \operatorname{supp} f_n)\backslash\{0\}\right).$$

The *Newton polygon* $\Gamma(f_1, \ldots, f_n)$ *of the system* (22.1) *at zero* is defined to be the union of the closed compact faces of the polyhedron $\Gamma_+(f_1, \ldots, f_n)$. We let $\Gamma_-(f_1, \ldots, f_n)$ be the union of all the segments with initial point at 0 and final point on $\Gamma(f_1, \ldots, f_n)$, and we call the number $n!v(\Gamma_-(f_1, \ldots, f_n))$ *the Newton number of the system* (22.1) *at zero*. For brevity we denote $\Gamma(f_1, \ldots, f_n)$ by Γ and call the polynomials $f_{1\Gamma}, \ldots, f_{n\Gamma}$ the *principal parts of the system* (22.1) *at zero*. We say that the system (22.1) is *nondegenerate at zero* if for any face σ of the Newton polygon the system is nondegenerate on the set σ (this is a condition on the principal parts of the system at zero).

It is assumed that the system (22.1) is *suitable*, i.e., such that for any i from 1 to n at least one of the functions $f_1, \ldots, f_n$ contains a monomial of the form $z_i^{k_i}$ ($k_i \geqslant 1$) with nonzero coefficient.

THEOREM 22.8 (Kushnirenko). 1. *The multiplicity of the isolated zero $z = 0$ of the system* (22.1) *is not less than the Newton number of the system at zero, and if the system is nondegenerate at zero, then they are equal.*

2. *The set of principal parts of all the systems that are nondegenerate at zero and have a given Newton polygon is an open dense subset of the manifold of all principal parts corresponding to the given Newton polygon.*

PROOF. 1. It suffices to prove the assertion when the supports coincide: $\operatorname{supp} f_1 = \cdots = \operatorname{supp} f_n$. Indeed, for small $\tau_j \neq 0$ ($j = 1, \ldots, n$) the systems (22.1) and $F_i = f_i + \Sigma_{j \neq i} \tau_j f_j$, $i = 1, \ldots, n$, have the zero $z = 0$ with the same multiplicity, are degenerate at the same time on faces on the Newton polygon $\Gamma = \Gamma(f_1, \ldots, f_n)$, and, moreover, have $\operatorname{supp} F_1 = \cdots = \operatorname{supp} F_n$. For simplicity let $n = 2$. Let $\sigma \subset \Gamma$ be a one-dimensional face, which is determined by its endpoints (α_1, α_2), $(\beta_1, \beta_2) \in \mathbf{N}^2$ (assume that $\beta_1 > \alpha_1$). To the face σ there correspond $\beta_1 - \alpha_1$ (counting multiplicity) local solutions of $f_1(z_1, z_2) = 0$, and they have the form

$$z_1^\nu(z_2) = a^\nu z_2^\lambda + o(z_2^\lambda), \qquad \nu = 1, \ldots, \beta_1 - \alpha_1,$$

where $\lambda = (\alpha_2 - \beta_2)/(\beta_1 - \alpha_1)$, $a^{(\nu)} \neq 0$, and satisfy the condition (see [198])

$$f_{1\sigma}(a^\nu z_2^\lambda, z_2) \equiv 0, \qquad \nu = 1, \ldots, \beta_1 - \alpha_1.$$

The product

$$\varphi_\sigma(z_2) = \prod_{\nu=1}^{\beta_1-\alpha_1} f_2(z_1^\nu(z_2), z_2)$$

$$= \prod_{\nu=1}^{b_1-\alpha_1} \left[f_{2\sigma}(a^\nu z_2^\lambda, z_2) + o(z_2^{\lambda\alpha_1+\alpha_2}) \right]$$

in the resultant $R(f_1, f_2)$ corresponds to the face σ. If the system (f_1, f_2) is nondegenerate on σ, then $f_{2\sigma}(a^{(\nu)}z_2^\lambda, z_2) \not\equiv 0$, $\nu = 1,\dots,\beta_1 - \alpha_1$; therefore, the smallest degree in the expansion of $\varphi_\sigma(z_2)$ in fractional powers of the variable z_2 is equal to $(\beta_1 - \alpha_1)(\lambda\alpha_1 + \alpha_2)$, and it is greater than this number if (f_1, f_2) is degenerate on σ. It is not hard to see that $(\beta_1 - \alpha_1)(\lambda\alpha_1 + \alpha_2)$ is double the area of the triangle with vertex at 0 and base σ. Taking into account the fact that $R(f_1, f_2) = \prod_{\sigma\subset\Gamma} \varphi_\sigma(z_2)$, we get what is required.

2. Let σ be a face of the Newton polygon Γ, and $f_1,\dots,f_n$ polynomials with variable coefficients and with $\operatorname{supp} f_i \subset \Gamma$. Obviously, $f_{1\sigma},\dots,f_{n\sigma}$ are weighted-homogeneous polynomials with respect to a common weight; therefore, the change of variables $z_j = \zeta_j^{s_j}$, $j = 1,\dots,n$ (see the proof of Theorem 22.7), takes the system $f_\sigma = (f_{1\sigma},\dots,f_{n\sigma})$ to a system $\tilde{f}(\zeta) = (\tilde{f}_1(\zeta),\dots,\tilde{f}_n(\zeta))$ of homogeneous polynomials. But for any i from 1 to n the system

$$\tilde{f}_j\left(\frac{\zeta_1}{\zeta_i},\dots,1,\dots,\frac{\zeta_n}{\zeta_i}\right), \qquad j = 1,\dots,n,$$

of polynomials in the $n-1$ variables $\zeta_1/\zeta_i,\dots,[i]\cdots\zeta_n/\zeta_i$ does not have zeros in $\mathbf{C}^{n-1}$ for almost all coefficients. Therefore, the systems $\tilde{f}(\zeta)$ and $f_\sigma(z)$ almost always do not have zeros in $(\mathbf{C}\setminus 0)^n$.

REMARK. We have actually proved that in the two-dimensional case the nondegeneracy condition for (22.1) on each face of the Newton polygon $\Gamma(f)$ is also necessary for $\mu_0(f)$ to equal the Newton number.

In Theorem 22.8 the Newton polygon $\Gamma(f)$ was constructed for the whole system f, namely, for the union of the supports $\operatorname{supp} f_i$, $i = 1,\dots,n$. However, the Newton polygon can be constructed for each support separately. Then the polynomials $f_{1\Gamma_1},\dots,f_{n\Gamma_n}$, where $\Gamma_i = \Gamma(f_i)$ is the Newton polygon constructed for $\operatorname{supp} f_i$, enter the roles of the principal parts for f. This approach to the computation of the multiplicity of a zero can be reduced to the case where the supports coincide, $\operatorname{supp} f_1 = \cdots = \operatorname{supp} f_n$, i.e., to Theorem 22.8.

PROPOSITION 22.9. *Let $S = (S_1,\dots,S_n)$ be an n-tuple of finite subsets of $\mathbf{N}^n\setminus\{0\}$ such that there exists a system of polynomials g with* $\operatorname{supp} g_i \subset S_i$ *that has an isolated zero at $z = 0$. Then the multiplicity $\mu_0(f)$ is one and the same and is minimal among all possible multiplicities for all systems f with* $\operatorname{supp} f_i \subset S_i$ *except for an analytic set in the coefficient space.*

PROOF. Let c be the variable in the space of coefficients $\mathbf{C}^m$ (m is equal to the sum of the cardinalities of the sets S_i, $i = 1,\dots,n$), and let $f_i(z, c) = \sum_{\alpha \in S_i} c_{i\alpha} z^\alpha$, $i = 1,\dots,n$, be a system with variable coefficients. Rouché's principle implies that the set M of c with $\mu_0(f(z, c))$ minimal is open in $\mathbf{C}^m$. Therefore, there exists a neighborhood U of the point $(0, c^{(0)}) \in \mathbf{C}^n \times M$ such that for the analytic set $E_f = \{(z, c) \in \mathbf{C}^n \times \mathbf{C}^m: f_i(z, c) = 0, i = 1,\dots,n\}$ we have $U \cap E_f = \{(z, c) \in U: z = 0\}$. If $\tilde{c}$ is a boundary point of M, then $V \cap E_f \setminus \{(z, c): z = 0\} = \varnothing$ for any neighborhood V of $(0, \tilde{c})$. Indeed, otherwise there would exist balls $B_1(0) \in \mathbf{C}^n$ and $B_2(\tilde{c}) \in \mathbf{C}^m$ about the respective points 0 and $\tilde{c}$ such that for any $c \in B_2(\tilde{c})$ the system $f(z, c)$ has in $B_1(0)$ the unique zero $z = 0$, i.e., $\tilde{c} \in M$. Further, the germ of the analytic set $\Pi = \{(z, c): z = 0\}$ at $(0, \tilde{c})$ does not belong to any other irreducible component of the germ of the analytic set E_f. Therefore, there exist functions $\varphi_1,\dots,\varphi_s$ holomorphic in the neighborhood $V(0, \tilde{c})$ such that $E_f = \Pi \cup \{(z, c) \in V(0, \tilde{c}): \varphi_1(z, c) = \dots = \varphi_s(z, c) = 0\}$.

Thus, the set of points $(0, c) \in \Pi$ in any neighborhood of which $E_f \setminus \Pi \neq \varnothing$ satisfies the system of equations $\psi_j(c) = \varphi_j(0, c) = 0$, $j = 1,\dots,s$, in a neighborhood of $(0, \tilde{c})$. Consequently, $\mathbf{C}^m \setminus M$ is an analytic set. □

For the tuple $S = (S_1,\dots,S_n)$ we let $\mu(S)$ stand for the minimal multiplicity of zero among all multiplicities $\mu_0(f)$ of systems f such that $\operatorname{supp} f_i \subset S_i$. The Rouché principle implies that for any S there exists a finite subtuple $S' = (S_1',\dots,S_n')$, $S_i' \subset S_i$ with the property that $\mu(S') = \mu(S)$. Therefore, for the computation of $\mu(S)$ it will be assumed that the S_i are finite.

We introduce an addition operation on subsets of $\mathbf{N}^n$ by setting $S_1 + S_2 = \{q_1 + q_2: q_1 \in S_1, q_2 \in S_2\}$. It is clear that to multiplication of the functions $g_1 g_2$ there corresponds addition of their supports: $\operatorname{supp} g_1 + \operatorname{supp} g_2$. Hence, by Proposition 22.1,

$$\mu(S_1 + S_1', S_2,\dots,S_n) = \mu(S_1,\dots,S_n) + \mu(S_1', S_2,\dots,S_n). \quad (22.10)$$

Moreover, $\mu(S)$ is symmetric with respect to its arguments, and is thus multilinear by (22.10). Consequently, $\mu(S)$ can be recovered from its values on the diagonal as follows:

$$\mu(S) = \frac{(-1)^n}{n!} \sum_{k=1}^{n} (-1)^k \sum_{1 \le i_1 < \dots < i_k \le n} \mu(S_{i_1} + \dots + S_{i_k},\dots,S_{i_1} + \dots + S_{i_k}).$$

For a set $S \subset \mathbf{N}^n \setminus \{0\}$ we use the following notation: $\Gamma_+(S)$ is the convex hull of the set $\{\alpha + \mathbf{R}_+^n : \alpha \in S\}$, $\Gamma(S)$ is the union of the closed compact faces of the polyhedron $\Gamma_+(S)$, and $\Gamma_-(S)$ is the union of all segments with initial point at 0 and final point on $\Gamma(S)$. In what follows we assume that for each i from 1 to n the set S_i in the n-tuple $S = (S_1,\dots,S_n)$ contains points of the form $\alpha = (0,\dots,\alpha_j,\dots,0)$, $j = 1,\dots,n$. Then, by Theorem 22.8,

$$\mu(S_{i_1} + \cdots + S_{i_k},\dots,S_{i_1} + \cdots + S_{i_k}) = n!v(\Gamma_-(S_{i_1} + \cdots + S_{i_k})),$$

$k = 1,\dots,n$, and

$$\mu(S) = (-1)^n \sum_{k=1}^{n} (-1)^k \sum_{1 \leqslant i_1 < \cdots < i_k \leqslant n} v(\Gamma_-(S_{i_1} + \cdots + S_{i_k})). \qquad (22.11)$$

We describe nondegeneracy conditions on a system f with $\operatorname{supp} f_i \subset S_i$ that ensure the equality $\mu_0(f) = \mu(S)$. Suppose that $S \subset \mathbf{N}^n$ and $l = (l_1,\dots,l_n)$, the $l_j > 0$ being rational numbers. We set

$$m(l, S) = \min\{\langle l, \alpha\rangle = l_1\alpha_1 + \cdots + l_n\alpha_n,\ \alpha \in S\}$$

and let $S_l = \{\alpha \in S\colon \langle l, \alpha\rangle = m(l, S)\}$ (the intersection of S with the supporting plane in the direction l). For a function $g = \sum_\alpha c_\alpha z^\alpha$ such that $\operatorname{supp} g \subset S$ let $g_l = \sum_{\alpha \in S_l} c_\alpha z^\alpha$. If $f = (f_1,\dots,f_n)$ is a system such that $\operatorname{supp} f_i \subset S_j$, $j = 1,\dots,n$, then let $f_l = (f_{1l},\dots,f_{nl})$.

THEOREM 22.10. *Suppose that* $f = (f_1,\dots,f_n)$ *is a system such that* $\operatorname{supp} f_i \subset S_i$, $i = 1,\dots,n$. *If for any positive rational vector l the system of f_l does not have zeros in* $(\mathbf{C}\setminus 0)^n$, *then* $\mu_0(f) = \mu(S)$, *where* $\mu(S)$ *can be computed by* (22.11).

It is important to note that the condition of the theorem has to be checked only for finitely many vectors l giving distinct tuples S_{il}, $i = 1,\dots,n$. Namely: it suffices to take those l for which the intersections of the polygons $\Gamma(S_i)$ $(i = 1,\dots,n)$ with the supporting planes in the direction l consist of more than one point, i.e., are faces and not vertices.

PROOF. Consider the system $f + tg = (f_1 + tg_1,\dots,f_n + tg_n)$, where g is a system such that $\operatorname{supp} g_i \subset S_i$ and $\mu_0(g) = \mu(S)$. Suppose that $\mu_0(f) > \mu(S) = \mu_0(g)$. Then for the curve $K = \{(z, t) \in \mathbf{C}^{n+1},\ f_j + tg_j = 0,\ j = 1,\dots,n\}$ there exists an analytic element $z(t) \not\equiv 0$ such that $z(0) = 0$. By Proposition 22.9, g can be chosen so that $z(t) = at^l + o(t^l)$, where $a = (a_1,\dots,a_n) \in (\mathbf{C}\setminus 0)^n$. Since $z(0) = 0$, it follows that $l_j > 0$; therefore, the conditions $f_j(z(t)) + tg_j(z(t)) = t^q f_{jl}(a) + o(t^q) \equiv 0$, $j = 1,\dots,n$, imply that $f_{jl}(a) = 0$, $j = 1,\dots,n$. This is a contradiction. $\square$

§23. Application of multiple residues for finding generating functions and computing combinatorial sums

The main idea for the application of one-dimensional and multidimensional residues in combinatorial analysis (an idea systematically developed by Egorychev) runs as follows. Suppose that we must find the generating function of some multiple combinatorial sequence $c_\alpha = c_{\alpha_1\cdots\alpha_n}$ or a combinatorial sum whose general term c_α can be expressed with the help of the addition and multiplication operations in terms of the coefficients of known generating functions. Substituting integral representations of the coefficients of these known functions (the Taylor coefficients in integral form) into the power series of the generating function $F(t) = F(t_1,\ldots,t_n) = \sum_{\alpha\geq 0} c_\alpha t^\alpha$ or, respectively, into the unknown combinatorial sum, then replacing products of integrals by multiple integrals, and then changing the order of integration and summation, we get a series or a finite sum under the sign of a multiple integral, and these, as a rule, are much simpler to convolve than the original expressions. Most frequently a geometric progression ends up under the integral sign. Computing the integral of the resulting sum with the help of residues, we find the desired generating function or combinatorial sum. It is sometmes useful to regard a combinatorial sum as a term in some combinatorial sequence. We get the desired sum by finding the generating function for the sequence by the indicated method and expanding this function in a Taylor series in the usual way.

The following are most often used as initial elementary integral formulas:

$$\binom{a}{m} = \frac{a(a-1)\cdots(a-m+1)}{m!} = \frac{1}{2\pi i}\int_{|z|=\varepsilon}\frac{(1+z)^a\,dz}{z^{m+1}},$$

$$\frac{a^n}{n!} = \int_{|z|=\varepsilon}\frac{e^{az}\,dz}{z^{n+1}}, \tag{23.1}$$

and so on.

The next result realizes the idea expounded above for a sufficiently general situation.

THEOREM 23.1 (Egorychev and Yuzhakov). *Suppose that the terms of an n-fold numerical sequence $c_\alpha = c_{\alpha_1\cdots\alpha_n}$ can be represented in the form*

$$c_\alpha = \frac{1}{(2\pi i)^n}\int_{\Gamma_\rho}\frac{\varphi(z)[f_1(z)]^{\alpha_1\beta_1}\cdots[f_n(z)]^{\alpha_n\beta_n}\,dz}{z_1^{\beta_1(\alpha_1+1)}\cdots z_n^{\beta_n(\alpha_n+1)}}, \tag{23.2}$$

where $\varphi, f_j \in A(\overline{U}_R)$, $\overline{U}_R = \{z \in \mathbf{C}^n\colon \|z\| \leq R\}$ is some polydisk, and $\Gamma_\rho = \{z\colon |z_1| = \cdots = |z_n| = \rho\}$, $0 < \rho < R$. Then the series

$$F(t) = \sum_{\alpha\geq 0} c_\alpha t^\alpha, \qquad t = (t_1,\ldots,t_n), \tag{23.3}$$

converges absolutely in some neighborhood $V_\varepsilon = \{t\colon \|t\| < \varepsilon\}$, and

$$F(t) = \frac{1}{(2\pi i)^n}\int_{\Gamma_\rho}\frac{\varphi(z)\,dz}{\prod_{j=1}^n\left[z_j^{\beta_j} - t_j f_j^{\beta_j}(z)\right]}. \tag{23.4}$$

In particular:

a) *if* $\beta_1 = \cdots = \beta_n = 1$, *then*

$$F(t) = \left(\varphi(z)\Big/\frac{\partial w}{\partial z}\right)\Big|_{z=z(t)} = \left[\varphi(z)\Big/\det\left\|\delta_{jk} - t\frac{\partial f_j(z)}{\partial z_k}\right\|\right]\Big|_{z=z(t)}, \quad (23.5)$$

where

$$w_j = z_j - t_j f_j(z), \qquad j = 1,\dots,n, \quad (23.6)$$

and $z = z(t)$ *is a zero of* (23.6) *lying in* $U_\rho = \{z: \|z\| < \rho\}$; *and*

b) *if* $f_j(0) \neq 0, j = 1,\dots,n$, *then*

$$F(t) = \sum_{\mu \in M} \left[\varphi(z)\Big/\frac{\partial w}{\partial z}\right]\Big|_{z=z^{(\mu)}(t)}, \quad (23.7)$$

where

$$w_j = z_j^{\beta_j} - t_j f_j^{\beta_j}(z), \qquad j = 1,\dots,n, \quad (23.8)$$

$\{z^{(\mu)}(t), \mu \in M\}$ *is the set of zeros of* (23.8) *in the polydisk* U_ρ, *and* $M = \{\mu = (\mu_1,\dots,\mu_n), \mu_j = 1,\dots,\beta_j, j = 1,\dots,n\}$.

For the indicated cases formulas (23.5) and (23.7) are expressions of the integral (23.4) in terms of multiple residues. In the general case we could not determine residue formulas for the integral (23.6). However, the computation of the integral (23.4) does not involve difficulties in concrete examples. It can sometimes be computed conveniently by iterated integration.

PROOF OF THEOREM 23.1. If $z \in \Gamma_\rho$ (ρ fixed), $t \in V_\varepsilon$, $0 < \varepsilon < (\rho/M_j)^{\beta_j}$, and $M_j = \max_{z\in\Gamma_\rho}|f_j(z)|, j = 1,\dots,n$, then the multiple geometric progression

$$\frac{\varphi(z)}{\prod_{j=1}^n \left[z_j^{\beta_j} - t_j f_j^{\beta_j}(z)\right]} = \sum_{\alpha \geqslant 0} \varphi(z) \prod_{j=1}^n \frac{t_j^{\alpha_j}[f_j(z)]^{\beta_j\alpha_j}}{z_j^{\beta_j(\alpha_j+1)}}$$

converges uniformly and absolutely. Integrating this series termwise over the cycle Γ_ρ, we get for $t \in V_\varepsilon$ formula (23.4) for the sum of the series (23.3) with coefficients (23.2).

By Lemma 4.9,

$$\Gamma_\rho \sim \Gamma_\delta = \left\{z: \left|z_j^{\beta_j} - t_j f_j^{\beta_j}(z)\right| = \delta, j = 1,\dots,n\right\},$$

for fixed $t \in V_\varepsilon$, where $\delta > 0$ is sufficiently small, $\gamma_\delta = \sum_{\mu\in M}\gamma_\mu$, $\gamma_\mu = \gamma_\delta \cap U_\mu$, and U_μ is a neighborhood of the zero $z^\mu(t)$ of (23.8). By the Cauchy-Poincaré

Theorem 0.6, the integral in (23.4) is equal to

$$\int_{\Gamma_\rho} \frac{\varphi\, dz}{\prod_{j=1}^n \left(z_j^{\beta_j} - t_j f_j^{\beta_j}(z)\right)} = \sum_{\mu \in M} \int_{\gamma_\mu} \frac{\varphi(z)\, dz}{\prod_{j=1}^n \left(z_j^{\beta_j} - t_j f_j^{\beta_j}(z)\right)}. \tag{23.9}$$

In case a) the system (23.8) takes the form (23.6) and has one simple zero in U_ρ. In case b) this system has $\beta_1 \cdots \beta_n$ distinct simple zeros in U_ρ, because the systems

$$z_j - t_j^{1/\beta_j} f_j(z) = 0, \qquad j = 1, \ldots, n,$$

each have a single simple zero in U_ρ, and these zeros are distinct for distinct values of the root t_j^{1/β_j} if $f_j(0) \neq 0$. Since $(\partial w/\partial z)|_{z^{(\mu)}(t)} \neq 0$, the mapping (23.8) is biholomorphic in a sufficiently small neighborhood U_μ of the zero $z^\mu(t)$. Performing the change (23.8) in U_μ, we get

$$\frac{1}{(2\pi i)^n} \int_{\gamma_\mu} \frac{\varphi(z)\, dz}{\prod_{j=1}^n \left[z_j^{\beta_j} - i_j f_j^{\beta_j}(z)\right]} = \frac{1}{(2\pi i)^n} \int_{\substack{|w_j| = \delta \\ j=1,\ldots,n}} \frac{\varphi(z(w))(\partial z/\partial w)\, dw}{w^{I}}$$

$$= \left[\varphi(z(w)) \frac{\partial z}{\partial w}\right]\Bigg|_{w=0} = \left[\varphi(z) \Big/ \frac{\partial w}{\partial z}\right]\Bigg|_{z=z^{(\mu)}(t)}. \tag{23.10}$$

From (23.10) and (23.9) we obtain (23.7), and then (23.5) for $\beta_1 = \cdots = \beta_n = 1$. □

The next theorem, which is important in combinatorics, is a special case of Theorem 23.1.

THEOREM 23.9 (main theorem of McMahon). *Let $X_j = \sum_{k=1}^n a_{kj} x_j$, $j = 1, \ldots, n$. Then the coefficient $b_\alpha = b_{\alpha_1 \cdots \alpha_n}$ of $x_1^{\alpha_1} \cdots x_n^{\alpha_n}$ in the expansion of $X_1^{\alpha_1} \cdots X_n^{\alpha_n}$ is equal to the coefficient of $t_1^{\alpha_1} \cdots t_n^{\alpha_n}$ in the expansion of*

$$F(t) = \begin{vmatrix} 1 - a_{11}t_1 & -a_{12}t_1 & \cdots & -a_{1n}t_1 \\ -a_{12}t_2 & 1 - a_{22}t_2 & \cdots & -a_{2n}t_2 \\ \cdots & \cdots & \cdots & \cdots \\ -a_{1n}t_n & -a_{2n}t_n & \cdots & 1 - a_{nn}t_n \end{vmatrix}^{-1}. \tag{23.11}$$

PROOF. The coefficient b_α can be represented in the form

$$b_\alpha = \frac{1}{(2\pi i)^n} \int_{|x_j| = \varepsilon,\, j=1,\ldots,n} \frac{X_1^{\alpha_1} \cdots X_n^{\alpha_n}\, dx_1 \wedge \cdots \wedge dx_n}{x_1^{\alpha_1+1} \cdots x_n^{\alpha_n+1}}.$$

By Theorem 23.1 (see case a)), the generating function $F(t) = \sum_{\alpha \geqslant 0} b_\alpha t^\alpha$ is determined by

$$F(t) = 1 \Big/ \left. \frac{\partial(w_1, \ldots, w_n)}{\partial(x_1, \ldots, x_n)} \right|_{x = x(t)},$$

where $w_j = x_j - t_j X_j$, $j = 1, \ldots, n$, or (23.11). □

EXAMPLE 1. Prove the identity

$$\sum_{k=0}^{N} \binom{m}{k} \binom{n}{k} \binom{m+n+p-k}{m+n} = \binom{m+p}{m} \binom{n+p}{n}, \qquad (23.12)$$

where $m, n, p > 0$ and $N = \min(m, n, p)$. We solve the problem of computing the combinatorial sum

$$A_{mn} = \sum_{k=0}^{N} \binom{m}{k} \binom{n}{k} \binom{m+n+p-k}{m+n}.$$

Using (23.1) and the fact that

$$\int_{|z| = \varepsilon} \frac{(1+z)^m \, dz}{z^{k+1}} = 0$$

for $k > m$, we repesent A_{mn} in the form

$$A_{mn} = \sum_{k=0}^{\infty} \frac{1}{(2\pi i)^3} \int_{|z_1| = \rho_1} \frac{(1+z_1)^m \, dz_1}{z_1^{k+1}} \int_{|z_2| = \rho_2} \frac{(1+z_2)^n \, dz_2}{z_2^{k+1}}$$
$$\times \int_{|z_3| = \rho_3} \frac{(1+z_3)^{m+n+p-k} \, dz_3}{z_3^{m+n+1}}.$$

Consider the generating function of the sequence A_{mn}:

$$F(u, v) = \sum_{m, n = 0}^{\infty} A_{mn} u^m v^n = \sum_{m, n, k = 0}^{\infty} \frac{1}{(2\pi i)^3} \int_\Gamma \left[\frac{u(1+z_1)(1+z_3)}{z_3} \right]^m$$
$$\times \left[\frac{v(1+z_2)(1+z_3)}{z_3} \right]^n \left[\frac{1}{z_1 z_2 (1+z_3)} \right]^k \frac{(1+z_3)^p}{z_1 z_2 z_3} dz_1 \wedge dz_2 \wedge dz_3, \qquad (23.13)$$

where $\Gamma = \{z : |z_j| = \rho_j,\ j = 1, 2, 3\}$. For $\rho_1 = \rho_2 = 2$, $\rho_3 = \frac{1}{2}$, $|u| < \frac{1}{9}$, and $|v| < \frac{1}{9}$ it is possible to interchange the summation and integration in (23.13), because the triple geometric progression then converges uniformly on Γ. An argument similar to that for (23.4) yields

$$F(u, v) = \frac{1}{(2\pi i)^3}\cdot$$

$$\times \int_\Gamma \frac{(1+z_3)^p z_3\, dz_1 \wedge dz_2 \wedge dz_3}{[z_3 - u(1+z_1)(1+z_3)][z_3 - v(1+z_2)(1+z_3)][z_1 z_2(1+z_3) - 1]}.$$

Computing the integral obtained as an iterated integral, we find that

$$F(u, v) = \frac{1}{(1-u)^{p+1}(1-v)^{p+1}}.$$

The Taylor expansion of this function has the form

$$F(u, v) = \sum_{m,n=0}^{\infty} \binom{m+p}{m}\binom{n+p}{n} u^m v^n.$$

From this and (23.13), we get (23.12).

EXAMPLE 2. Compute the sum

$$S_m = \sum_{k=0}^{m} (-1)^k \binom{m}{k}^3.$$

With the help of (23.1), taking account of the fact that

$$\int_{|z|=\varepsilon} (1+z)^m\, dz/z^{k+1} = 0$$

for $k > m$, we find the integral representation

$$S_m = \sum_{k=0}^{\infty} \frac{(-1)^k}{(2\pi i)^3} \int_\Gamma \frac{(1+z_1)^m}{z_1^{m-k+1}} \frac{(1+z_2)^m}{z_2^{m-k+1}} \frac{(1+z_3)^m}{z_3^{k+1}}\, dz_1 \wedge dz_2 \wedge dz_3,$$

where $\Gamma = \{z : |z_j| = \rho_j, j = 1, 2, 3\}$. To assure that not all the variable powers end up in the denominator, we use the the identity $\binom{m}{k} = \binom{m}{m-k}$. Interchanging the summation and integration, we get for $\rho_1 = \rho_2 = 1$ and $\rho_3 = 2$ that

$$S_m = \frac{1}{(2\pi i)^3} \int_\Gamma \frac{(1+z_1)^m(1+z_2)^m(1+z_3)^m}{z_1^{m+1} z_2^{m+1}(z_3 + z_1 z_2)}\, dz_1 \wedge dz_2 \wedge dz_3.$$

The integration with respect to z_3 yields

$$S_m = \frac{1}{(2\pi i)^2} \int_{|z_1|=|z_2|=1} \frac{(1+z_1)^m(1+z_2)^m(1-z_1 z_2)^m\, dz_1 \wedge dz_2}{z_1^{m+1} z_2^{m+1}}.$$

If we now seek the generating function immediately for S_m, we get a double integral whose direct computation leads either to the original power series or to the period of an elliptic integral. Therefore, it is expedient to consider the double sequence

$$R_{mn} = \frac{1}{(2\pi i)^2} \int_{|z_1|=|z_2|=1} \frac{(1+z_1)^m(1+z_2)^m(1-z_1z_2)^n \, dz_1 \wedge dz_2}{z_1^{m+1}z_2^{n+1}},$$

for which $R_{mm} = S_m$. Theorem 23.1 (see case a)) can now be applied to the sequence R_{mn}. By (23.3) and (23.5), elementary computations give us

$$\begin{aligned} F(t_1, t_2) &= \sum_{m,n=0}^{\infty} R_{mn} t_1^m t_2^n \\ &= \frac{1}{(2\pi i)^2} \int_{|z_1|=|z_2|=1} \frac{dz_1 \wedge dz_2}{[z_1 - t_1(1+z_1)(1+z_2)][z_2 - t_2(1 - z_1 z_2)]} \\ &= \left[(1-t_1)^2 + 4t_1t_2^2\right]^{-1/2}. \end{aligned}$$

Expanding this function in a binomial series first with respect to t_2 and then with respect to t_1, we find that

$$R_{mn} = \begin{cases} 0 & \text{for } n \text{ odd}, \\ \dfrac{(-1)^p(m+p)!}{(p!)^2(m-p)!} & \text{for } n = 2p. \end{cases}$$

Hence,

$$S_m = \sum_{k=0}^{m} (-1)^k \binom{m}{k}^3 = \begin{cases} 0 & \text{for } m \text{ odd}, \\ \dfrac{(-1)^p(3p)!}{(p!)^3} & \text{for } m = 2p. \end{cases}$$

CHAPTER V

Some Applications in Multidimensional Complex Analysis

§24. The Hartogs-Bochner Theorem on Required Analytic Extension. Approximation of Holomorphic Functions on Weil Polyhedra and Linearly Convex Compact Sets

1°. We discuss the question of holomorphic extension into a domain D of a function given on the boundary ∂D. Let us consider first a simple corollary to Theorem 1.2.

THEOREM 24.1 (Aronov and Kytmanov). *Let D be a bounded domain with $\partial D \in C^{(3)}$, and let $f \in C^{(2)}(\partial D)$. A function $F \in C^{(1)}(\bar{D})$ holomorphic in D and such that $F|_{\partial D} = f$ exists if and only if*

$$\int_{\partial D} f(\zeta)\omega(\zeta - z, \overline{\zeta - z}) = 0 \tag{24.1}$$

for $z \in \mathbf{C}^n \setminus \bar{D}$. If $\mathbf{C}^n \setminus \bar{D}$ is connected, then it suffices that (24.1) holds for $z \in \{w : |w_i| > R,\ i = 1, \ldots, n\}$, where R is a sufficiently large positive number.

PROOF. The first part of the theorem follows at once from Theorem 1.2, Lemma 1.4 and the fact that

$$\int_{\partial D} \frac{f(\zeta)\, d\bar{\zeta}_{[k]} \wedge d\zeta}{|\zeta - z|^{2n-2}}, \qquad z \in D, \tag{24.2}$$

extends to $\bar{D}$ as a function of class $C^{(2)}(\bar{D})$ (see Proposition 0.10). It remains only to observe that the Martinelli-Bochner type integral of f is a linear combination of derivatives of integrals of the form (24.2) and, consequently, is in the class $C^{(1)}(\bar{D})$. To prove the second assertion of the theorem, recall that a Martinelli-Bochner type integral is harmonic outside ∂D. □

THEOREM 24.2 (Bochner). *Suppose that D is a bounded domain, $\partial D \in C^{(1)}$, $\mathbf{C}^n \setminus \bar{D}$ is connected, $n > 1$, and $f \in C^{(1)}(\partial D)$. A function $F \in A(D) \cap C^{(1)}(\bar{D})$ such that $F|_{\partial D} = f$ exists if and only if the Severi conditions (the tangential Cauchy-Riemann equations) hold:*

$$df \wedge d\zeta|_{\partial D} = 0. \tag{24.3}$$

PROOF. The necessity follows at once from the Cauchy-Riemann conditions $\partial F/\partial \bar{\zeta}_k = 0$, $k = 1, \ldots, n$. For simplicity we prove the sufficiency under the assumptions that $\partial D \in C^{(3)}$ and $f \in C^{(2)}(\partial D)$, which enable us to use the preceding theorem. Note that

$$\omega(\zeta - z, \overline{\zeta - z}) = d_\zeta \Omega(\zeta, z), \tag{24.4}$$

where

$$\Omega(\xi, z) = \frac{(n-2)!}{(2\pi i)^n} \sum_{k=2}^{n} \frac{(-1)^k}{|\zeta - z|^{2n-2}} \frac{\bar{\zeta}_k - \bar{z}_k}{\zeta_1 - z_1} d\bar{\zeta}_{[1,k]} \wedge d\zeta.$$

Let $R > 0$ be such that $D \subset C(0, R)$. For $z \in \{w : |w_1| > R\}$ and $\zeta \in \partial D$ the form Ω is regular. By Stokes' formula, (24.3) and (24.4) now give us that

$$\int_{\partial D} f(\zeta)\omega = \int_{\partial D} f(\zeta)\, d_\zeta \Omega = \int_{\partial D} d_\zeta[f(\zeta)\Omega] = 0.$$

It remains to refer to Theorem 24.1. □

THEOREM 24.3 (Hartogs, Osgood and Braun). *Suppose that D is a bounded domain, $\mathbf{C}^n \setminus \bar{D}$ is connected, and $n > 1$. Every function $f(z)$ holomorphic on ∂D (i.e., in some neighborhood $U(\partial D)$) extends holomorphically to the whole of D.*

PROOF. We choose a domain D' such that $\partial D' \in C^{(1)}$, $D' \subset D$, $\partial D' \subset U(\partial D)$ and $\mathbf{C}^n \setminus \bar{D}'$ is, as before, connected. The function $f(z)$ is holomorphic on $\partial D'$; therefore, the tangential Cauchy-Riemann conditions hold on $\partial D'$. By theorem 24.2, there exists a function $F \in A(D') \cap C^{(1)}(\bar{D}')$ such that $F|_{\partial D'} = f$. It remains to show that $F = f$ in $U(\partial D) \cap D'$. This follows from the fact that they coincide on $\partial D'$ and the fact that the zeros of a holomorphic function that is not identically zero cannot fill a $(2n-1)$-dimensional continuous real surface (see, for example, [173], Theorem 4.7). □

In proving this theorem it can be assumed that $\partial D' \in C^{(3)}$ in order to use only the case of Theorem 24.2 we actually proved.

If Theorem 1.6 is used instead of Theorem 1.2, then the following result can be obtained in a way similar to that for Theorem 24.1.

THEOREM 24.4 (Kytmanov and Aĭzenberg). *In each of the two cases in Theorem* 1.6 *condition* (24.1) *is necessary and sufficient for every* $f \in C(\partial D)$ *to have an extension* $F \in A_C(D)$.

As a corollary to this assertion we give another criterion for the existence of a holomorphic extension. Let $Z_{k,l}(\zeta)$ be spherical (homogeneous harmonic) polynomials forming an orthonormal basis in the space $L^2(\partial B_1(0))$ on the unit sphere $\partial B_1(0) \subset \mathbf{C}^n$. Here k is the degree of the polynomial, and l is its index among the polynomials of degree k appearing in the basis, $k = 0, 1, \ldots, l = 1, \ldots, \sigma(k)$, where

$$\sigma_k = \frac{2(n+k-1)!\,(2n+k-3)!}{k!\,(2n-2)!}.$$

COROLLARY 24.5. *Suppose that* $\partial D \subset C^{(2)}$ *is connected and* D *is bounded. A function* $f \in C(\partial D)$ *is the trace on* ∂D *of a function in* $A_C(D)$ *if and only if*

$$\int_{\partial D} f(\zeta) \sum_{j=1}^{n} (-1)^{j-1} \frac{\partial}{\partial \zeta_j} Z_{k,l}(\zeta)\, d\bar{\zeta}_{[j]} \wedge d\zeta = 0,$$

$l = 1, \ldots, \sigma(k)$, $k = 0, 1, \ldots$.

For $n = 1$ this corollary is the familiar classical criterion that $f(\zeta)$ be orthogonal to all the monomials ζ^k, $k = 0, 1, \ldots$.

We need the following concept to get another corollary to Theorem 24.4. Suppose that D is bounded, $\partial D \in C^{(1)}$, and $f \in C(\partial D)$. We say that f has the one-dimensional holomorphic extension property if for any complex line l the restriction $f_l = f|_{\partial D \cap l}$ has an extension $F_l \in C(\overline{D} \cap l)$ that is holomorphic at interior (with respect to l) points of $\overline{D} \cap l$.

LEMMA 24.6. *If a function* f *has the one-dimensional holomorphic extension property, then* (24.1) *holds for* $z \in \mathbf{C}^n \setminus \overline{D}$.

PROOF. Suppose that z belongs to the unbounded component of $\mathbf{C}^n \setminus \overline{D}$. Let us perform an orthogonal linear transformation and a translation in $\mathbf{C}^n$ such that $z = 0$ and the plane $\{z : z_1 = 0\}$ does not intersect $\overline{D}$. This can be done for points z sufficiently far from D, in which case the kernel $\omega(\overline{\zeta - z}, \zeta - z)$ becomes $\omega(\bar{\zeta}, \zeta)$. Next, we make the change of variables $\zeta_1 = v_1^{-1}$, $\zeta_j = v_j v_1^{-1}$, $j = 2, \ldots, n$, and get

$$d\zeta = d\zeta_1 \wedge \cdots \wedge d\zeta_n = \begin{vmatrix} -\dfrac{1}{v_1^2} & 0 & \cdots & 0 \\ -\dfrac{v_2}{v_1^2} & \dfrac{1}{v_1} & \cdots & 0 \\ \cdots & \cdots & \cdots & \cdots \\ -\dfrac{v_n}{v_1^2} & 0 & \cdots & \dfrac{1}{v_1} \end{vmatrix} dv = -\frac{1}{v_1^{n+1}}\, dv.$$

Then $\mathbf{C}^n_z \setminus \{z : z_1 = 0\}$ is mapped biholomorphically onto $\mathbf{C}^n_v \setminus \{v : v_1 = 0\}$. Note that

$$\sum_{k=1}^{n} (-1)^{k-1} \bar{\zeta}_k\, d\bar{\zeta}_{[k]} = \sum_{k=1}^{n} \bar{\Delta}_k\, d\bar{v}_{[k]}.$$

where

$$\Delta_k = \begin{vmatrix} \zeta_1 & \zeta_2 & \cdots & \zeta_n \\ \zeta'_{1v_1} & \zeta'_{2v_1} & \cdots & \zeta'_{nv_1} \\ \cdots & \cdots & \cdots & \cdots \\ \zeta'_{1v_n} & \zeta'_{2v_n} & \cdots & \zeta'_{nv_n} \end{vmatrix} = \begin{vmatrix} \frac{1}{v_1} & \frac{v_2}{v_1} & \cdots & \frac{v_n}{v_1} \\ -\frac{1}{v_1^2} & -\frac{v_2}{v_1^2} & \cdots & \frac{v_n}{v_1^2} \\ 0 & \frac{1}{v_1} & \cdots & 0 \\ [k] & & \cdots & \cdot \\ \cdot & & \cdots & \cdot \\ 0 & 0 & \cdots & \frac{1}{v_1} \end{vmatrix} = \begin{cases} 1/v_1^n, & k = 1, \\ 0, & k \neq 1. \end{cases}$$

Therefore,

$$\omega(\zeta, \bar\zeta) = (-1)^n \frac{dv_1}{v_1} \wedge \frac{d\bar v_{[1]} \wedge dv_{[1]}}{(1 + |v_2|^2 + \cdots + |v_n|^2)^n} = \frac{dv_1}{v_1} \wedge \lambda(v'),$$

where $v' = (v_2, \ldots, v_n)$. Under our change of variables the lines $\{\zeta : \zeta_1 = t, \zeta_2 = c_2 t, \ldots, \zeta_n = c_n t\}$ pass into the lines $\{v : v_2 = c_2, \ldots, v_n = c_n\}$, and the boundary ∂D passes into some cycle Γ. If f has the one-dimensional holomorphic extension property, then after the change of variables this function (denote it now by $f(v)$) has the one-dimensional holomorphic extension property along the complex lines $\{v : v_2 = c_2, \ldots, v_n = c_n\}$. Further,

$$\int_{\partial D} f\omega(\zeta, \bar\zeta) = c_n \int_\Gamma f(v) \frac{dv_1}{v_1} \wedge \lambda(v') = c_n \int_{\pi(\Gamma)} \lambda(v') \int_{\pi^{-1}(v') \cap \Gamma} f(v) \frac{dv_1}{v_1},$$

where π is the projection $v \to v'$. By Sard's theorem, $\pi^{-1}(v') \cap \Gamma$ is a smooth curve for almost all $v' \in \pi(\Gamma)$. The point 0 lies outside the domain bounded by Γ; therefore, the inside integral is equal to zero for almost all v', i.e., (24.1) holds for z sufficiently far from D. It is also valid for all points in the unbounded component of $\mathbf{C}^n \setminus \bar D$, because an integral of the Martinelli-Bochner type in (24.1) is harmonic.

Suppose now that z is in a bounded component $Q \subset \mathbf{C}^n \setminus \bar D$, and assume that $z = 0$ (this situation can always be attained by a translation of the origin). Consider a cone Π_ε with center at 0 formed by complex lines passing through 0, where ε is the area of the part of the hypersurface ∂Q interior to Π_ε. After the same change of variables the cone passes into a cylinder formed by lines $\{v : v_2 = c_2, \ldots, v_n = c_n\}$, and Π_ε can be chosen so that: 1) the area of $\partial(Q \cap \Pi_\varepsilon) \cap \partial D$ tends to zero as $\varepsilon \to 0$, whence

$$\int_{\partial D} f\omega(\zeta, \bar\zeta) = \lim_{\varepsilon \to 0} \int_{\partial D \setminus \partial(Q \cap \Pi_\varepsilon)} f\omega(\zeta, \bar\zeta);$$

and 2) $\partial D \setminus \partial(Q \cap \Pi_\varepsilon)$ passes into a compact set Γ_ε under the change of variables. As above, it can be shown that

$$\int_{\Gamma_\varepsilon} f(v) \frac{dv_1}{v_1} \wedge \lambda(v') = 0,$$

consequently, (24.1) holds also for $z \in Q$. □

This lemma and Theorem 24.4 give us

Theorem 24.7 (Agranovskiĭ, Val'skiĭ and Stout). *If the domain D is bounded, $\partial D \in C^{(2)}$, and the function $f \in C(\partial D)$ has the one-dimensional holomorphic extension property, then f extends holomorphically into D.*

We remark that the proof of Theorem 12.9 amounted to an application of Theorem 12.2 and the fact that in the case under consideration linear combinations of polynomials of the form (12.16) are dense in $[Ph_C(D)]_\mu^\perp$ in the metric of $C(S(D))$. These considerations (and analogous considerations for holomorphic functions) led to the following characterization of traces of pluriharmonic (respectively, holomorphic) functions on the Shilov boundary.

THEOREM 24.8 (Aĭzenberg and Dautov). *Suppose that D is a circular strongly star-like bounded domain, μ is a nonnegative rotation-invariant measure that is massive on $S(D)$ and for which there is a Szegö kernel, and each point of $S(D)$ is a peak point for the algebra $A_C(D)$. If $\varphi \in C(S(D))$, then:*

1) $\varphi = f|_{S(D)}$ *for some $f \in A_C(D)$ if and only if*

$$\int_{S(D)} \varphi P \, d\mu = 0 \tag{24.3}$$

for every polynomial of the form

$$\begin{cases} z^J \bar{z}^{\mathfrak{J}} - P_{J\mathfrak{J}}(z), & |J| \leq |\mathfrak{J}|, \\ z^J \bar{z}^{\mathfrak{J}}, & |J| > |\mathfrak{J}|, \end{cases} \tag{24.4}$$

2) $\varphi = u|_{S(D)}$ *for some $u \in Ph_C(D)$ if and only if* (24.3) *holds for any polynomial of the form* (12.16).

If, moreover, D is an n-circular domain, and the measure is invariant with respect to n-rotations, then (24.4) *can be replaced by*

$$\begin{cases} z^J \bar{z}^{\mathfrak{J}}, & J \not< \mathfrak{J}, \\ z^J \bar{z}^{J+\mathfrak{J}}\left(a^2_{J+\mathfrak{J}+I_k} - a^2_{J+\mathfrak{J}}|z_k|^2\right), \end{cases}$$

and (12.16) *by* (12.20).

2°. Let us consider a Weil polyhedron $\Delta = \{z : z \in D, \chi_i(z) < 1, i = 1,\ldots,N\}$, $\chi_i(z) \in A(D)$, $i = 1,\ldots,N$, $\Delta \Subset D$, and a function $f(z) \in A(\Delta)$. For every r with $0 < r < 1$ we have $f \in A(\overline{\Delta(r)})$, where $\Delta(r) = \{z : z \in D, |\chi_i(z)| < r, i = 1,\ldots,N\}$. The function $f(z)$ can be represented in $\Delta(r)$ by the Bergman-Weil formula (9.1). Note that if $z \in \Delta(r)$ and $\zeta \in \gamma_{i_1\cdots i_n}$, then

$$\frac{|\chi_{i_k}(z)|}{|\chi_{i_k}|} < 1;$$

therefore, (9.1) gives us

$$f(z) = \sum_{i_1 < \cdots < i_n} \sum_{k_1,\ldots,k_n = 0} \left[\frac{1}{(2\pi i)^n} \int_{\gamma_{i_1\cdots i_n}} \frac{f(\zeta)\Omega_{i_1\cdots i_n}(\zeta, z)\, d\zeta}{\chi_{i_1}^{k_1+1}(\zeta)\cdots\chi_{i_n}^{k_n+1}(\zeta)} \right] \times \chi_{i_1}^{k_1}(z)\cdots\chi_{i_n}^{k_n}(z). \tag{24.5}$$

By the Cauchy-Poincaré Theorem 0.6, the series (24.5) does not depend on $r < 1$ for r sufficiently close to 1. Thus, we have

THEOREM 24.9 (Weil). *Every function $f \in A(\Delta)$ can be expanded for $z \in \Delta$ in a series* (24.5) *uniformly convergent on compact subsets of Δ.*

A polyhedron Δ is called a *polynomial Weil domain* if all the $\chi_i(z)$ are polynomials. A domain D such that every $f \in A(D)$ can be approximated uniformly on compact subsets of D by polynomials is called a *Runge domain.*

COROLLARY 24.10. *Every polynomial Weil domain is a Runge domain.*

The series (24.5) is a generalization of a multiple power series. It differs from a power series not just in the presence of the sum over $i_1, \ldots, i_n$, nor just in the fact that the monomials $z_1^{k_1} \cdots z_n^{k_n}$ are replaced by powers of the functions $\chi_i(z)$, $i = 1, \ldots, N$, but principally in the fact that the coefficients of (24.5) are functions of z and that an expansion in this series is, generally speaking, not unique.

3°. Consider the following problem: find conditions under which a compact set $M \subset \mathbf{C}^n$ has the property that every function $f \in A(M)$ can be represented in some neighborhood (depending on f) of M by a uniformly convergent series of "partial fractions":

$$f(z) = \sum_{p=1}^{\infty} \frac{A_p}{\prod_{j=1}^{n}(a_{pj1}z_1 + \cdots + a_{pjn}z_n + b_{pj})}, \tag{24.6}$$

where $\sum_{m=1}^{n} |a_{pjm}|^2 = 1, j = 1, \ldots, n, p = 1, 2,$ and $\sum_1^{\infty} |A_p| < \infty$.

Without loss of generality we can assume that the compact set M is connected. Certain concepts will be needed in what follows. A compact set M is said to be *approximated from the outside* by a sequence of domains D_p, $p = 1, 2, \ldots,$ if $D_{p+1} \Subset D_p$ and $M = \bigcap_p D_p$. We introduce the concept of the *envelope of holomorphy of a compact set*. Suppose that there exists a sequence of domains D_p, $p = 1, 2, \ldots,$ approximating M from the outside such that each D_p has a one-sheeted envelope of holomorphy $\mathcal{H}(D_p)$; then we say that *the compact set M has a one-sheeted envelope of holomorphy*, and define $\mathcal{H}(M) = \bigcap_p \mathcal{H}(D_p)$.

The following assertions are easy to see:

1) $\mathcal{H}(M)$ *is compact and does not depend on the choice of the sequence* D_p, *and each function holomorphic on M is holomorphic also on* $\mathcal{H}(M)$.

2) *If every function holomorphic on M is also holomorphic on a set E, then* $E \subset \mathcal{H}(M)$.

3) *On* $\mathcal{H}(M)$ *each function in* $A(M)$ *takes only the values it takes on M.*

In the solution of the problem posed above we assume that M has a one-sheeted envelope of holomorphy. This restriction seems natural in the case at hand. Indeed, it is natural to assume that M can be approximated from the outside by a sequence of domains D_p, $p = 1, 2, \ldots,$ such that each $f \in A(D_p)$

can be approximated uniformly on compact subsets of D_p by rational functions, and that the domains have one-sheeted envelopes of holomorphy $\mathcal{H}(D_p)$ (see [174], Chapter I, §3.2).

A *domain* $D \subset \mathbf{C}^n$ is said to be *linearly convex* if for each $\zeta \in \partial D$ there is an analytic plane of complex dimension $n-1$ passing through ζ and not intersecting D. Through each point in a $(2n-1)$-dimensional hyperplane α it is possible to pass an analytic plane of complex dimension $n-1$ lying in α, so every convex domain in $\mathbf{C}^n$ is also linearly convex. Further, a topological product $D = D_1 \times \cdots \times D_n$ of domains in $\mathbf{C}^1$ is a linearly convex domain, and, in particular, any domain $D \subset \mathbf{C}^1$ is linearly convex.

A *compact set* M is said to be *linearly convex* if there exists a sequence of linearly convex domains approximating it from the outside.

Suppose that $E \subset \mathbf{C}^n$ and $0 \in E$. The *conjugate set* is defined to be $\tilde{E} = \{w : \langle w, z\rangle \neq 1 \text{ for all } z \in E\}$. The set conjugate to $\tilde{E}$ is denoted by $\tilde{\tilde{E}}$. We always have $E \subset \tilde{\tilde{E}}$. It is not hard to prove the following properties:

4) *If E is open, then $\tilde{E}$ is compact, and, conversely, if E is compact, then $\tilde{E}$ is open.*

5) *If $E_1 \Subset E$, then $\tilde{E} \Subset \tilde{E}_1$ and $\tilde{\tilde{E}}_1 \Subset \tilde{\tilde{E}}$.*

6) *A domain $D \ni 0$ is linearly convex if and only if $(\tilde{\tilde{D}})_C = D$, where $(\tilde{\tilde{D}})_C$ is the connected component of $\tilde{\tilde{D}}$ containing D.*

7) *A compact set E is linearly convex if and only if there exists a sequence of linearly convex Weil polyhedra defined by means of functions of the form*

$$\frac{c}{(a_1 z_1 + \cdots + a_n z_n + b)^m}, \tag{24.7}$$

and approximating E from the outside (here $a_1, \ldots, a_n$, b, and c are complex numbers and m is a positive integer).

We can now formulate a result solving our problem.

THEOREM 24.11 (Aĭzenberg). *Suppose that the compact set M has a one-sheeted envelope of holomorphy $\mathcal{H}(M)$. Then every $f \in A(M)$ can be repesented in some neighborhood of M (depending on f) by a uniformly convergent series* (24.6) *if and only if the envelope of holomorphy $\mathcal{H}(M)$ is a linearly convex compact set.*

If $n = 1$, then $\mathcal{H}(D) = D$ for every domain D, and every connected compact set is linearly convex. Therefore, Theorem 24.11 contains the classical result of Wolff that the approximation under consideration is always valid when $n = 1$.

Some auxiliary propositions are needed to prove Theorem 24.11. Let $E(U(M))$ be the set of points $\eta = (\eta_1, \ldots, \eta_{n+1}) \in \mathbf{C}^{n+1}$, such that $\Sigma_1^n |\eta_i|^2 = 1$

and the plane $\alpha(\eta)=\{z: z_1\eta_1+\cdots+z_n\eta_n+\eta_{n+1}=0\}$ does not intersect the neighborhood $U(M)$ of the compact set M, and let $F=\{(z,\eta): z\in M, \eta\in E(U(M))\}\subset \mathbf{C}^{2n+1}$.

LEMMA 24.12. *There exists a constant $\delta>0$ such that for all $(z,\eta)\in F$*

$$|z_1\eta_1+\cdots+z_n\eta_n+\eta_{n+1}|\geqslant\delta. \tag{24.8}$$

PROOF. The function $|z_1\eta_1+\cdots+z_n\eta_n|$ is continuous on the compact set $F_1=\{(z,\eta_1,\ldots,\eta_n): z\in M, \Sigma_1^n|\eta_i|^2=1\}$. Therefore, there exists a $\delta_1>0$ such that $|z_1\eta_1+\cdots+z_n\eta_n|\leqslant\delta_1$ for all points in F_1. We break up the set $E(U(M))$ into two parts: $E_1=\{\eta:\eta\in E(U(M)), |\eta_{n+1}|>2\delta_1\}$ and $E_2=\eta:\eta\in E(U(M)), |\eta_{n+1}|\leqslant 2\delta_1\}$. For all points $(z,\eta)\in M\times E_1$ we get

$$|z_1\eta_1+\cdots+z_n\eta_n+\eta_{n+1}|\geqslant|\eta_{n+1}|-|z_1\eta_1+\cdots+z_n\eta_n|>\delta_1.$$

Therefore, it suffices to establish an inequality of the form (24.8) for $\eta\in E_2$.

The set E_2 is bounded, since it is contained in the compact set $\{\eta:\Sigma_1^n|\eta_i|^2=1, |\eta_{n+1}|\leqslant 2\delta_1\}$. We show that E_2 is closed. Suppose that $\eta^{(0)}=\lim_{p\to 0}\eta^{(p)}$, $\eta^{(p)}\in E_2$, $p=1,2,\ldots$. All the planes $\alpha(\eta^{(p)})$ are disjoint from $U(M)$, because $\eta^{(p)}\in E(U(M))$. Obviously, the limit plane $\alpha(\eta^{(0)})$ is also disjoint from $U(M)$. Consequently, $\eta^{(0)}\in E(U(M))$. Finally, observe that $|\eta_{n+1}^{(0)}|=\lim_{p\to\infty}|\eta_{n+1}^{(p)}|\leqslant 2\delta_1$, and so $\eta^{(0)}\in E_2$. The positive continuous function $|z_1\eta_1+\cdots+z_n\eta_n+\eta_{n+1}|$ attains its infimum $\delta_2>0$ on the compact set $F_2=\{(z,\eta): z\in M, \eta\in E_2\}$. Thus, an inequality of the form (24.8) holds also for $(z,\eta)\in F_2$. □

LEMMA 24.13. *Suppose that a closed bounded domain $\overline{D}$ can be approximated from the outside by a sequence of Weil polyhedra given by means of functions of the form* (24.7). *Then every function $f\in A(\overline{D})$ can be represented by a series* (24.6) *that converges uniformly on compact subsets of D.*

The proof of this lemma does not differ in principle from that of the analogous theorem of Wolff for the case $n=1$ (see [171], Chapter I, Theorem 13), but is considerably more cumbersome because the Weil integral is employed instead of the Cauchy integral, so we omit it.

PROOF OF THEOREM 24.11. *Sufficiency*. If $\mathcal{H}(M)$ is a linearly convex compact set, then it can be approximated from the outside (see property 7)) by a sequence of linearly convex bounded Weil polyhedra Δ_p, $p=1,2,\ldots$, given by means of functions of the form (24.7). Every function $f\in A(M)$ is in $A(\mathcal{H}(M))$, by property 1), so there is a p_0 such that $f\in A(\overline{\Delta}_{p_0})$. The polyhedron $\overline{\Delta}_{p_0}$ can in turn be approximated from the outside by Weil polyhedra of the same form. It remains to apply Lemma 24.13.

Necessity. Without loss of generality it can be assumed that $0\in M$ (this situation can always be attained by a translation of the origin, whereupon the series (24.6) passes into a series of the same form, the linearly convex compact sets pass into linearly convex compact sets, etc.). We take any sequence of domains D_p, $p = 1, 2, \ldots$, that approximates $\mathcal{H}(M)$ from the outside and consider the sequence $(\tilde{\tilde{D}}_p)_C$, $p = 1, 2, \ldots$, of linearly convex (see property 6)) domains, which approximate the compact set $M_1 = \bigcap_p (\tilde{\tilde{D}}_p)_C$, $M_1 \supset \mathcal{H}(M)$, from the outside (see 5)). Every $f \in A(M)$ can be assumption be represented in some neighborhood U of M by a uniformly convergent series (24.6). It is well known that the domains of uniform convergence of series of holomorphic functions are (see [174], Chapter I, §1.2) domains of holomorphy; therefore, the series (24.6) converges uniformly on compact sets lying in the envelope of holomorphy $\mathcal{H}(U) \supset \mathcal{H}(M)$. Hence, there exists a p_0 such that this series converges uniformly in D_{p_0}. The denominator of each term in the series is nonzero in D_{p_0}, i.e., $(-a_{pj1}b_{pj}^{-1}, \ldots, -a_{pjn}b_{pj}^{-1}) \in \tilde{D}_{p^0}$ (note that $0 \in M \subset \mathcal{H}(M) \subset D_{p_0}$, and so $b_{pj} \neq 0$); hence, this denominator is also nonzero in $(\tilde{\tilde{D}}_{p_0})_C$. By Lemma 24.12 and the convergence of $\Sigma_1^\infty |A_p|$, this gives us the uniform convergence of (24.6) in $\tilde{\tilde{D}}_{p_0} \supset M_1$. Thus, every $f \in A(M)$ is also in $A(M_1)$, so (see property 2)) $M_1 \subset \mathcal{H}(M)$. It has thus been shown that $\mathcal{H}(M) = M_1$, and the compact set M_1 is linearly convex. □

4°. The integral representation (8.5) enabled us to describe the general form of the continuous linear functionals on the spaces $A(D)$ (respectively, $A(M)$) of functions holomorphic in linearly convex domains (on compact sets) that can be approximated from the inside (from the outside) by bounded regular linearly convex domains. Such an approximation is not always possible. A topology is introduced in $A(D)$ and $A(M)$ in the usual way. This description is a generalization of the familiar results of Köthe, Sebastião e Silva, and Grothendieck for the case $n = 1$. The conjugate set plays the role of the exterior. Let $D_m = \{z : \Phi_m(z) < 0\}$ be the approximating domains indicated above, with $0 \in D_m$, $m = 1, 2, \ldots$. We set $\tau(\Phi) = (\tau_1(\Phi), \ldots, \tau_n(\Phi))$, where $\tau_i(\Phi) = \Phi'_{z_i} \langle \operatorname{grad} \Phi, z \rangle^{-1}$, $i = 1, \ldots, n$.

THEOREM 24.14 (Martineau and Aĭzenberg). *Every continuous linear functional $F(f)$ in $A(D)$ (in $A(M)$) has the form*

$$F(f) = \int_{\partial D_m} f(z) \varphi(\tau(\Phi_m)) \omega(z, \operatorname{grad} \Phi_m(z)), \tag{24.9}$$

where $\varphi \in A(\tilde{D})$ (respectively, $\varphi \in A(\tilde{M})$) and m depends only on φ. This formula gives an isomorphism between the topological linear spaces $A'(D)$ and $A(\tilde{D})$ (respectively, $A'(M)$ and $A(\tilde{M})$).

Suppose that $F \in A'(D)$ (respectively, $F \in A'(M)$). The Fantappiè indicatrix is defined to be the function $F_z[(1 - \langle z, w \rangle)^{-n}]$, where $z \in D$, $w \in \tilde{D}$ and $0 \in D$ (respectively, $z \in M$, $w \in \tilde{M}$ *and* $0 \in M$). In (24.9) the function φ is the indicatrix of the corresponding functional F. A domain D (a compact set M) is said to be *strongly linearly convex* if it is linearly convex, and the mapping $F \to F_z[(1 - \langle z, w \rangle)^{-n}]$ is an isomorphism between the spaces $A'(D)$ and $A(\tilde{D})$ (respectively, $A'(M)$ and $A(\tilde{M})$). It follows from Theorem 24.14 that the approximation mentioned before this theorem is a sufficient condition for strong linear convexity. The next result is based on Theorem 24.11.

THEOREM 24.15 (Znamenskiĭ). *A domain (compact set) is strongly linearly convex if and only if it intersects every complex line in a connected and simply connected set.*

It is now possible to formulate a result on expansion of holomorphic functions in fractions of a simpler form than in (24.6):

$$f(z) = \sum_{p=1}^{\infty} \frac{A_p}{(a_{p1}z_1 + \cdots + a_{pn}z_n + b_n)^n}, \tag{24.10}$$

where $\sum_{m=1}^{n} |a_{pm}|^2 = 1$ and $\sum_{p=1}^{\infty} |A_p| < \infty$. As in 3°, it is assumed that the compact set M is connected and has a one-sheeted envelope of holomorphy $\mathcal{H}(M)$.

THEOREM 24.16 (Trutnev). *Every $f \in A(M)$ can be represented in a neighborhood of M by a uniformly convergent series* (24.10) *if and only if the envelope of holomorphy $\mathcal{H}(D)$ is a strongly linearly convex compact set.*

COROLLARY 24.17 (Aĭzenberg). *For every $f \in A(M)$ to be representable in a neighborhood of M by a uniformly convergent series* (24.10) *it is necessary that $\mathcal{H}(M)$ be a linearly convex compact set, and sufficient that $\mathcal{H}(M)$ be approximable from the outside by regular linearly convex domains.*

§25. The $\bar{\partial}$-problem and Oka's theorems

1°. The following result occupies an important place in complex analysis.

THEOREM (Oka). *If at each point of its boundary a domain $D \subset \mathbf{C}^n$ is not holomorphically extendible, then D is a domain of holomorphy.*

Recall that a domain *D is not holomorphically extendible at a point $\zeta \in \partial D$* if there exist a neighborhood U of ζ and a function $f \in A(D \cap U)$ that cannot be extended holomorphically to ζ.

The existing accounts of this theorem, which solves the famous *Levi problem*, make use either of algebraic methods (sheaf theory) or of the theory of differential operators (solvability of the $\bar{\partial}$-Neumann problem).

Here we give a proof of Oka's theorem, due to Khenkin, that uses (as in the original work of Oka) only function-theoretic methods, mainly integral representations.

The following two assertions can be found in many complex analysis courses (see, for example, [51], Chapter III, §16.7, and Chapter II, §12.10).

I. *If at each point of its boundary the domain D is not holomorphically extendible, then D is pseudoconvex (i.e., there exists a function plurisubharmonic in D that tends to $+\infty$ as ∂D is approached).*

II. *If D is pseudoconvex, then $D = \bigcup_1^{\infty} D_k$, where the D_k are strictly pseudoconvex domains and $D_k \Subset D_{k+1}$, $k = 1, 2, \ldots$.*

To prove Oka's theorem it now suffices to prove the following.

III. *If D is a strictly pseudoconvex domain, then D is a domain of holomorphy.*

IV. *If* $D = \bigcup_1^\infty D_k$, $D_k \Subset D_{k+1}$, *and the* D_k *are domains of holomorphy, then* D *is a domain of holomorphy.*

It is easy to see that III follows from the next assertion (see, for example, [51], Chapter III, §16.6).

THEOREM 25.1 (Oka). *Suppose that* D *is a strictly pseudoconvex domain. Then for any* $\zeta \in \partial D$ *there exists a* $g \in A(D)$ *such that* $g(z) \to \infty$ *as* $z \to \zeta$, $z \in D$.

Next, assertion IV is the Behnke-Stein theorem (see [51], Chapter III, §16.10). The Oka-Hefer theorem can be used in the following formulation to prove this theorem.

THEOREM 25.2 (Oka and Hefer). *Suppose that* D *is a pseudoconvex domain in* $\mathbf{C}^n$ *and* $f \in A(D)$. *Then for any domain* $G \Subset D$ *there exist functions* $P_j(\zeta, z) \in A(G \times G)$, $j = 1, \ldots, n$, *such that*

$$f(\zeta) - f(z) = P_1(\zeta, z) \cdot (\zeta_1 - z_1) + \cdots + P_n(\zeta, z) \cdot (\zeta_n - z_n)$$

for $\zeta, z \in G$.

Thus, to solve the Levi problem the main thing is to prove Theorems 25.1 and 25.2. Moreover, the following assertions are needed to construct a barrier function for a strictly pseudoconvex domain (Lemma 10.1).

THEOREM 25.2′. *Suppose that* D *is a domain of holomorphy in* $\mathbf{C}^n$ *and* $G \Subset D$. *There exist linear operators* $P_j \colon A(D) \to A_C(G \times G)$, $j = 1, \ldots, n$, *such that for* $\zeta, z \in \bar G$

$$f(\zeta) - f(z) = (P_1 f)(\zeta, z) \cdot (\zeta_1 - z_1) + \cdots + (P_n f)(\zeta, z) \cdot (\zeta_n - z_n).$$

THEOREM 25.3. *Suppose that* D *is a domain of holomorphy in* $\mathbf{C}^n$, *and let* $A_{0,1}(\bar D)$ *be the space of* $\bar\partial$*-closed continuous forms of type* $(0, 1)$ *in the closed domain* $\bar D$. *Then for any domain* $G \Subset D$ *there exists a linear operator* $\tau \colon A_{0,1}(\bar D) \to C(\bar G)$, *such that* $\bar\partial \tau g = g$ *in* G *for any form* $g \in A_{0,1}(\bar D)$.

2°. The $\bar\partial$-problem for strictly convex domains. Let $A_{p,q}(\bar D)$ denote the space of continuous forms f of type (p, q) ;on $\bar D$ such that $\bar\partial f = 0$, and let $C_{p,q}^{(\tau)}(\bar D)$ denote the space of forms of type (p, q) whose coefficients satisfy a Hölder condition with exponent τ, $0 < \tau \leqslant 1$. It is assumed that these spaces are equipped with the natural topology. Here D is a bounded domain in $\mathbf{C}^n$.

Suppose now that D is a strictly convex domain, i.e., $D = \{z, \rho(z) < 0\}$, where $\rho \in C^{(2)}(\Omega)$, Ω is a domain $\bar D \subset \Omega$, $\operatorname{grad} \rho|_{\partial D} \neq 0$, and the second differential $d^2\rho$ is positive definite in Ω. In this subsection we prove

THEOREM 25.4. *Suppose that* D *is a strictly convex domain in* $\mathbf{C}^n$. *There exists a linear operator* $G_{p,q} \colon A_{p,q}(\bar D) \to C_{p,q-1}^{(1/2)}(\bar D)$, $0 \leqslant p \leqslant n$, $1 \leqslant q \leqslant n$, *such that the equation* $\bar\partial u = f$ *is satisfied by the form* $u = G_{p,q} f$.

In other words, $G_{p,q}$ is a right inverse of the operator $\bar\partial$. The construction([1]) of $G_{p,q}$ is given by (25.12) for any $k \geqslant 2$. Theorem 25.4 is a corollary to Theorems 25.6 and 25.9.

Let

$$P(\zeta) = (P_1(\zeta),\dots,P_n(\zeta)) = \left(\frac{\partial\rho}{\partial\zeta_1}(\zeta),\dots,\frac{\partial\rho}{\partial\zeta_n}(\zeta)\right),$$
$$\Phi(\zeta, z) = \sum_{j=1}^{n} P_j(\zeta)\cdot(\zeta_j - z_j), \qquad \tilde\Phi(\zeta, z) = \Phi(\zeta, z) - \rho(\zeta). \tag{25.1}$$

From Taylor's formula and the strict convexity of the function ρ it follows that there exists a constant $\sigma > 0$ such that if $\zeta, z \in \Omega'$ and $\bar D \subset \Omega' \Subset \Omega$, then

$$2\operatorname{Re}\Phi(\zeta, z) \geqslant \rho(\zeta) - \rho(z) + \sigma|\zeta - z|^2,$$
$$2\operatorname{Re}\tilde\Phi(\zeta, z) \geqslant -\rho(\zeta) - \rho(z) + \sigma|\zeta - z|^2. \tag{25.2}$$

We now get from (25.2) that for $\zeta, z \in \Omega'$

$$|\zeta - z|^2 \leqslant \frac{2}{\gamma}|\tilde\Phi(\zeta, z)|, \tag{25.3}$$

$$|\Phi(\zeta, z)| \leqslant \tilde\Phi(\zeta, z)|, \tag{25.4}$$

$$|\tilde\Phi(\zeta, z)| \leqslant \sigma_1\left[|\rho(\zeta) - \rho(z)| + |\operatorname{Im}\Phi(\zeta, z)| - \rho(\zeta) - \rho(z) + |\zeta - z|^2\right], \tag{25.5}$$

where $\sigma_1 > 0$ is a constant depending only on D.

Let $w(\zeta, z, \lambda) = (w_1,\dots,w_n)$ be a vector-valued function defined on a set $U \subset (\mathbf{C}^n_\zeta \times \mathbf{C}^n_z \times \mathbf{R}^1_\lambda)\setminus\{(\zeta, z, \lambda): z = \zeta\}$ and such that $\langle w, \zeta - z\rangle = 1$. For $w \in C^{(1)}(U)$ let

$$\tilde W_{p,q}(w, \zeta, z, \lambda) = \frac{(-1)^{q+p(n-q-1)}\binom{n-1}{q}}{(2\pi i)^n p!\,(n-p)!}$$
$$\times D_{1,q,n-q-1}\left(w, \bar\partial_z w, \left(\bar\partial_\zeta + d_\lambda\right)w\right) \wedge D_{p,n-p}(\partial z, \partial\zeta),$$
$$W_{p,-1} = W_{p,n} = 0.$$

If w does not depend on λ, then $\tilde W_{p,q}(w, \zeta, z, \lambda) = W_{p,q}(w, \zeta, z)$ (see §7). We introduce the vector-valued functions

$$u(\zeta, z) = \frac{P(\zeta)}{\Phi(\zeta, z)} \quad \text{and} \quad \eta(\zeta, z, \lambda) = (1 - \lambda)t + \lambda u(\zeta, z), \tag{25.6}$$

([1]) The construction of $G_{p,q}$ given in (25.12) is made with its use in §26 also in mind. Simpler assertions suffice to prove Theorem 25.4.

and the function

$$\alpha_k(\zeta, z) = \left[1 - \left(\frac{\Phi(\zeta, z)}{\tilde\Phi(\zeta, z)}\right)^{2n}\right]^k = \left(-\frac{\rho(z)}{\tilde\Phi(\zeta, z)}\right)^k \left[\sum_{j=0}^{2n-1}\left(\frac{\Phi(\zeta, z)}{\tilde\Phi(\zeta, z)}\right)^j\right]^k.$$

Here $\zeta, z \in \bar D$, $\Phi(\zeta, z) \neq 0$, $\lambda \in [0, 1]$, and $t = (\bar\zeta - \bar z)/|\zeta - z|^2$.

LEMMA 25.5. *If $\Phi(\zeta, z) \neq 0$, then the form*

$$\bar\partial_\zeta \alpha_k(\zeta, z) \wedge \left(\tilde W_{p,q}(\eta, \zeta, z, \lambda) - W_{p,q}(\eta, \zeta, z)\right)$$

can be represented as a linear combination of forms like

$$\left(\frac{\rho(\zeta)}{\tilde\Phi(\zeta, z)}\right)^{k-1}\left(\frac{\Phi(\zeta, z)}{\tilde\Phi(\zeta, z)}\right)^s \frac{(\bar\zeta_j - \bar z_j)\varphi(\zeta, z, \lambda) \wedge \bar\partial\rho(\zeta)}{|\zeta - z|^{2(q+r+1)}\tilde\Phi^{n-q-r}(\zeta, z)} \tag{25.8}$$

and

$$\left(\frac{\rho(\zeta)}{\tilde\Phi(\zeta, z)}\right)^{k}\left(\frac{\Phi(\zeta, z)}{\tilde\Phi(\zeta, z)}\right)^s \frac{(\bar\zeta_j - \bar z_j)(\zeta_m - z_m)\Psi(\zeta, z, \lambda)}{|\zeta - z|^{2(q+r+1)}\tilde\Phi^{n-q-r}(\zeta, z)}, \tag{25.9}$$

where $1 \leqslant j, m \leqslant n$, $0 \leqslant r \leqslant n - q - 2$, $s = 0, 1, \ldots$, and the forms φ and Ψ have type (p, q) in z, degree 1 in λ, and type $(n - p, n - q - 2)$ and $(n - p, n - q - 1)$ in ζ. The coefficients of the forms φ and Ψ are derivatives of $\rho(\zeta)$ of order at most two, multiplied by a polynomial in λ.

PROOF. From (25.6), the properties of a determinant, and the fact that the vector-valued function u is holomorphic in z when $\Phi(\zeta, z) \neq 0$, we get

$$\begin{aligned} D_{1,q,n-q-1}&\left(\eta, \bar\partial_z\eta, (\bar\partial_\zeta + d_\lambda)\eta\right) - D_{1,q,n-q-1}\left(\eta, \bar\partial_z\eta, \bar\partial_\zeta\eta\right) \\ &= (n-q-1)D_{1,q,1,n-q-2}\left(t, \bar\partial_z\eta, d_\lambda\eta, \bar\partial_\zeta\eta\right) \\ &= (-1)^{n-q}(n-q-1)D_{1,q,1,n-q-2}\left(t, (1-\lambda)\bar\partial_z t, u - t,\right. \end{aligned} \tag{25.10}$$

$$\begin{aligned} &\left.\left((1-\lambda)\bar\partial_\zeta t + \lambda\bar\partial_\zeta u\right) \wedge d\lambda\right) \\ &= \sum_{r=0}^{n-q-2} (-1)^n (n-q-1)\binom{n-q-2}{r} D_{1,1q,r,n-q-r-2}\left(t, u, \bar\partial_z t, \bar\partial_\zeta t, \bar\partial_\zeta u\right) \\ &\quad \wedge \lambda^{n-q-r-2}(1-\lambda)^{q+r}\, d\lambda. \end{aligned}$$

Just as in the proof of (7.2),

$$\begin{aligned} D_{1,1q,r,n-q-r-2}\left(t, u, \bar\partial_z t, \bar\partial_\zeta t, \bar\partial_\zeta u\right) &= \frac{(-1)^q}{|\zeta - z|^{2(q+r+1)}\Phi^{n-q-r-1}(\zeta, z)} \\ &\quad \times D_{1,1q,r,n-q-r-2}\left(\bar\zeta - \bar z, P, \bar\partial_z \bar z, \bar\partial_\zeta \bar\zeta, \bar\partial P\right). \end{aligned}$$

It now follows from (25.10) and the definitions of $\tilde{W}_{p,q}$ and $W_{p,q}$ that $\tilde{W}_{p,q}(\eta) - W_{p,q}(\eta)$ is a linear combination of forms like

$$\frac{(\bar{\zeta}_{j'} - \bar{z}_j)\varphi(\zeta, z, \lambda)}{|\zeta - z|^{2(q+r+1)}\Phi^{n-q-r-1}(\zeta, z)}, \tag{25.11}$$

where $1 \leqslant j \leqslant n$, $0 \leqslant r \leqslant n-q-2$; φ has type (p, q) in z, type $(n-p, n-q-2)$ in ζ, and degree 1 in λ, and the coefficients of φ are derivatives of $\rho(\zeta)$ of order at most two, multiplied by a polynomial in λ.

On the other hand,

$$\bar{\partial}_\zeta \alpha_k(\zeta, z) = -k \cdot 2n\alpha_{k-1}(\zeta, z)\left(\frac{\Phi}{\tilde{\Phi}}\right)^{2n-1}\left(\frac{\Phi}{\tilde{\Phi}}\frac{\bar{\partial}\rho(\zeta)}{\tilde{\Phi}(\zeta, z)} - \frac{\rho(\zeta)\bar{\partial}_\zeta\Phi}{\tilde{\Phi}^2}\right).$$

The assertion of the lemma follows from this equality, (25.11), and the equality $\bar{\partial}_\zeta\Phi = \langle \bar{\partial}_\zeta P, \zeta - z\rangle$. $\square$

We define the operators

$$(G^k_{p,q}f)(z) = -\int_{D_\zeta} f(\zeta) \wedge \alpha_k(\zeta, z) U_{p,q-1}(\zeta, z)$$
$$+ (-1)^{p+q-1}\int_{D_\zeta \times [0,1]_\lambda} f(\zeta) \wedge \bar{\partial}_\zeta \alpha_k(\zeta, z) \wedge \tilde{W}_{p,q-1}(\eta, \zeta, z, \lambda), \tag{25.12}$$

where $0 \leqslant p \leqslant n$, $1 \leqslant q \leqslant n$, $f \in C_{p,q}(\bar{D})$.

REMARK. The form $f(\zeta) \wedge \bar{\partial}_\zeta \alpha_k(\zeta, z) \wedge W_{p,q-1}(\eta)$ is of type $(n, n+1)$ in ζ; therefore, it vanishes. Consequently, the integrand in the second integral in (25.12) actually has the form

$$f(\zeta) \wedge \bar{\partial}_\zeta \alpha_k \wedge (\tilde{W}_{p,q}(\eta) - W_{p,q}(\eta)).$$

Since $\tilde{\Phi}(\zeta, z) \neq 0$ and $\Phi(\zeta, \zeta) = 0$ for $z \in D$ and $\zeta \in \bar{D}$, the form $\bar{\partial}_\zeta \alpha_k \wedge (\tilde{W}_{p,q}(\eta) - W_{p,q}(\eta))$ can be extended continuously to $\bar{D}_\zeta \times D_z \times [0, 1]_\lambda$ (this follows from Lemma 25.5). Thus, the second integral in (25.12) is a proper integral.

THEOREM 25.6. *If $f \in A_{p,q}(\bar{D})$, $q \geqslant 1$, then the form $u = G^k_{p,q}f$ is a solution of the equation $\bar{\partial}u = f$ for any $k \geqslant 2$.*

PROOF. Let us take the function $\alpha_k(\zeta, z)$ as φ in (7.15). Since α_k is holomorphic in z and $\alpha_k(\zeta, z) \equiv 0$ for $\zeta \in \partial D$ (see (25.7)), the equality $\bar{\partial}f = 0$

gives us that for $z \in D$

$$f(z) = (-1)^{p+q-1}\int_{D_\zeta} f(\zeta) \wedge \bar\partial_\zeta \alpha_k(\zeta, z) \wedge U_{p,q}(\zeta, z)$$

$$-\bar\partial_z \int_{D_\zeta} f(\zeta) \wedge \alpha_k(\zeta, z) U_{p,q-1}(\zeta, z). \tag{25.13}$$

By Stokes' formula,

$$\int_{D_\zeta \times [0,1]_\lambda} d\big(f(\zeta) \wedge \bar\partial_\zeta \alpha_k(\zeta, z) \wedge \tilde W_{p,q}(\eta, \zeta, z, \lambda)\big)$$

$$= \int_{D_\zeta} f(\zeta) \wedge \bar\partial_\zeta \alpha_k(\zeta, z) \wedge W_{p,q}(u, \zeta, z)$$

$$-\int_{D_\zeta} f(\zeta) \wedge \bar\partial_\zeta \alpha_k(\zeta, z) \wedge U_{p,q}(\zeta, z)$$

$$+\int_{\partial D_\zeta \times [0,1]_\lambda} f(\zeta) \wedge \bar\partial_\zeta \alpha_k(\zeta, z) \wedge \tilde W_{p,q}(\eta, \zeta, z, \lambda).$$

The vector-valued function u is holomorphic in $z \in D$, so $W_{p,q}(u, \zeta, z) \equiv 0$ for $q \geqslant 1$. For $k \geqslant 2$ and $\zeta \in \partial D$ we have $\bar\partial_\zeta \alpha_k = 0$; consequently, the preceding equality gives us

$$\int_{D_\zeta} f(\zeta) \wedge \bar\partial_\zeta \alpha_k(\zeta, z) \wedge U_{p,q}(\zeta, z)$$

$$= (-1)^{p+q} \int_{D_\zeta \times [0,1]_\lambda} f(\zeta) \wedge \bar\partial_\zeta \alpha_k(\zeta, z) \wedge \big(\bar\partial_\zeta + d_\lambda\big) \tilde W_{p,q}(\eta, \zeta, z, \lambda)$$

$$= \bar\partial_z \int_{D_\zeta \times [0,1]_\lambda} f(\zeta) \wedge \bar\partial_\zeta \alpha_k(\zeta, z) \wedge \tilde W_{p,q-1}(\eta, \zeta, z, \lambda). \tag{25.14}$$

Here we have used the condition $\bar\partial f = 0$, the equality

$$\big(\bar\partial_\zeta + d_\lambda\big) \tilde W_{p,q} = (-1)^{p+q} \bar\partial_z \tilde W_{p,q-1}$$

(which can be proved in the same way as Lemma 7.2), and the fact that α_k is holomorphic in z. Substituting (25.14) into (25.12), we get the theorem. □

Two lemmas are needed for estimating the solution of $\bar\partial u = f$ given in Theorem 25.6.

LEMMA 25.7. *Let*

$$J_{\alpha,k,l}(\varepsilon) = \int_{\substack{|t|<1 \\ t_1+\varepsilon>0}} \frac{(t_1 + \varepsilon)^\alpha dt_1 \cdots dt_{2n}}{\big(|t_1| + |t_2| + \varepsilon + |t|^2\big)^k |t|^l},$$

where α, k, and l are real, and $\alpha > -1$. Then:

a)

$$J_{\alpha,k,2n-1}(\varepsilon) = \begin{cases} O(\varepsilon^{1-k+\alpha}), & k-\alpha > 1, \\ O(\ln\varepsilon), & k-\alpha = 1. \end{cases}$$

b) *For* $m \geqslant 0$

$$J_{\alpha,k,2n-3-2m}(\varepsilon) = \begin{cases} O(\varepsilon^{-k+\alpha+m+5/2}), & k-\alpha-m \geqslant 5/2, \\ O(\ln\varepsilon), & k-\alpha-m = 5/2. \end{cases}$$

PROOF. Let us pass to the spherical coordinates $t_1 = r\cos\varphi_1$, $t_2 = r\sin\varphi_1\cos\varphi_2, \ldots,$ in the integral defining $J_{\alpha,k,l}$ and make the change of variables $s_i = \cos\varphi_i$, $i = 1, 2$:

$$J_{\alpha,k,l}(\varepsilon) = O(1)\int_{\substack{0<r,s_2<1\\ -1<s<1\\ rs_1+\varepsilon>0}} \frac{(rs_1+\varepsilon)^\alpha r^{2n-1-l}\,dr\,ds_1\,ds_2}{\left[r\left(|s_1|+s_2\sqrt{1-s_1^2}\right)+\varepsilon+r^2\right]^k}.$$

We use the inequality $|s_1|+s_2\sqrt{1-s_1^2} \geqslant \frac{1}{4}(|s_1|+s_2)$ for $-1 \leqslant s_1 \leqslant 1$, $0 \leqslant s_2 \leqslant 1$. The change of variables $s_1 = su$, $s_2 = s(1-|u|)$ leads to

$$J_{\alpha,k,l}(\varepsilon) = O(1)\int_{\substack{0<r,s<2\\ -1<u<1\\ rsu+\varepsilon>0}} \frac{(rsu+\varepsilon)^\alpha r^{2n-1-l} s\,dr\,ds\,du}{(rs+\varepsilon+r^2)^k}.$$

Let $l = 2n-1$. We pass to the coordinates (v, s, u), where $v = \varepsilon^{-1}rs$, and make use of the inequality $rs+\varepsilon+r^2 \geqslant rs+\varepsilon$:

$$\begin{aligned} J_{\alpha,k,2n-1}(\varepsilon) &= O(\varepsilon^{1-k+\alpha})\left[\int_0^1 \frac{dv}{(v+1)^k}\int_{-1}^1 (vu+1)^\alpha\,du \right.\\ &\qquad \left. + \int_1^{4\varepsilon^{-1}} \frac{dv}{(v+1)^k}\int_{-v^{-1}}^1 (vu+1)^\alpha\,du\right] \\ &= O(\varepsilon^{1-k+\alpha})\left[\int_0^1 \frac{(1+v)^{1+\alpha}-(1-v)^{1+\alpha}}{v}\,dv + \int_1^{4\varepsilon^{-1}} \frac{dv}{v(v+1)^{k-\alpha+1}}\right]. \end{aligned}$$

This gives us assertion a).

Suppose now that $l = 2n - 3 - 2m$, $m \geqslant 0$. Then

$$J_{\alpha,k,2n-3-2m} = O(1)\int_0^2 r^{2m}dr\int_0^2 \frac{r\,ds}{(rs+\varepsilon+r^2)^k}\int_{-\varepsilon(rs)^{-1}}^1 (rsu+\varepsilon)^\alpha rs\,du$$

$$= O(1)\int_0^2 r^{2m}dr\int_0^2 \frac{r\,ds}{(rs+\varepsilon+r^2)^{k-\alpha-1}}$$

$$= O(1)\int_0^2 \frac{r^{2m}\,dr}{(\varepsilon+r^2)^{k+\alpha-2}} = O(1)\int_0^2 \frac{dr}{\left(r+\sqrt{\varepsilon}\right)^{2(k-\alpha-m-2)}}.$$

Assertion b) is obtained by direct integration. □

LEMMA 25.8. *Suppose that $D = \{x,\ \rho(x) < 0\}$ is a bounded domain in $\mathbf{R}^n$ with smooth boundary and the function $F\colon D \to \mathbf{C}^1$ satisfies the condition*

$$|F(x) - F(y)| = |x-y|\lambda(|\rho(x)|) \quad \textit{for } \rho(x) > \rho(y),$$

where $\lambda(t)$ is a monotonically decreasing nonnegative function on $(0,\infty)$ such that $\int_0^1 \lambda(t)\,dt < \infty$. Then

$$|F(x) - F(y)| = O(1)(\Lambda(|x-y|) + |x-y|\lambda(|x-y|)), \qquad x,\ y \in D,$$

where $\Lambda(t) = \int_0^t \psi(\tau)\,d\tau$, and the constant in $O(1)$ does not depend on F.

In particular,

$$|F(x) - F(y)| = O(1)|x-y|^{1-\tau},$$

if $\lambda(t) = t^{-\tau}$, $0 \leqslant \tau < 1$, and

$$|F(x) - F(y)| = O(1)|x-y|\ln|x-y|$$

if $\lambda(t) = -\ln t$.

This assertion is a slight generalization of a result of Hardy and Littlewood for the case where D is the disk in the plane and $\lambda(t) = t^{-\tau}$ (see, for example, [65]).

THEOREM 25.9. *If $f \in C_{p,q}(\bar D)$, then $G^k_{p,q}f \in C^{(1/2)}_{p,q-1}(\bar D)$, and $G^k_{p,q}$ is a linear operator from $C_{p,q}(\bar D)$ to $C^{(1/2)}_{p,q-1}(\bar D)$.*

PROOF. Let $\tilde G^k_{p,q}$ and $\tilde{\tilde G}{}^k_{p,q}$ be the operators defined, respectively, by the first and second integrals in (25.12). Consider the function

$$H(z) = \int_{D_\zeta} h(\zeta)\alpha_k(\zeta,z)\frac{\bar\zeta_j - \bar z_j}{|\zeta - z|^{2n}}\,dv, \quad \textit{where } h \in C(\bar D).$$

Then for $z, w \in D$

$$|H(w) - H(z)| \leqslant \int_{D_\zeta} |h(\zeta)||\alpha_k(\zeta, w)| \left| \frac{\bar{\zeta}_j - \bar{w}_j}{|\zeta - w|^{2n}} - \frac{\bar{\zeta}_j - \bar{z}_j}{|\zeta - z|^{2n}} \right| dv$$

$$+ \int_{D_\zeta} |h(\zeta)||\alpha_k(\zeta, w) - \alpha_k(\zeta, z)| \frac{dv}{|\zeta - z|^{2n-1}}. \qquad (25.15)$$

From (25.2), (25.4), and (25.7),

$$|\operatorname{grad}_z \alpha_k(\zeta, z)| = 2kn|\alpha_{k-1}(\zeta, z)| \left| \frac{\Phi(\zeta, z)}{\tilde{\Phi}(\zeta, z)} \right|^{2n-1} \left| \frac{\rho(\zeta) \operatorname{grad}_z \Phi(\zeta, z)}{\tilde{\Phi}^2(\zeta, z)} \right|$$

$$= O(1) \frac{1}{|\tilde{\Phi}(\zeta, z)|}.$$

Suppose that $\rho(z) \geqslant \rho(w)$. By the mean value theorem, there is a θ, $0 < \theta < 1$, such that

$$\alpha_k(\zeta, w) - \alpha_k(\zeta, z)| \leqslant |\operatorname{grad}_z \alpha_k(\zeta, w + \theta(z - w))| \cdot |w - z|$$

$$= O(1) \frac{|w - z|}{|\tilde{\Phi}(\zeta, w + \theta(z - w))|} = O(1) \frac{|w - z|}{|\tilde{\Phi}(\zeta, z)|}.$$

If $\mathcal{J}(z, w)$ denotes the second integral in (25.15), we get that

$$\mathcal{J}(z, w) = O(1)\|h\|_{C(\bar{D})} \left(1 + \int_{(D \cap B_\delta(z))_\zeta} \frac{dv(\zeta)}{|\tilde{\Phi}(\zeta, z)||\zeta - z|^{2n-1}} \right) |w - z|.$$

Now let σ and δ be chosen in accordance with Lemma 10.6, and suppose that $z \in D$ and $\rho(z) > -\delta$. We pass to the variables $t_{2j-1} = \operatorname{Re} \eta_j(\zeta)$, $t_{2j} = \operatorname{Im} \eta_j(\zeta)$ (the functions η_j are defined in Lemma 10.6), use (25.5), and find for $\rho(z) \geqslant \rho(w)$ that

$$\mathcal{J}(z, w) = O(1)\|h\|_{C(\bar{D})}(1 + J_{0,1,2n-1}(|\rho(z)|))|w - z|.$$

By Lemma 25.7,

$$\mathcal{J}(z, w) = O(1)\|h\|_{C(\bar{D})}|w - z| \ln|\rho(z)| \qquad (25.16)$$

when $\rho(z) > -\delta$. Obviously, (25.16) holds also when $\rho(z) \leqslant -\delta$. On the other hand, (25.7) and (25.4) show that $\alpha_k(\zeta, w) = O(1)$ for $\zeta, w \in D$. Moreover, it is known that

$$\int_{D_\zeta} |h(\zeta)| \left| \frac{\bar{\zeta}_j - \bar{w}_j}{|\zeta - w|^{2n}} - \frac{\bar{\zeta}_j - \bar{z}_j}{|\zeta - z|^{2n}} \right| dv(\zeta) = O(1)\|h\|_{C(D)}|w - z| \ln|w - z|.$$

Consequently, (25.15) and (25.16) imply that

$$|H(w) - H(z)| = O(1)\|h\|_{C(\bar{D})}|w - z|(\ln|\rho(z) + \ln|w - z|)$$

for $\rho(z) \geqslant \rho(w)$. By Lemma 25.8,

$$|H(w) - H(z)| = O(1)\|h\|_{C(\bar D)}|w - z|\ln|w - z|.$$

From the definition of the operator $\tilde G^k_{p,q}$ and the kernel $U_{p,q-1}$ it now follows that $\tilde G^k_{p,q}$ satisfies the conditions of the theorem.

It remains to prove that $\tilde{\tilde G}{}^k_{p,q}$ also satisfies these conditions. Lemma 25.5 and the remark after (25.12) show that to do this it suffices to prove that

$$|F_1(w) - F_1(z)| = O(1)\|h\|_{C(\bar D)}|w - z|^{1/2} \tag{25.17}$$

and

$$|F_2(w) - F_2(z)| = O(1)\|h\|_{C(\bar D)}|w - z|\ln|w - z|, \tag{25.18}$$

where

$$F_1(z) = \int_{D_\zeta} h(\zeta)\left(\frac{\rho(\zeta)}{\tilde\Phi(\zeta, z)}\right)^{k-1}\left(\frac{\Phi(\zeta, z)}{\tilde\Phi(\zeta, z)}\right)^s \frac{(\bar\zeta_j - \bar z_j)\varphi(\zeta)}{|\zeta - z|^{2r}\tilde\Phi^{n-r+1}(\zeta, z)}\,dv(\zeta),$$

$$F_2(z) = \int_{D_\zeta} h(\zeta)\left(\frac{\rho(\zeta)}{\tilde\Phi(\zeta, z)}\right)^{k}\left(\frac{\Phi(\zeta, z)}{\tilde\Phi(\zeta, z)}\right)^s \frac{(\bar\zeta_j - \bar z_j)(\zeta_m - z_m)\varphi(\zeta)}{|\zeta - z|^{2r}\tilde\Phi^{R-r+1}(\zeta, z)}\,dv(\zeta).$$

Here $\varphi(\zeta)$ is a product of derivatives of $\rho(\zeta)$ of order at most two, $1 \leqslant j$, $m \leqslant n$, and $s = 0, 1, \ldots$.

We can differentiate under the integral sign in the integrals defining F_1 and F_2. A direct computation with use of (25.2)–(25.4) gives us that

$$\begin{aligned}|\operatorname{grad} F_1(z)| &= O(1)\left(\int_{D_\zeta}\frac{dv}{|\zeta - z|^{2r-1}|\tilde\Phi(\zeta, z)|^{n-r+2}}\right.\\ &\qquad\left.+\int_{D_\zeta}\frac{dv}{|\zeta - z|^{2r}|\tilde\Phi(\zeta, z)|^{n-r+1}}\right)\\ &= O(1)\left(\int_{D_\zeta}\frac{dv}{|\zeta - z|^{2n-3}|\tilde\Phi(\zeta, z)|^{3}} + \int_{D_\zeta}\frac{dv}{|\zeta - z|^{2n-1}|\tilde\Phi(\zeta, z)|^{3/2}}\right).\end{aligned}$$

Let σ and δ be chosen according to Lemma 10.6. Then for $\rho(z) > -\delta$

$$\begin{aligned}|\operatorname{grad} F_1(z)| &= O(1)\|h\|_{C(\bar D)}\left(1 + \int_{(D\cap B_\delta(z))_\zeta}\frac{dv}{|\zeta - z|^{2n-3}|\tilde\Phi(\zeta, z)|^{3}}\right.\\ &\qquad\left.+\int_{(D\cap B_\delta(z))_\zeta}\frac{dv}{|\zeta - z|^{2n-1}|\tilde\Phi(\zeta, z)|^{3/2}}\right)\\ &= O(1)\|h\|_{C(\bar D)}\big(1 + J_{0,3,2n-3}(|\rho(z)|) + J_{0,3/2,2n-1}(|\rho(z)|)\big).\end{aligned}$$

Here we have used the substitution $t_{2j-1} = \operatorname{Re}\eta_j(\zeta)$, $t_{2j} = \operatorname{Im}\eta_j(\zeta)$ and (25.5). Lemma 25.7 now shows that for $\rho(z) > -\delta$

$$|\operatorname{grad} F_1(z)| = O(1)\|h\|_{C(\bar{D})}|\rho(z)|^{-1/2}.$$

The mean value theorem and Lemma 25.8 then give us (25.17). The proof of (25.18) is similar. □

3°. The $\bar{\partial}$-problem for strictly pseudoconvex domains. Let $B_{p,q}(\bar{D})$ denote the subspace of forms $f \in A_{p,q}(\bar{D})$, $q \geq 1$, such that $\bar{\partial}u = f$ for some form $u \in C_{p,q}^{(1/2)}(\bar{D})$.

THEOREM 25.10. *Let D be a bounded strictly pseudoconvex domain in $\mathbf{C}^n$. Then $B_{p,q}$ is a subspace of finite codimension in $A_{p,q}(\bar{D})$.*

PROOF. Lemma 10.2 implies the existence of a finite covering $\{U_j\}$, $j = 1,\ldots,k$, of the compact set $\bar{D}$ such that for each $j = 1,\ldots,k$ there is a mapping $\varphi_j(z)$ biholomorphic in a neighborhood of $\bar{D}_j = \bar{U}_j \cap \bar{D}$ for which $\varphi_j(\bar{D}_j)$ is the closure of a strictly convex domain. If $f \in A_{p,q}(D)$, then $f(\varphi_j^{-1}(w)) \in A_{p,q}(\varphi_j(\bar{D}))$. By Theorem 25.4, there are linear operators $G_{p,q}^{(j)}\colon A_{p,q}(\varphi_j(D_j)) \to C_{p,q-1}^{(1/2)}(\varphi_j(D_j))$ such that

$$\bar{\partial}G_{p,q}^{(j)} f\big(\varphi_j^{(-1)}(w)\big) = f\big(\varphi_j^{-1}(w)\big).$$

Thus, we can define linear operators $G_{p,q}^{(j)}\colon A_{p,q}(\bar{D}) \to C_{p,q-1}^{(1/2)}(\bar{D}_j)$, $j = 1,\ldots,k$, for which $\bar{\partial}G_{p,q}^{(j)} f = f$ on $\bar{D}_j$.

Let $\{\chi_j\}$ be a partition of unity subordinate to the covering $\{U_j\}$, and define operators

$$N_{p,q}\colon A_{p,q}(\bar{D}) \to C_{p,q-1}^{(1/2)}(\bar{D}) \quad \text{and} \quad T_{p,q}\colon A_{p,q}(\bar{D}) \to C_{p,q}(\bar{D})$$

by the equalities

$$N_{p,q}f = \sum_{j=1}^{k} \chi_j G_{p,q}^{(j)} f, \qquad T_{p,q}f = \sum_{j=1}^{k} \bar{\partial}\chi_j \wedge G_{p,q}^{(j)} f.$$

We have

$$\bar{\partial}N_{p,q}f = T_{p,q}f + f. \tag{25.19}$$

This implies that $T_{p,q}$ carries $A_{p,q}(D)$ into $A_{p,q}(\bar{D})$. Moreover, the definition of $T_{p,q}$ implies that it is continuous, regarded as an operator from $A_{p,q}(\bar{D})$ to $C_{p,q}^{(1/2)}(\bar{D})$. By Ascoli's theorem (see [80], Chapter 7, §5), $T_{p,q}\colon A_{p,q}(\bar{D}) \to A_{p,q}(\bar{D})$ is a compact operator. The Riesz-Fredholm theorem (see [80], Chapter 11, §3) tells us that the set $\tilde{B}_{p,q}(\bar{D})$ of forms $T_{p,q}f + f$ with $f \in A_{p,q}(\bar{D})$ is a closed subspace of $A_{p,q}(\bar{D})$ with finite codimension. From (25.19) it follows that $\tilde{B}_{p,q}(\bar{D}) \subset B_{p,q}(\bar{D})$, and so $B_{p,q}(\bar{D})$ is a closed subspace of finite codimension in $A_{p,q}(\bar{D})$. □

We note that results in this section are used to prove Lemma 10.1, which enables us to construct operators analogous to the operators $G^k_{p,q}$ for a strictly psueudoconvex domain. A repetition of all the arguments in 2° of this section gives us

THEOREM 25.11 (Romanov and Khenkin). *Let D be a strictly pseudoconvex domain in $\mathbf{C}^n$. Then there exist linear operators $G_{p,q}: A_{p,q}(\bar D) \to C^{(1/2)}_{p,q-1}(\bar D)$, $0 \leqslant p \leqslant n$, $1 \leqslant q \leqslant n$, such that the form $u = G_{p,q}f$ satisfies the equation $\bar\partial u = f$.*

This theorem is stronger than Theorem 25.10, but its proof uses consequences of the latter.

4°. PROOF OF THEOREM 25.1. Suppose that ζ is a fixed point of ∂D, where D is a bounded strictly pseudoconvex domain. In Lemma 10.2 it is actually proved that there is a $\delta > 0$ such that $\operatorname{Re} F(\zeta, z) > 0$ when $0 < |\zeta - z| < 2\delta$ and $z \in \bar D$, where

$$F(\zeta, z) = 2\sum_{j=1}^{n} \frac{\partial\rho}{\partial\zeta_j}(\zeta)(z_j - \zeta_j) + \sum_{j,k=1}^{n} \frac{\partial^2\rho}{\partial\zeta_j\partial\zeta_k}(\zeta)(z_j - \zeta_j)(z_k - \zeta_k).$$

Let χ be a compactly supported infinitely differentiable function on $\mathbf{C}^n$ with support in $B_{2\delta}(\zeta)$ and with $\chi(z) \equiv 1$ for $z \in B_\delta(\zeta)$. We construct the forms

$$f_k(z) = \begin{cases} \bar\partial\left(\dfrac{\chi(z)}{F^k(\zeta, z)}\right) = \dfrac{\bar\partial\chi(z)}{F^k(\zeta, z)} & \text{for } z \in \bar D \cap (B_{2\delta}(\zeta)\setminus B_\delta(\zeta)), \\ 0 & \text{for } z \in \bar D\setminus(B_{2\delta}(\zeta)\setminus B_\delta(\zeta)), \end{cases}$$

$k = 1, 2, \ldots$. It is not hard to verify that the f_k are in $A_{0,1}(\bar D)$ and are linearly independent; therefore, by Theorem 25.10, there exist a positive integer k and complex numbers $a_1, \ldots, a_{k-1}$ such that $f_k + a_{k-1}f_{k-1} + \cdots + a_1 f_1 = \bar\partial u$, where $u \in C(\bar D)$. The function

$$g(z) = \chi(z)\left(\frac{1}{F^k(\zeta, z)} + \frac{a_{k-1}}{F^{k-1}(\zeta, z)} + \cdots + \frac{a_1}{F(\zeta, z)}\right) - u(z)$$

is holomorphic in D, and

$$\lim_{\substack{z\to\zeta \\ z\in D}} g(z) = \infty. \quad \square$$

5°. PROOFS OF THEOREM 25.2 AND 25.2′. We need

LEMMA 25.12. *Suppose that D and G satisfy the conditions of Theorem 25.2, $\Pi_{n-1} = \{z_n = 0\}$, where $\Pi_{n-1} \cap G \neq \varnothing$, and $f('z)$ is a holomorphic function on $D \cap \Pi_{n-1}$, where $'z = (z_1, \ldots, z_{n-1})$. Then there exists an $F \in A(G)$ such that $F|_{G\cap\Pi_{n-1}} = f$.*

PROOF. Part II in the scheme for proving Oka's theorem (see 1°) shows that there exists a strictly pseudoconvex domain Ω such that $G \Subset \Omega \Subset D$. Let $\Omega_\varepsilon = \{z \in \Omega : |z_n| < \varepsilon\}$. Since $\Omega \Subset D$, f can be extended to $\Omega_{2\varepsilon}$ for sufficiently small ε by setting $f(\,) = f('z)$.

Let χ be a compactly supported infinitely differentiable function on the plane with support in $B_{2\varepsilon}(0)$ with $\chi \equiv 1$ on $B_\varepsilon(0)$. We introduce the forms

$$f_k(z) = \begin{cases} \bar{\partial}\left(\dfrac{\chi(z_n)f(z)}{z_n^k}\right) = \bar{\partial}_\chi(z_n)\dfrac{f(z)}{z_n^k} & \text{for } z \in \Omega_{2\varepsilon}\backslash\Omega_\varepsilon, \\ 0 & \text{for } z \in \Omega_\varepsilon \cup (\Omega\backslash\Omega_{2\varepsilon}), \end{cases}$$

$k = 1, 2, \ldots$. Just as in the preceding subsection, we get that $f_k + a_{k-1}f_{k-1} + \cdots + a_1 f_1 = \bar{\partial}u$, where $u \in C(\Omega)$ for some positive integer k and complex numbers $a_1, \ldots, a_{k-1}$. The function

$$F(z) = \chi(z_n)f(z)(1 + a_{k-1}z_n + \cdots + a_1 z_n^{k-1}) - z_n^k u(z)$$

gives the desired extension of the function f in Ω. □

The proof of Theorem 25.2 now goes through in the usual way, for example, as in [201] (p. 260), except that again "domain of holomorphy" must be replaced by "pseudoconvex domain". □

If Theorem 25.3 is used in the proof of Lemma 25.12 instead of Theorem 25.10, then a strengthened form of Lemma 25.12 is obtained in which F is constructed from f with the help of a linear operator $P: A(D) \to A_C(G)$, and this, in turn, implies Theorem 25.2′. We remark that the proof of Theorem 25.3 uses Theorem 25.2 and not 25.2′. □

6°. PROOF OF THEOREM 25.3. Since D is a domain of holomorphy and $G \Subset D$, there is a Weil polyhedron Δ such that $G \Subset \Delta \Subset D$. This polyhedron has the form (see §9)

$$\Delta = \{z \in D : |\chi_j(z)| < 1, j = 1, \ldots, N\},$$

where the χ_j are holomorphic in D, and the nonempty faces $\sigma_{j_1\cdots j_k} = \{z \in D: |\chi_{j_1}(z)| = \cdots = |\chi_{j_k}(z)| = 1,\ |\chi_j(z)| \leq 1,\ j = 1, \ldots, N,\ j \neq j_1, \ldots, j_k\}$ have dimension $2n - k$ for all $j_1, \ldots, j_k$, $1 \leq j_1 << \cdots < j_k \leq N$.

We construct a chain Γ in $\mathbf{C}_\zeta^n \times \mathbf{R}_\lambda^{N+1}$ as follows: if $j_1, \ldots, j_k$ are such that $1 \leq j_1 < \cdots < j_k \leq N$ and $\sigma_{j_1, \ldots, j_k} \neq \varnothing$, then

$$\Gamma_{j_1, \ldots, j_k} = \{(\zeta, \lambda) : \zeta \in \sigma_{j_1, \ldots, j_k};\ \lambda = (\lambda_0, \lambda_1, \ldots, \lambda_N), \lambda_m \geq 0 \\ \text{for } m = 0, 1, \ldots, N,\ \lambda_m = 0 \text{ for } m \neq j_1, \ldots, j_k, 0 \text{ and} \\ \lambda_0 + \lambda_{j_1} + \cdots + \lambda_{j_k} = 1\};$$

$$\Gamma = \sum_{k=1}^{N} \sum_{1 \leq j_1 < \cdots < j_k \leq N} \Gamma_{j_1, \ldots, j_k}.$$

It is not hard to verify that $\partial\Gamma = \Gamma \cap \{\lambda_0 = 1\} - \Gamma \cap \{\lambda_0 = 0\}$ and $\Gamma \cap \{\lambda_0 = 1\} = \partial\Delta$.

Suppose now that $w(\zeta, z, \lambda)$ is defined in the set $U \subset \mathbf{C}^n_\zeta \times \mathbf{C}^n_z \times \mathbf{R}^{N+1}_\lambda$. The forms $\tilde W_{p,q}(w, \zeta, z, \lambda)$ are defined as in 1°, the vector-valued functions u_j as in §9 (here is where Theorem 25.2 is used), $t = (\bar\zeta - \bar z)/|\zeta - z|^2$, and $\eta(\zeta, z, \lambda) = \lambda_0 t + \Sigma_1^N \lambda_j u^j$. We define the following operator for $f \in A_{p,q}(\bar\Delta)$:

$$(\tau_{p,q} f)(z) = \int_{\Gamma_{\zeta,\lambda}} f(\zeta) \wedge \tilde W_{p,q-1}(\eta, \zeta, z, \lambda)$$

$$- \int_{\Delta_\zeta} f(\zeta) \wedge U_{p,q-1}(\zeta, z). \tag{25.20}$$

THEOREM 25.13 (Khenkin-Polyakov). *If $f \in A_{p,q}(\bar\Delta)$, $0 \leqslant p \leqslant n$, $1 \leqslant q \leqslant n$, then $\tau_{p,q} f \in C_{p,q-1}(\bar G)$, where $G \Subset \Delta$, where $G \Subset \Delta$. The operator $\tau_{p,q}: A_{p,q}(\bar\Delta) \to C_{p,q-1}(\bar G)$ is continuous, and*

$$\bar\partial \tau_{p,q} f = f. \tag{25.21}$$

Theorem 25.3 follows at once from Theorem 25.13.

PROOF OF THEOREM 25.13. The first integral in (25.20) is proper for $z \in G$, and the second converges uniformly; therefore, only (25.21) needs to be checked. By Stokes' formula,

$$\int_{\Gamma_{\zeta,\lambda}} d_{\zeta,\lambda}\big(f(\zeta) \wedge \tilde W_{p,q}(\eta, \zeta, z, \lambda)\big) = \int_{(\Gamma \cap \{\lambda_0 = 1\})_{\zeta,\lambda}} f(\zeta) \wedge \tilde W_{p,q}(\eta, \zeta, z, \lambda)$$

$$- \int_{(\Gamma \cap \{\lambda_0 = 0\})_{\zeta,\lambda}} f(\zeta) \wedge \tilde W_{p,q}(\eta, \zeta, z, \lambda) = \int_{\partial\Delta_\zeta} f(\zeta) \wedge U_{p,q}(\zeta, z)$$

$$- \sum_{k=1}^{N} \sum_{1 \leqslant j_1 < \cdots < j_k \leqslant N} \int_{\substack{\zeta \in \sigma_{j_1 \cdots j_k} \\ \lambda_{j_1} \geqslant 0, \ldots, \lambda_{j_k} \geqslant 0 \\ \lambda_{j_1} + \cdots + \lambda_{j_k} = 1}} f(\zeta) \wedge \tilde W_{p,q}\left(\sum_{m=1}^{k} \lambda_{j_m} u^{j_m}, \zeta, z, \lambda \right).$$

Since the u^j are holomorphic in $z \in D$, it follows that

$$\tilde W_{p,q}\left(\sum_{m=1}^{k} \lambda_{j_m} u^{j_m}, \zeta, z, \lambda \right) = 0$$

for $q \geqslant 1$. Consequently,

$$\int_{\partial\Delta_\zeta} f(\zeta) \wedge U_{p,q}(\zeta, z) = (-1)^{p+q} \int_{\Gamma_{\zeta,\lambda}} f(\zeta) \wedge \left(\bar\partial_\zeta + d_\lambda\right) \tilde W_{p,q}(\eta, \zeta, z, \lambda)$$

$$= \bar\partial_z \int_{\Gamma_{\zeta,\lambda}} f(\zeta) \wedge \tilde W_{p,q-1}(\eta, \zeta, z, \lambda).$$

Now (25.21) follows from the Martinelli-Bochner-Koppelman formula (7.1), applied to Δ. $\square$

7°. In this subsection we give some estimates for the solution of the $\bar{\partial}$-equation and applications of them. Let $f \in L^1_{p,q}(D)$. We say that $u \in L^1_{p,q-1}(D)$ and $g \in L^1_{p,q-1}(\partial D)$ have the properties that $\bar{\partial} u = f$ and $u|_{\partial D} = g$ if for any smooth form φ of type $(n-p, n-q)$ on $\bar{D}$

$$\int_{\partial D} g \wedge \varphi = \int_D f \wedge \varphi + (-1)^{p+q} \int_D u \wedge \bar{\partial}\varphi.$$

THEOREM 25.14 (Khenkin and Skoda). *Suppose that D is a strictly convex domain in $\mathbf{C}^n$ and $f \in L^1_{p,q}(D)$, $0 \leq p \leq n$, $1 \leq q \leq n$. If $\bar{\partial} f = 0$ and $|f(z)| + |\rho(z)|^{-1/2} |f(z) \wedge \bar{\partial}\rho(z)| \in L^1(D)$, then there exist forms $u \in L^1_{p,q-1}(D)$ and $g \in L^1_{p,q-1}(\partial D)$ with the properties that $\bar{\partial} u = f$ and $u|_{\partial D} = g$.*

As usual, the *modulus of a form* is understood to be the sum of the moduli of its coefficients.

Theorem 25.14 enables us to describe the set of zeros of functions in the *Nevanlinna class*, i.e., the functions f holomorphic in D such that

$$\sup_{\varepsilon>0} \int_{\zeta \in \partial D} \log^+ |f(\zeta - \varepsilon\nu(\zeta))| \, d\sigma_{2n-1}(\zeta) < \infty.$$

Here $\log^+ t = \max(0, \log t)$, $t > 0$, $\nu(\xi)$ is the outward unit vector normal to the boundary of D at the point ζ, and $d\sigma_k$ is the k-dimensional Lebesgue volume element.

THEOREM 25.15 (Khenkin-Skoda). *Suppose that D is a strictly convex domain in $\mathbf{C}^n$. An $(n-1)$-dimensional analytic subset M of D is the set of zeros of a function in the Nevanlinna class $N(D)$ if and only if M satisfies the Blaschke condition*

$$\int_M |\rho(z)| \, d\sigma_{2n-2}(z) = \sum_j \gamma_j \int_{M_j} |\rho(z)| \, d\sigma_{2n-2}(z) < \infty,$$

where the M_j are the irreducible components of the set M, and γ_j is the multiplicity of the component M_j.

From the form of the function α_k (see (25.7)) it follows that the operator $G^k_{p,q}$ defined by (25.12) can be applied to forms f such that $f(z)\rho^{k-1}(z) \in L^1_{p,q}(D)$. Using Theorem 25.6 and Lemmas 25.5 and 25.7, we can prove the following result.

THEOREM 25.16 (Dautov and Khenkin). *Suppose that D is a strictly convex domain in $\mathbf{C}^n$, and $f \in L^1_{p,q}(D, \mathrm{loc})$. If $\bar{\partial} f = 0$ and the number $\alpha > 0$ is such that*

$$|\rho(z)|^{\alpha} \left(|f(z)| + |\rho(z)|^{-1/2} |f(z) \wedge \bar{\partial}\rho(z)|\right) \in L^1(D),$$

then the form $u_k = G^k_{p,q} f$ satisfies for any $k > \alpha + \frac{1}{2}$ the equation $\bar{\partial} u = f$ in D, and

$$|(G^k_{p,q} f)(z)| |\rho(z)|^{\alpha-1} \in L^1_{p,q}(D).$$

A function f holomorphic in a bounded domain $D \subset \mathbf{C}^n$ is called a function of *Nevanlinna-Dzhrbashyan class $N_\alpha(D)$* ($\alpha > 0$) if

$$\int_D |\rho(z)|^{\alpha-1} \ln^+ |f(z)| \, d\sigma_{2n}(z) < \infty.$$

THEOREM 25.17 (Dautov and Khenkin). *Suppose that D is a strictly convex doamin in $\mathbf{C}^n$. An $(n-1)$-dimensional analytic subset M of D is the set of zeros of a function of Nevanlinna-Dzhrbashjan class $N_\alpha(D)$, $\alpha > 0$, if and only if*

$$\int_M |\rho(z)|^{\alpha+1} d\sigma_{2n-2}(z) < \infty.$$

Theorems 25.14 and 25.15 are limiting cases (as $\alpha \to 0$) of Theorems 25.16 and 25.17.

§26. Forms orthogonal to holomorphic functions. Lewy's equation. General form of integral representation formulas for holomorphic functions

1°. Let D be a bounded domain with smooth boundary in $\mathbf{C}^n$. We are interested in forms $\alpha \in C_{p,n-1}(\partial D)$ orthogonal to holomorphic forms in the sense that

$$\int_{\partial D} f \wedge \alpha = 0, \tag{26.1}$$

for all $f \in A_{n-p}(\bar{D})$ ($A_{n-p}(\bar{D})$ is the space of forms of type $(p, 0)$ with coefficients holomorphic on $\bar{D}$).

For $n = 1$ a function $\alpha \in C(\partial D)$ satisfying condition (26.1) for all forms $f(z)\,dz$ with $f \in A(\bar{D})$ extends to a function $\alpha \in A_C(D)$ (see [132], §29.3).

In this section we present an analogue of this assertion for $n > 1$ along with a consequence of it that gives a general form for the logarithmic residue formula, in particular, a general form for integral representations of holomorphic functions of several complex variables. At the end of the section a necessary and sufficient condition is given for the (local) solvability of Lewy's equation.

2°. We need some notation and definitions for what follows. A *modulus of continuity* is defined to be a continuous increasing concave function θ on $[0, \infty)$ that equals 0 at 0.

Let F be a compact subset of $\mathbf{R}^n$, and m an integer, $m \geqslant 0$. If $g \in C^{(m)}(F)$, then a modulus of continuity θ is called *a modulus of continuity of the function* g if for $x, y \in F$ and $|J| \leqslant m$

$$|D^J g(x) - D^J g(y)| = O(1)\theta(|x - y|).$$

Here $J = (i_1, \ldots, i_n)$ is a multi-index, and

$$D^J = \frac{\partial^{|J|}}{\partial x_1^{i_1} \cdots dx_n^{i_n}}.$$

If D is a bounded domain in $\mathbf{R}^n$ and m is an integer, then $\tilde{C}^{(m)}(D)$ denotes the space of complex-valued infinitely differentiable functions g on D for which there exists a modulus of continuity θ such that for $x, y \in D$

$$|D^J g(x) - D^J g(y)| = O(1)\theta(|x - y|) \quad \text{for } |J| \leqslant m \tag{26.2}$$

and

$$|D^J g(x)| \leqslant \frac{O(1)\theta(\rho(x, \partial D))}{\rho(x, \partial D)^{|J| - m}} \quad \text{for } |J| > m. \tag{26.3}$$

Here $\rho(x, \partial D) = \inf_{\xi \in \partial D} |\xi - x|$ is the Euclidean distance from x to ∂D.

A modulus of continuity θ appearing in (26.2) and (26.3) is called *a modulus of continuity of the function* $g \in \tilde{C}^{(m)}(D)$. It is not hard to show that for $m \geqslant 0$ any function $g \in \tilde{C}^{(m)}(D)$ extends to a function $g_1 \in \tilde{C}^{(m)}(\bar{D})$ with the same modulus of continuity as g. We regard $\tilde{C}^{(m)}(D)$ as imbedded in $C^{(m)}(\bar{D})$ if $m \geqslant 0$. If $D \subset \mathbf{C}^n$, then $\tilde{C}^{(m)}_{p,q}(D)$ denotes the space of forms of type (p,q) with coefficients in $\tilde{C}^{(m)}(D)$. *A modulus of continuity of a form* in $\tilde{C}^{(m)}_{p,q}(D)$ or $\tilde{C}^{(m)}_{p,q}(F)$ is defined to be a common modulus of continuity for its coefficients. Since for any finite set $\{\theta_1,\dots,\theta_k\}$ of moduli of continuity there is a modulus of continuity θ (see [119], Chapter I, §2) such that $\theta_j = O(1)\cdot\theta$, any form in $\tilde{C}^{(m)}_{p,q}(D)$ or $\tilde{C}^{(m)}_{p,q}(F)$ has a modulus of continuity.

If $m = 0$ and a form has modulus of continuity t^τ, where $0 < \tau \leqslant 1$, then we say that *the form satisfies a Hölder condition with exponent* τ.

Let D be a domain in $\mathbf{R}^n$ with boundary of class $C^{(l)}$, $l \geqslant 1$, i.e., $D = \{x \in \Omega : \rho(x) < 0\}$, where $\bar{D} \Subset \Omega$, $\rho \in \tilde{C}^{(l)}(\Omega)$, and $\operatorname{grad}\rho|_{\partial D} \neq 0$. By Whitney's theorem (see Theorem 3.2 and [119], Supplement 3.3 to Chapter I), there is a function $\tilde{\rho} \in C^{(l)}(\Omega) \cap \tilde{C}^{(l)}(D)$ coinciding with ρ on ∂D up to derivatives of order l; therefore, we assume in what follows that D is given by a function $\rho \in C^{(l)}(\Omega) \cap \tilde{C}^{(l)}(D)$.

Since $\operatorname{grad}\rho|_{\partial D} \neq 0$, we have $\rho(x,\partial D) = O(1)|\rho(x)| = O(1)\rho(x,\partial D)$, so (26.3) is equivalent to the condition

$$|D^J g(x)| \leqslant \frac{O(1)\theta(|\rho(x)|)}{|\rho(x)|^{|J|-m}} \quad \text{for } |J| > m. \tag{26.3'}$$

3°. Description of forms orthogonal to holomorphic forms.

THEOREM 26.1 (Dautov). *Suppose that $D = \{z : \rho(z) < 0\}$ is a strictly pseudoconvex domain in $\mathbf{C}^n$ $(n \geqslant 2)$ with boundary of class C^{m+2} $(m \geqslant 0)$, and $\alpha \in C_{p,n-1}(\partial D)$ satisfies* (26.1). *Suppose also that*

1) $\alpha \in C^{(m)}_{p,n-1}(\partial D)$ *for* $m \geqslant 1$, *or*

2) α *satisfies a Hölder condition of order* τ, $0 < \tau \leqslant 1$, *for* $m = 0$.

Then there exists a form $\beta = \beta_1 + \bar\partial\rho \wedge \sqrt{|\rho|}\beta_2$, with $\beta_1 \in \tilde{C}^{(m)}_{p,n-1}(D)$ and $\beta_2 \in \tilde{C}^{(m-1)}_{p,n-2}(D)$, such that $\bar\partial\beta = 0$ and $\beta_1|_{\partial D} = \alpha$. Here a modulus of continuity of the form α is also a modulus of continuity of the forms β_1 and β_2.

REMARK. If $m > 1$, then $\beta_2 \in C_{p,n-2}(\bar{D})$ and, therefore, $\beta|_{\partial D} = \beta_1|_{\partial D} = \alpha$. But if $m = 0$ and $\tau > \frac{1}{2}$, then it follows from (26.3′) that

$$\sqrt{|\rho(z)|}\,|\beta_2(z)| = O(1)\frac{|\rho(z)|^{1/2+\tau}}{|\rho(z)|} = O(1)|\rho(z)|^{\tau-1/2}.$$

Therefore, $\sqrt{|\rho|}\beta_2$ extends continuously to $\bar{D}$ and $\sqrt{|\rho|}\beta_2|_{\partial D} = 0$; consequently, $\beta|_{\partial D} = \alpha$ also in this case. Thus, the form α is the restriction to ∂D of a continuous $\bar\partial$-closed form when $m \geq 1$ or when $m = 0$ and $\tau > \frac{1}{2}$. In the general case, for any $f \in C^{(1)}_{n-p,0}(\bar D)$ we have

$$(-1)^{p+n-1}\int_D \beta \wedge \bar\partial f = \lim_{\delta \to 0}\int_{D_{-\delta}} d(\beta \wedge f) = \lim_{\delta \to 0}\int_{\partial D_{-\delta}} \beta \wedge f$$

$$= \int_{\partial D} \beta_1 \wedge f + \lim_{\delta \to 0}\int_{\partial D_{-\delta}} \bar\partial\rho \wedge \sqrt{|\rho|}\,\beta_2 \wedge f$$

$$= \lim_{\delta \to 0}\int_{\partial D_{-\delta}} \partial\rho \wedge \sqrt{|\rho|}\,\beta_2 \wedge f = \int_{\partial D} \alpha \wedge f. \qquad (26.4)$$

Here $D_{-\delta} = \{z : \rho(z) < -\delta\}$. We have used the equality $\bar\partial\rho|_{\partial D} + \partial\rho|_{\partial D} = d\rho|_{\partial D} = 0$ and the fact that the form $\partial\rho \wedge \beta_2 \wedge f$ has type $(n+1, n-2)$; consequently, it vanishes.

Equality (26.4) can be taken as the definition of the $\bar\partial$-closed extension of a form α in the case when this extension cannot be made a continuous form on D. Hence, in the general case as well α is the restriction (in the generalized sense) of the form β to ∂D.

We give without proof an assertion containing information about $\bar\partial$-closed extension of forms in the case $m = 0$ and $\tau = 0$. Some needed definitions are considered first. Let $\alpha \in L^1_{p,n-1}(\partial D)$. Represent α in the form

$$\alpha = \sum_{J,\mathcal{J}}{}' \alpha_{J,\mathcal{J}}(z)\,dz_J \wedge d\bar z_{\mathcal{J}}$$

and let

$$|\alpha(z)| = \sum_{J,\mathcal{J}}{}' |\alpha_{J,\mathcal{J}}(z)|, \qquad \tilde f(z,t) = \sum_{J,\mathcal{J}}{}' \tilde f_{J,\mathcal{J}}(z,t),$$

where $\tilde f_{J,\mathcal{J}}(z,t)$ is the *Hörmander maximal function*:

$$\tilde f_{J,\mathcal{J}}(z,t) = \sup_{\delta \geq t} \frac{1}{\sigma_{2n-1}(B_{z,\delta})}\int_{B_{z,\delta}} |f_{J,\mathcal{J}}(\zeta)|\,d\sigma_{2n-1}(\zeta).$$

Here $B_{z,\delta}$ is a *Hörmander "ellipsoid"*, i.e., the set of points $\zeta \in \partial D$ whose distance to the complex plane through z tangent to ∂D is at most δ.

THEOREM 26.2 (Khenkin). *Suppose that D is a strictly pseudoconvex domain in $\mathbf{C}^n$, and $\alpha \in L^1_{p,n-1}(\partial D)$ satisfies condition* (26.1). *Then there exists a form $\beta \in C_{p,n-1}(D)$, that is a $\bar\partial$-closed extension of the form α and has the property that*

$$\lim_{\varepsilon \to 0}\int_{\partial D_{-\varepsilon}} \beta(z) \wedge \varphi(z) = \int_{\partial D} \alpha(z) \wedge \varphi(z)$$

for all $\varphi \in C_{n-p,0}(\overline{D})$. *Moreover,*

$$\int_{\partial D_{-r}} (|\beta(z) \wedge \bar{\partial}\rho(z)| + |\rho(z)|^{1/2}|\beta(z)|)d\sigma_{2n-1}(z) = O(1)\|f\|_{L^1(D)}$$

and

$$|F(z) \wedge \bar{\partial}\rho(z)| + |\rho(z)|^{1/2}|F(z)| = O(1)\tilde{f}(\tilde{z}(z), |\rho(z)|),$$

where $\tilde{z}(z)$ *is a point of* ∂D *closest to* z.

Let us prove Theorem 26.1 for strictly convex domains. In the general case there are additional technical difficulties which we discuss at the end of this subsection.

LEMMA 26.3. *If* D *is a domain in* $\mathbf{R}^n$ *with boundary of class* $C^{(l)}$, $l \geq 1$, *and* $g \in \tilde{C}^{(m)}(D)$, $m \leq l-2$, *then* $\rho g \in \tilde{C}^{(m+1)}(D)$. *If* $m \geq -1$, *then* $\rho g|_{\partial D} = 0$. *Moreover, a modulus of continuity of* g *is a modulus of continuity of* ρg.

PROOF. Condition (26.2) for $|J| \leq m$ and condition (26.3) for $|J| > m+1$ follow at once for ρg from the Leibniz formula and conditions (26.2) and (26.3) for the functions g and ρ. Suppose that $|J| = m+1$. By the Leibniz formula,

$$D^J(g\rho)(x) = \sum_{\mathcal{J} \leq J} \binom{J}{\mathcal{J}} D^{\mathcal{J}}\rho(x) \cdot D^{J-\mathcal{J}}g(x).$$

Here, as usual, $\mathcal{J} \leq J$ means that $j_l \leq j_m$ for all $m = 1, \dots, n$, and $\binom{J}{\mathcal{J}} = \binom{i_1}{j_1} \cdots \binom{i_n}{j_n}$.

For terms with index $\mathcal{J} \neq 0$ condition (26.2) for g and ρ gives an estimate of the type (26.2). It remains to prove that

$$|\rho(x)h(x) - \rho(y)h(y)| = O(1)\theta(|x-y|), \tag{26.5}$$

where θ is a modulus of continuity of g, and $h = D_g^J$. It can be assumed that $\rho(x, \partial D) \geq \rho(y, \partial D)$ Denote $\rho(x, \partial D)$ by d. Assume first that $d \leq 2|x-y|$. We note that the concavity of θ and the equality $\theta(0) = 0$ imply that $\theta(\lambda t) \leq \lambda\theta(t)$ for $\lambda \geq 1$. Since θ is monotone and concave, (26.3) and (26.3′) imply that

$$\begin{aligned}|\rho(x)h(x) - \rho(y)h(y) &= O(1)(\theta(d) + \theta(\rho(y, \partial D))) \\ &= O(1)\theta(d) = O(1)\theta(2|x-y|) = O(1)\theta(|x-y|),\end{aligned}$$

i.e., (26.5) holds in this case.

If $d > 2|x-y|$, then the segment Γ joining x and y lies in the ball $B_{d/2}(x)$; hence, $\frac{3}{2}d \geq \rho(\tilde{x}, \partial D) \geq \frac{1}{2}d$ for any $\tilde{x} \in \Gamma$. From (26.3) for $\tilde{x} \in \Gamma$ and from the monotonicity of θ we get

$$|\operatorname{grad}(\rho h)(\tilde{x})| = |\rho(\tilde{x}) \operatorname{grad} h(\tilde{x}) + h(\tilde{x}) \operatorname{grad} \rho(\tilde{x})|$$
$$= O(1)\frac{\theta(\rho(\tilde{x}, \partial D))}{\rho(\tilde{x}, \partial D)} = O(1)\frac{\theta(d)}{d}.$$

Furthermore,

$$|\rho(x)h(x) - \rho(y)h(y)| = \left|\int_0^1 \frac{d}{dt}[(\rho h)(x + t(x-y))]\, dt\right|$$
$$= \left|\int_0^1 (\operatorname{grad}(\rho h)(x + t(x-y)), (x-y))\, dt\right|$$
$$= O(1)\frac{|x-y|}{d}\theta(d) = O(1)\theta(|x-y|).$$

At the last step we used the inequality $\theta(\lambda t) \leqslant \lambda\theta(t)$ for $t = |x-y|$ and $\lambda = d/|x-y| \geqslant 2$. □

LEMMA 26.4. *Suppose that D is a bounded domain in $\mathbf{C}^n$ with boundary of class $C^{(l)}$, $l \geqslant 1$, and $\alpha \in C^{(m)}_{p,n-1}(\partial D)$, $0 \leqslant m < l$. Then there exists a form $\tilde{\alpha} \in \tilde{C}^{(m)}_{p,n-1}(D)$, such that $\tilde{\alpha}|_{\partial D} = \alpha$, $\bar{\partial}\tilde{\alpha}(\zeta) = \rho^{m-1}(\zeta) \cdot \kappa(\zeta)$, $\kappa \in \tilde{C}^{(0)}_{p,n}(D)$, and $\kappa(\zeta) = 0$ for $\zeta \in \partial D$. Moreover, a modulus of continuity of α is a modulus of continuity of the forms $\tilde{\alpha}$ and κ.*

PROOF. Since $\bar{\partial}\rho \in \tilde{C}^{(l-1)}_{0,1}(D)$ and $\bar{\partial}\rho \neq 0$ on ∂D, there is a $\delta > 0$ such that any form $\mu \in \tilde{C}^{(m-1)}_{p,n}(\bar{D})$ vanishing for $\rho(\zeta) < -\delta$ satisfies $\mu(\zeta) = \bar{\partial}\rho(\zeta) \wedge \nu(\zeta)$, for some $\nu \in \tilde{C}^{(m-1)}_{p,n-1}(D)$.

By Whitney's theorem (see Theorem 3.2 and [119], Supplement 3.3 to Chapter I), there is a form $\alpha_0 \in \tilde{C}^{(m)}_{p,n-1}(\bar{D})$ with the same modulus of continuity as α such that $\alpha_0|_{\partial D} = \alpha$. Multiplying (if necessary) by an infinitely differentiable function identically equal to 1 in a neighborhood of ∂D and identically equal to 0 for $\rho(\zeta) \leqslant -\delta$, we obtain the equality $\alpha_0 \equiv 0$ for $\rho(\zeta) < -\delta$.

Lemma 26.3 and the equality $\bar{\partial}\alpha_0(\zeta) = \rho^{-1}(\zeta)\kappa$, where $\kappa(\zeta) = \rho(\zeta)\bar{\partial}\alpha_0(\zeta)$, conclude the proof of the lemma when $m = 0$.

Suppose that $m \geqslant 1$. Then $\bar{\partial}\alpha_0 = \bar{\partial}\rho(\zeta) \wedge \kappa_0(\zeta)$ for $\zeta \in \bar{D}$ and $\kappa_0 \in \tilde{C}^{(m-1)}_{p,n-1}(\bar{D})$. Let $\alpha_1 = \alpha_0 + \rho\kappa_0$. By Lemma 26.3, $\alpha_1 \in \tilde{C}^{(m)}_{p,n-1}(D)$. We have

$$\bar{\partial}\alpha_1 = \bar{\partial}\alpha_0 - \bar{\partial}\rho \wedge \kappa_0(\zeta) - \rho\bar{\partial}\kappa_0 = -\rho\bar{\partial}\kappa_0.$$

By Lemma 26.3 again, $\rho\bar{\partial}\kappa_0 \in \tilde{C}^{(0)}_{p,n}(D)$ and $\rho(\zeta)\bar{\partial}\kappa_0(\zeta) = 0$ for $\zeta \in \partial D$. For $m = 1$ the proof is finished. If $m \geqslant 2$, we represent $\bar{\partial}\kappa_0$ in the form $\bar{\partial}\rho \wedge \kappa_1$ and set $\alpha_2 = \alpha_1 - \frac{1}{2}\rho^2\kappa_1$. Application of this device m times gives us the lemma. □

Suppose now that D is a strictly convex domain with boundary of class $C^{(m+2)}$ $(m \geq 0)$, and the functions P_j $(j = 1,\dots,n)$, Φ, $\tilde{\Phi}$, and α_k are those defined in §25. We introduce the functions

$$\Phi^*(\zeta, z) = -\Phi(z, \zeta), \qquad \tilde{\Phi}^*(\zeta, z) = -\tilde{\Phi}(z, \zeta),$$
$$\alpha_k^*(\zeta, z) = \alpha_k(\zeta, z),$$

and the vector-valued functions

$$u^*(\zeta, z) = \frac{P(z)}{\Phi^*(\zeta, z)},$$
$$\eta^*(\zeta, z, \lambda) = -\eta(z, \zeta, \lambda) = (1-\lambda)t + \lambda u^*,$$

where, as earlier, $t(\zeta, z) = (\bar{\zeta} - \bar{z})/|\zeta - z|^2$.

Consider the following operators for $\mu \in L^1_{p,n}(D)$:

$$\left(G_p^{*k}\mu\right)(z) = -\int_{D_\zeta} \mu(\zeta) \wedge \alpha_k^*(\zeta, z) U_{p,n-1}(\zeta, z)$$
$$+ (-1)^{p+n} \int_{D_\zeta \times [0,1]_\lambda} \mu(\zeta) \wedge \bar{\partial}_z \alpha_k^* \wedge \tilde{W}_{p,n-2}(\eta^*, \zeta, z, \lambda). \quad (26.6)$$

The kernel $U_{p,n-1}$ was defined in §7, and $\tilde{W}_{p,n-2}$ in §25.

It follows from Lemma 25.5 that the form

$$\bar{\partial}_z \alpha_k^* \wedge \left(\tilde{W}_{p,n-2}(\eta^*, \zeta, z, \lambda) - W_{p,n-2}(\eta^*, \zeta, z)\right)$$

is a linear combination of forms like

$$\bar{\partial}\rho \wedge \rho^{k-1}(z)\left(\frac{\Phi^*}{\tilde{\Phi}^*}\right)^s \frac{(\bar{\zeta}_j - \bar{z}_j)\varphi(\zeta, z, \lambda)}{|\zeta - z|^{2r}(\tilde{\Phi}^*)^{n-r+k}} \quad (26.7)$$

and

$$\rho^k(z)\left(\frac{\Phi^*}{\tilde{\Phi}^*}\right)^s \frac{(\bar{\zeta}_j - \bar{z}_j)(\zeta_i - z_i)\Psi(\zeta, z, \lambda)}{|\zeta - z|^{2r}(\tilde{\Phi}^*)^{n-r+k+1}}, \quad (26.8)$$

where $s = 0, 1, \dots, j = 1, \dots, n$, $r = 0, \dots, n-1$, and φ and Ψ are forms whose coefficients are derivatives of $\rho(z)$ of order at most two, multiplied by a polynomial in λ.

Just as in the remark after (25.12), we get that the second integral in (26.6) is proper.

LEMMA 26.5. *Suppose that* $\mu \in L^1_{p,n}(D)$ *and*

$$\int_D \mu \wedge f = 0 \quad (26.9)$$

for any $f \in A_{n-p}(\bar{D})$. *Then* $\gamma_k = G_p^{*k}\mu$ *is a solution of the equation* $\bar{\partial}\gamma_k = \mu$ *for any* $k \geq 0$.

PROOF. Using (7.15) with $\varphi = \alpha_k^*$, we get

$$\mu(z) = -\bar{\partial}_z \int_{D_\zeta} \mu(\zeta) \wedge \alpha_k^* U_{p,n-1}(\zeta, z)$$
$$+ (-1)^{p+n} \int_{D_\zeta} \mu(\zeta) \wedge \bar{\partial}_z \alpha_k^* \wedge U_{p,n-1}(\zeta, z). \qquad (26.10)$$

By Stokes' formula,

$$\int_{D_\zeta \times [0,1]_\lambda} d_{\zeta,\lambda}\big(\mu(\zeta) \wedge \bar{\partial}_z \alpha_k^* \wedge \tilde{W}_{p,n-1}(\eta^*, \zeta, z, \lambda)\big)$$
$$= -\int_{D_\zeta} \mu(\zeta) \wedge \bar{\partial}_z \alpha_k^*(\zeta, z) \wedge U_{p,n-1}(\zeta, z)$$
$$+ \int_{D_\zeta} \mu(\zeta) \wedge \bar{\partial}_z \alpha_k^* \wedge W_{p,n-1}(u^*, \zeta, z). \quad (26.11)$$

It is not hard to see that $\bar{\partial}_z \alpha_k^*(\zeta, z) \wedge W_{p,n-1}(u^*, \zeta, z)$ is holomorphic in $\zeta \in \bar{D}$ for $z \in D$; therefore, (26.9) implies that the last integral in (26.11) is equal to 0. Then

$$\int_{D_\zeta} \mu(\zeta) \wedge \bar{\partial}_z \alpha_k^* \wedge U_{p,n-1}(\zeta, z)$$
$$= (-1)^{p+n} \int_{D_\zeta \times [0,1]_\lambda} \mu(\zeta) \wedge \bar{\partial}_z \alpha_k^* \wedge \big(\bar{\partial}_\zeta + d\lambda\big) \tilde{W}_{p,n-1}(\eta^*, \zeta, z, \lambda)$$
$$= \bar{\partial}_z \int_{D_\zeta \times [0,1]_\lambda} \mu(\zeta) \wedge \bar{\partial}_z \alpha_k^* \wedge \tilde{W}_{p,n-2}(\eta^*, \zeta, z, \lambda).$$

Here, as in the proof of Theorem 25.6, we have used the equality

$$\bar{\partial}_z \tilde{W}_{p,q} = (-1)^{p+q}\big(\bar{\partial}_\zeta + d_\lambda\big) \tilde{W}_{p,q+1}.$$

It now follows from (26.10) that $\bar{\partial} G_p^{*k}\mu = \mu$. □

LEMMA 26.6. *Suppose that* $h \in \tilde{C}^{(0)}(D)$, $h|_{\partial D} = 0$; m, k, r, *and* s *are nonnegative integers with* $r \leq n - 1$, *the function* $\varphi(z)$ *is a product of derivatives of* $\rho(z)$ *or order at most two, and for* $m = 0$ *the function* h *satisfies a Hölder condition of order* τ *for some* τ, $0 < \tau \leq 1$. *Let*

$$H_1(z) = \rho^k(z) \int_{D_\zeta} \rho^{m-1}(\zeta) h(\zeta) \frac{\bar{\zeta}_j - \bar{z}_j}{|\zeta - z|^{2n} (\bar{\Phi}^*)^k} \left(\frac{\Phi^*}{\bar{\Phi}^*}\right)^s dv(\zeta);$$

$$H_2(z) = \rho^k(z)\int_{D_\zeta} \rho^{m-1}(\zeta)h(\zeta)\frac{(\bar{\zeta}_j - \bar{z}_j)(\zeta_l - z_l)\varphi(z)}{|\zeta - z|^{2r}(\tilde{\Phi}^*)^{n-r+k+1}}\left(\frac{\Phi^*}{\tilde{\Phi}^*}\right)^s dv(\zeta);$$

$$H_3(z) = |\rho(z)|^{k-3/2}\int_{D_\zeta} \rho^{m-1}(\zeta)h(\zeta)\frac{(\bar{\zeta}_j - \bar{z}_j)\varphi(z)}{|\zeta - z|^{2r}(\tilde{\Phi}^*)^{n-r+k}} \times \left(\frac{\Phi^*}{\tilde{\Phi}^*}\right)^s dv(\zeta).$$

If $k > m + 2$, *then* H_1, $H_2 \in \tilde{C}^{(m)}(D)$, $H_1|_{\partial D} = H_2|_{\partial D} = 0$ *and* $H_3 \in \tilde{C}^{(m-1)}(D)$. *Moreover, a modulus of continuity of* h *is a modulus of continuity of* H_1, H_2, *and* H_3.

The proof is carried out for H_3; the arguments are similar in the remaining cases. The fact that $h|_{\partial D} = 0$ and condition (26.2) give us that

$$|h(\zeta)| = O(1)\theta(|\rho(\zeta)|),$$

where θ is a modulus of continuity of h. By assumption, $\theta(t) = t^\tau$ when $m = 0$; therefore, the integral defining H_3 converges.

Suppose that $\chi \in C^{(\infty)}(\mathbf{C}^n)$, $\operatorname{supp}\chi \subset B_2(0)$, and $\chi(\zeta) \equiv 1$ in $B_1(0)$. Fix $z^0 \in D$, let $a = \rho(z^0, \partial D)/3$, and let H' and H'' be defined in the same way as H_3, except that the integrand must be multiplied by $\chi(a^{-1}(\zeta - z^0))$ and $1 - \chi(a^{-1}(\zeta - z^0))$, respectively. It is clear that $H = H' + H''$. Moreover, $a = O(1)\rho(z^0) = O(1)a$.

In the integral defining H'' we can differentiate under the integral sign for $z \in B_a(z^0)$, because $1 - \chi(a^{-1}(\zeta - z^0)) \equiv 0$ for $\zeta \in B_a(z^0)$. Let D be a linear differential operator of order $s \geqslant 0$ with constant coefficients. Using Leibniz' formula, inequalities (25.3) and (25.4), and conditions (26.2) and (26.3) for the function ρ, we get

$$D\left(\frac{H''(z)}{|\rho(z)|^{k-3/2}}\right)\Bigg|_{z=z^0} = O(1)\int_{D_\zeta\setminus B_a(z^0)} |\rho(\zeta)|^{m-1}|h(\zeta)|$$
$$\times \sum_{i_1+i_2+i_3=s} \frac{dv(\zeta)}{|\zeta - z^0|^{2n-3+i_1}|\tilde{\Phi}^*(\zeta, z^0)|^{k+1+i_2}a^{i_3}}$$
$$\times O(1) \sum_{i_1+i_2+i_3=s} \frac{1}{a^{i_1+i_3}}\int_{D_\zeta} \frac{|\beta(\zeta)|^{m-1}\theta(|\rho(\zeta)|)\, dv(\zeta)}{|\zeta - z^0|^{2n-3}|\tilde{\Phi}^*(\zeta, z^0)|^{k+1+i_2}}.$$

Let $m \geqslant 1$. From the fact that θ is monotone and concave and from the equality $\theta(0) = 0$ we get

$$\theta(|\rho(\zeta)|) \leqslant (|\rho(\zeta)| \cdot |\rho(z^0)|^{-1} + 1) \cdot \theta(|\rho(z^0)|)$$
$$= O(1)(|\rho(\zeta)|a^{-1} + 1) \cdot \theta(a).$$

Therefore,

$$D\left(\frac{H''(z)}{|\rho(z)|^{k-3/2}}\right)\Bigg|_{z=z^0} = O(1)\sum_{i_1+i_2=s}\frac{\theta(a)}{a_1^{i_1+1}}\int_{D_\zeta}\frac{(|\rho(z)|^m + a|\rho(\zeta)|^{m-1})\,dv(\zeta)}{|\zeta - z^0|^{2n-3}|\tilde{\Phi}^*(\zeta, z^0)|^{k+1+i_2}}.$$

Just as in the proof of Theorem 25.9, we break up each integral into two parts: over $D_\zeta\setminus B_\delta(z^0)$ and over $D\cap B_\delta(z^0)$, where δ is chosen according to Lemma 10.6. In the integral over $D\cap B_\delta(z^0)$ we make a change of variables (see Lemma 10.6) and use (25.5). By Lemma 25.7,

$$\begin{aligned}
&D\left(\frac{H''(z)}{|\rho(z)|^{k-3/2}}\right)\Bigg|_{z=z^0}\\
&\quad= O(1)\sum_{i_1+i_2=s}\frac{\theta(a)}{a^{i_1+1}}\Big(J_{m,k+1+i_2,2n-3}(|\rho(z^0)|)\\
&\qquad+aJ_{m-1,k+1+i_2,2n-3}(|\rho(z^0)|) + O(1)\Big)\\
&\quad= O(1)\sum_{i_1+i_2=s}\frac{\theta(a)}{a^{i_1+1}}\Big(|\rho(z^0)|^{m-k-1-i_2+5/2}\\
&\qquad+a|\rho(z^0)|^{m-1-k-1-i_2+5/2} + O(1)\Big) = O(1)\frac{\theta(a)}{a^{s+k-m-1/2}}.
\end{aligned}$$

We have $\theta(t) = t^\tau$ for $m = 0$; therefore,

$$\begin{aligned}
D\left(\frac{H''(z)}{|\rho(z)|^{k-3/2}}\right)\Bigg|_{z=z^0} &= O(1)\sum_{i_1+i_2=s}\frac{1}{a^{i_1}}\int_{D_\zeta}\frac{|\rho(\zeta)|^{\tau-1}dv(\zeta)}{|\zeta - z^0|^{2n-3}|\tilde{\Phi}^*(\zeta, z)|^{k+1+i_2}}\\
&= O(1)\sum_{i_1+i_2=s}\frac{1}{a^{i_1}}J_{\tau-1,k+1+i_2,2n-3}(|\rho(z^0)|) + O(1) = O(1)\frac{a^\tau}{a^{s+k-1/2}}.
\end{aligned}$$

The change of variables $\zeta' = \zeta - z$ in the integral defining H' gives us that

$$\begin{aligned}
H'(z) = {}&|\rho(z)|^{k-3/2}\int_{B_{2a}(z^0-z)}\rho^{m-1}(\zeta + z)h(\zeta + z)\chi(a^{-1}(\zeta + z - z^0))\\
&\times\frac{\bar{\zeta}_j\varphi(z)}{|\zeta|^{2r}[\tilde{\Phi}^*(\zeta + z, z)]^{n-r+k}}\left(\frac{\Phi^*(\zeta + z, z)}{\tilde{\Phi}^*(\zeta + z, z)}\right)^s dv(\zeta).
\end{aligned}$$

Differentiating under the integral sign, we have

$$\begin{aligned}
D\left(\frac{H'(z)}{|\rho(z)|^{k-3/2}}\right) = {}&O(1)\int_{B_{2a}(z^0-z)}\sum_{i_1+\cdots+i_4=s}\rho^{m-i_1-1}(\zeta + z)\\
&\times\frac{\theta(\rho(\zeta + z))}{|\rho(\zeta + z)|^{i_2}}\cdot\frac{1}{a^{i_3}}\cdot\frac{dv(\zeta)}{|\zeta|^{2n-3}|\tilde{\Phi}^*(\zeta + z, z)|^{k+1+i_4}}.
\end{aligned}$$

Here we have used Leibniz' formula, inequalities (25.3) and (25.4), and conditions (26.2) and (26.3) for the functions ρ and h. Let us make the inverse substitution:

$$D\left(\frac{H'(z)}{|\rho(z)|^{k-3/2}}\right)\Bigg|_{z=z_0}$$

$$= O(1)\int_{B_{2a}(z^0)}\sum_{i_1+\cdots+i_4=s}\frac{|\rho(\zeta)|^{m-1-i_1-i_2}\theta(|\rho(\zeta)|)\,dv(\zeta)}{a^{i_3}|\zeta-z^0|^{2n-3}|\tilde{\Phi}^*(\zeta,z^0)|^{k+1+i_4}}.$$

Since $\rho(z^0,\partial D)=3a$, it follows that $a\leqslant\rho(\zeta,\partial D)\leqslant 5a$ for $\zeta\in B_{2a}(z^0)$; therefore,

$$|\rho(\zeta)|^{m-1-i_1-i_2}\theta(|\rho(\zeta)|)=O(1)a^{m-1-i_1-i_2}\theta(a),$$

and so

$$D\left(\frac{H'(z)}{|\rho(z)|^{k-3/2}}\right)\Bigg|_{z=z_0}=\sum_{i_1+i_2=s}\theta(a)a^{m-i_1-1}\int_{D_\zeta}\frac{dv(\zeta)}{|\zeta-z^0|^{2n-3}|\tilde{\Phi}^*(\zeta,z^0)|^{k+1+i_2}}$$

$$=O(1)\theta(a)\sum_{i_1+i_2=s}a^{m-1-i_1}J_{0,k+1+i_2,2n-3}(|\rho(z^0)|)$$

$$=O(1)\frac{\theta(a)}{a^{s+k-m-1/2}}.$$

Thus, for any $z^0\in D$

$$D\left(\frac{H_3(z)}{|\rho(z)|^{k-3/2}}\right)\Bigg|_{z=z_0}=D\left(\frac{H'(z)}{|\rho(z)|^{k-3/2}}\right)\Bigg|_{z=z_0}+D\left(\frac{H''(z)}{|\rho(z)|^{k-3/2}}\right)\Bigg|_{z=z_0}$$

$$=O(1)\frac{\theta(a)}{a^{s+k-m-1/2}}=O(1)\frac{\theta(|\rho(z^0)|)}{|\rho(z^0)|^{s+k-m-1/2}}.$$

This and Leibniz' formula give us that

$$D\left(\frac{H_3(z)}{|\rho(z)|^{k-1}}\right)=D\left(|\rho(z)|^{-1/2}\frac{H_3(z)}{|\rho(z)|^{k-3/2}}\right)=O(1)\frac{\theta(|\rho(z)|)}{|\rho(z)|^{s+k-m}},$$

i.e. (compare with (26.3′)), $H_3\cdot|\rho|^{1-k}\in\tilde{C}^{(m-k)}(D)$, and θ is a modulus of continuity of $H_3\cdot|\rho|^{1-k}$; applying 26.3 $k-1$ times, we now get that $H_3\in\tilde{C}^{(m-1)}(D)$, and θ is a modulus of continuity of H_3. □

PROOF OF THEOREM 26.1. Suppose that D is a strictly convex domain and α satisfies the conditions of the theorem. By Lemma 26.4, $\alpha=\tilde{\alpha}|_{\partial D}$, where

$\tilde{\alpha} \in \tilde{C}^{(m)}_{p,n-1}(D)$ and $\bar{\partial}\tilde{\alpha}(\zeta) = \rho^{m-1}(\zeta)\kappa(\zeta)$, $\kappa \in \tilde{C}^{(0)}_{p,n}(D)$, $\kappa(\zeta) = 0$ for $\zeta \in \partial D$, and a modulus of continuity of α is a modulus of continuity of the forms $\tilde{\alpha}$ and κ. As in the proof of Lemma 26.6, we get that $\bar{\partial}\tilde{\alpha} \in L^1_{p,n}(D)$. By (26.1),

$$\int_D \bar{\partial}\tilde{\alpha} \wedge f = \int_D d(\tilde{\alpha} \wedge f) = \int_{\partial D} \alpha \wedge f = 0 \tag{26.12}$$

for all $f \in A_{n-p}(\bar{D})$. Lemma 26.5 gives us that $\bar{\partial}\tilde{\alpha} = \bar{\partial}\gamma_k$, where $\gamma_k = G^{*k}_p(\bar{\partial}\tilde{\alpha})$ for any $k \geqslant 0$. From the definition of the operator G^{*k}_p (see (26.6)), the definitions of $U_{p,n-1}$ and the function α^*_k, and from (26.7) and (26.8) it follows that $\gamma_k = \gamma'_k + \bar{\partial}\rho \wedge \sqrt{|\rho|} \cdot \gamma''_k$, where γ'_k and γ''_k are forms of type $(p, n-1)$ and $(p, n-2)$, respectively, the coefficients of γ'_k are linear combinations of functions of the kind H_1 and H_2 in Lemma 26.6, and the coefficients of γ''_k are linear combinations of functions of the kind H_3. By Lemma 26.6,

$$\gamma'_k \in \tilde{C}^{(m)}_{p,n-1}(D), \qquad \gamma''_k \in C^{(m-1)}_{p,n-2}(D)$$

and $\gamma'_k|_{\partial D} = 0$ for $k > m+2$. Any form of the kind $\tilde{\alpha} - \gamma_k$ for $k > m+2$ can now be taken as β, with $\beta_1 = \tilde{\alpha} - \gamma'_k$ and $\beta_2 = -\gamma''_k$, since $\bar{\partial}\beta = \bar{\partial}(\tilde{\alpha} - \gamma_k) = \bar{\partial}\tilde{\alpha} - \bar{\partial}\gamma_k = 0$ and $\beta_1|_{\partial D} = \tilde{\alpha}|_{\partial D} - \gamma'_k|_{\partial D} = \alpha$. □

To prove the theorem in the general case it is necessary to construct an operator analogous to G^{*k}_p. It can be constructed just as G^{*k}_p was, except that as P_j and Φ we must take the functions obtained in Lemmas 10.1 and 10.3, and as $\tilde{\Phi}(\zeta, z)$ the function $(F(\zeta, z) - \rho(\zeta))G(\zeta, z)$. The functions P_j, $\tilde{\Phi}$, and Φ are defined for $(\zeta, z) \in U_{\varepsilon,\delta} = \{(\zeta, z): |\rho(\zeta)| < \delta;\ \rho(z) < \delta;\ |\zeta - z| < \varepsilon\}$ for some $\varepsilon > 0$ and $\delta > 0$. Repeating the arguments of Lemmas 26.5 and 26.6, we find a $\bar{\partial}$-closed form $\tilde{\beta}$ defined in $U \cap D$, where U is a neighborhood of ∂D satisfying the conditions of the theorem. We smooth out $\tilde{\beta}$ near the boundary of U and get a form, again denoted by $\tilde{\beta}$, defined in D and satisfying all the conditions of the theorem, except that $\bar{\partial}\tilde{\beta} = 0$ in some (one-sided) neighborhood of ∂D. It is then necessary to modify $\tilde{\beta}$ in roughly the same way the form $\tilde{\alpha}$ was modified in the proof of the theorem to get the form β. This is done as follows.

In some neighborhood U_1 of ∂D the form

$$\begin{aligned}\tau(z) &= \int_{D_\zeta \times [0,1]_\lambda} \bar{\partial}\tilde{\beta}(\zeta) \wedge \tilde{W}_{p,n-2}(\eta^*, \zeta, z, \lambda) \\ &= \int_{(D\setminus U)_\zeta \times [0,1]_\lambda} \bar{\partial}\tilde{\beta}(\zeta) \wedge \tilde{W}_{p,n-2}(\eta^*, \zeta, z, \lambda)\end{aligned} \tag{26.13}$$

is defined, where, as before

$$\eta^*(\zeta, z) = (1-\lambda)t + \lambda u^* \quad \text{and} \quad u^* = P^*(\zeta, z)/\Phi^*(\zeta, z).$$

Since $\rho \in C^{(\infty)}(D)$, it follows from the proofs of Lemmas 10.1 and 10.3 that $\tilde{W}_{p,n-1}(\eta^*, \zeta, z, \lambda)$ is infinitely differentiable for $(\zeta, z, \lambda) \in (D\setminus U) \times U_1 \times [0, 1]$; therefore, $\tau \in C^{(\infty)}_{p,n-2}(U_1 \cap D)$. Smoothing out τ near the boundary of U_1, we find a form, again denoted by τ, that is in $C^{(\infty)}_{p,n-2}(D)$ and is given by (26.13) in some neighborhood U_2 of ∂D.

For $z \in U_2 \cap D$

$$\begin{aligned}\bar\partial\tau(z) &= \int_{D_\zeta\times[0,1]_\lambda} \bar\partial\tilde\beta(\zeta) \wedge \bar\partial_z \tilde W_{p,n-2}(\eta^*, \zeta, z, \lambda)\\ &= (-1)^{p+n-1}\int_{D_\zeta\times[0,1]_\lambda} \bar\partial\tilde\beta(\zeta) \wedge (\bar\partial_\zeta + d_\lambda)\tilde W_{p,n-1}(\eta^*, \zeta, z, \lambda)\\ &= -\int_{D_\zeta\times[0,1]_\lambda} d_{\zeta,\lambda}(\bar\partial\tilde\beta(\zeta) \wedge \tilde W_{p,n-1}(\eta^*, \zeta, z, \lambda))\\ &= \int_{D_\zeta} \bar\partial\tilde\beta(\zeta) \wedge U_{p,n-1}(\zeta, z) - \int_{D_\zeta} \bar\partial\tilde\beta \wedge W_{p,n-1}(u^*, \zeta, z).\end{aligned}$$

If z is close to ∂D, then the vector-valued function $u^*(\zeta, z)$ is holomorphic in $\zeta \in D_{\rho(z)} = \{w : \rho(w) < \rho(z)\}$. It can be shown that the functions in $A(D_{\rho(z)})$ can be uniformly approximated on $D\backslash U$ by functions in $A(\bar D)$. From this and from the equality for $\bar\partial\tilde\beta$ analogous to (26.12), we get that

$$\int_{D_\zeta} \bar\partial\tilde\beta \wedge W_{p,n-1}(u^*, \zeta, z) = 0$$

for z close to ∂D. Therefore, the form

$$\tilde\gamma(z) = \int_{D_\zeta} \bar\partial\tilde\beta(\zeta) \wedge U_{p,n-1}(\zeta, z) - \bar\partial\tau(z)$$

is in $C^{(\infty)}_{p,n-1}(D)$ and vanishes near ∂D. The Martinelli-Bochner-Koppelman formula (7.1) now gives us that $\beta = \tilde\beta - \tilde\gamma$ is the desired form.

4°. It was shown in the preceding subsection that any form orthogonal to holomorphic forms when integrated over the boundary of a strictly pseudoconvex domain D can be extended to a form that is $\bar\partial$-closed in D. Since the $\bar\partial$-problem is solvable in a strictly pseudoconvex domain (see Theorem 25.10), the following result looks natural.

THEOREM 26.7. *Suppose that D is a strictly pseudoconvex domain in $\mathbf{C}^n$ and* (26.1) *holds for the form $\alpha \in C_{p,n-1}(\partial D)$. If α satisfies a Hölder condition with exponent $\tau > \frac{1}{2}$, then $\alpha = \bar\partial\gamma|_{\partial D}$, where $\gamma \in C^{(1)}_{p,n-2}(\bar D)$.*

The proof will be carried out for strictly convex domains. To do this it suffices to solve the equation $\bar\partial\gamma = \beta$, where $\beta = \beta_1 + \bar\partial\rho \wedge \sqrt{|\rho|}\cdot\beta_2$ in Theorem 26.1. The solution of this equation gives the operator $G^k_{p,n-1}$ (see (25.12)), but significant difficulties arise in estimating the derivatives. Therefore, we construct another solution.

Let Ω be a domain containing D in which the estimates (25.2)–(25.5) are valid. Using Lemma 26.4, we find a form $\tilde\beta_1 \in \tilde C^{(0)}_{p,n-1}(\Omega\backslash D)$ such that $\tilde\beta_1|_{\partial D} = \beta_1|_{\partial D}$ and $\bar\partial\tilde\beta_1 = \rho^{-1}(\zeta)\kappa(\zeta)$, where $\kappa \in \tilde C^{(0)}_{p,n}(\bar D)$ and $\kappa(\zeta) = 0$ for $\zeta \in \partial D$. By Theorem 26.1, $\theta(t) = t^\tau$ is a modulus of continuity of the form β_1 for some τ; consequently, the forms β_1 and κ also have this function as a modulus of continuity. It can be assumed that $\tilde\beta_1|_{\partial\Omega} = 0$.

From the Martinelli-Bochner-Koppelman formula (7.1) and the remark after Theorem 26.1, for $z \in D$ we get

$$\beta(z) = \int_{\partial D_\zeta} \beta_1(\zeta) \wedge U_{p,n-1}(\zeta, z) - \bar{\partial}_z \int_{D_\zeta} \beta(\zeta) \wedge U_{p,n-2}(\zeta, z),$$

$$0 = -\int_{\partial D_\zeta} \tilde{\beta}_1(\zeta) \wedge U_{p,n-1}(\zeta, z) - \bar{\partial}_z \int_{(\Omega\setminus D)_\zeta} \tilde{\beta}_1(\zeta) \wedge U_{p,n-2}(\zeta, z)$$

$$- \int_{(\Omega\setminus D)_\zeta} \bar{\partial}\tilde{\beta}_1(\zeta) \wedge U_{p,n-1}(\zeta, z).$$

Adding these equalities and taking the equality $\beta_1|_{\partial D} = \tilde{\beta}_1|_{\partial D}$ into account, we get that for $z \in D$

$$\beta(z) = -\int_{(\Omega\setminus D)_\zeta} \bar{\partial}\tilde{\beta}_1(\zeta) \wedge U_{p,n-1}(\zeta, z)$$
$$-\bar{\partial}_z \left[\int_{D_\zeta} \beta(\zeta) \wedge U_{p,n-2}(\zeta, z) + \int_{(\Omega\setminus D)_\zeta} \tilde{\beta}_1(\zeta) \wedge U_{p,n-2}(\zeta, z) \right]. \tag{26.14}$$

A transformation of the first integral in (26.14) gives us

$$\int_{(\Omega\setminus D)_\zeta} \bar{\partial}\tilde{\beta}_1(\zeta) \wedge U_{p,n-1}(\zeta, z)$$
$$= -\int_{(\Omega\setminus D)_\zeta\times[0,1]_\lambda} d_{\zeta,\lambda}\left(\bar{\partial}\tilde{\beta}_1(\zeta) \wedge \tilde{W}_{p,n-1}(\eta, \zeta, z, \lambda)\right)$$
$$+ \int_{(\Omega\setminus D)_\zeta} \bar{\partial}\tilde{\beta}_1(\zeta) \wedge W_{p,n-1}(u, \zeta, z).$$

Here, as before, $u(\zeta, z) = P(\zeta)/\Phi(\zeta, z)$; and $\eta = (1-\lambda)t + \lambda u$.

Since u is holomorphic in $z \in D$ for $\zeta \notin D$, $W_{p,n-1}(u, \zeta, z) = 0$. Further,

$$\int_{(\Omega\setminus D)_\zeta} \bar{\partial}\tilde{\beta}_1(\zeta) \wedge U_{p,n-1}(\zeta, z)$$
$$= (-1)^{p+n+1} \int_{(\Omega\setminus D)_\zeta\times[0,1]_\lambda} \bar{\partial}\tilde{\beta}_1(\zeta) \wedge \left(\bar{\partial}_\zeta + d_\lambda\right) \tilde{W}_{p,n-1}(\eta, \zeta, z, \lambda)$$
$$= \bar{\partial}_z \int_{(\Omega\setminus D)_\zeta\times[0,1]_\lambda} \bar{\partial}\tilde{\beta}_1(\zeta) \wedge \tilde{W}_{p,n-2}(\eta, \zeta, z, \lambda).$$

This and (26.14) give us that $\bar{\partial}\gamma(z) = \beta$ for $z \in D$, where

$$\gamma(z) = -\int_{D_\zeta} \beta(z) \wedge U_{p,n-2}(\zeta, z) - \int_{(\Omega \setminus D)_\zeta} \tilde{\beta}_1(\zeta) \wedge U_{p,n-2}(\zeta, z)$$
$$- \int_{(\Omega \setminus D)_\zeta \times [0,1]_\lambda} \bar{\partial}\beta_1(\zeta) \wedge \tilde{W}_{p,n-2}(\eta, \zeta, z, \lambda). \tag{26.15}$$

We show that γ is the desired form. In the first two integrals in (26.15) the coefficients of the kernel are of the kind $(\bar{\zeta}_j - \bar{z}_j)/|\zeta - z|^{2n}$, while the forms β and β_1 satisfy a Hölder condition; therefore, the forms defined by these integrals are in $C^{(1)}_{p,n-2}(\bar{D})$ (see Proposition 0.10).

Consider the third integral in (26.15). Using (25.11), we integrate with respect to $\lambda \in [0, 1]$ and get that this integral is a linear combination of forms with coefficients like

$$F(z) = \int_{(\Omega \setminus D)_\zeta} \frac{(\bar{\zeta}_j - \bar{z}_j)\varphi(\zeta)}{|\zeta - z|^{2(n-1)}\Phi(\zeta, z)} \cdot \frac{\kappa(\zeta)}{\rho(\zeta)}, \qquad z \in D,$$

where $1 \leqslant j \leqslant n$, and $\varphi(z)$ is a product of derivatives of $\rho(\zeta)$ of order at most two. Note that $|\kappa(\zeta)| = O(1)\theta(|\rho(\zeta)|) = O(1)|\rho(\zeta)|^\tau$ (compare with the proof of Lemma 26.6).

Let D be a second-order differential operator with constant coefficients. Differentiation under the integral sign gives us that for $z \in D$

$$|DF(z)| = O(1)\left[\int_{(\Omega \setminus D)_\zeta} \frac{|\rho(\zeta)|^{\tau-1} dv(\zeta)}{|\zeta - z|^{2n-1}|\Phi(\zeta, z)|}\right.$$
$$\left. + \int_{(\Omega \setminus D)_\zeta} \frac{|\rho(\zeta)|^{\tau-1} dv(\zeta)}{|\zeta - z|^{2n-2}|\Phi(\zeta, z)|^2} + \int_{(\Omega \setminus D)_\zeta} \frac{|\rho(\zeta)|^{\tau-1} dv(\zeta)}{|\zeta - z|^{2n-3}|\Phi(\zeta, z)|^3}\right].$$

After using (25.2) and the change of variables mentioned in Lemma 10.6 just as in the proof of Theorem 25.9 (and Lemma 26.6), we get from Lemma 25.7 that

$$\begin{aligned} |DF(z)| &\\ &= O(1)\left[J_{\tau-1,1,2n-1}(|\rho(z)|) + J_{\tau-1,3/2,2n-1}(|\rho(z)|) + J_{\tau-1,3,2n-3}(|\rho(z)|)\right] \\ &= O(1)\left(|\rho(z)|^{\tau-1} + |\rho(z)|^{\tau-3/2} + |\rho(z)|^{\tau-3/2}\right) = O(1)|\rho(z)|^{-1+(\tau-1/2)}. \end{aligned}$$

Since $\tau > \frac{1}{2}$, the first derivatives of F satisfy a Hölder condition with exponent $\tau - \frac{1}{2}$, by Lemma 25.8. Thus, the third integral in (26.15) also belongs to $C^{(1)}_{p,n-2}(\bar{D})$. $\square$

5°. We present without proof some more results on forms orthogonal to holomorphic forms.

THEOREM 26.8 (Dautov). *If $D = \Omega \setminus \bigcup_1^k \Omega_j$, where Ω and the Ω_j are strictly pseudoconvex domains in $\mathbf{C}^n$ such that $\Omega_j \Subset \Omega$ and $\bar{\Omega}_j \cap \bar{\Omega}_l = \varnothing$ for $j \neq l$, and if $\alpha \in C^{(\tau)}_{(p,n-1)}(\partial D)$, $\frac{1}{2} < \tau \leq 1$, then α satisfies (26.1) if and only if there exists a $\bar{\partial}$- closed form*

$$\beta = \beta_1 + \bar{\partial}\rho \wedge \sqrt{|\rho|} \cdot \beta_2, \qquad \beta_1 \in \tilde{C}^{(0)}_{p,n-1}(D), \qquad \beta_2 \in \tilde{C}^{(-1)}_{p,n-2}(D),$$

such that $\beta_1|_{\partial D} = \alpha$ and t^τ is a modulus of continuity of the form β_2.

As in the remark after Theorem 26.1, it can be shown that β extends to a continuous form on $\bar{D}$.

THEOREM 26.9 (Khenkin). *If $D = \Omega \setminus \bigcup_1^n \Omega_j$, where Ω and the Ω_j are pseudoconvex domains in $\mathbf{C}^n$ with boundaries of class $C^{(\infty)}$ such that $\Omega_j \Subset \Omega$ and $\bar{\Omega}_j \cap \bar{\Omega}_l = \varnothing$ for $j \neq l$, and if $\alpha \in C^{(\infty)}_{p,n-1}(\partial D)$, then α satisfies (26.1) if and only if there exists a $\bar{\partial}$-closed form $\beta \in C^{(\infty)}_{p,n-1}(\bar{D})$ such that $\beta|_{\partial D} = \alpha$.*

The "exotic" shape of the domain in which this description is given is to a certain extent (and almost completely when $n = 2$) justified by the essence of the matter. The next result has to do with this.

THEOREM 26.10 (Dautov). *Let D be a bounded domain in $\mathbf{C}^2$ with boundary of class $C^{(2)}$. If for any form $\alpha \in C^{(\infty)}_{p,n-1}(\partial D)$ satisfying (26.1) there exists a $\bar{\partial}$-closed form $\beta \in C_{p,n-1}(\bar{D})$ such that $\beta|_{\partial D} = \alpha$, then $D = \Omega \setminus \bigcup_1^k \Omega_j$, where Ω and the Ω_j are pseudoconvex domains, $\Omega_j \Subset \Omega$, and $\bar{\Omega}_j \cap \bar{\Omega}_l = \varnothing$ for $j \neq l$. And if $\beta = \bar{\partial}\gamma$, where $\gamma \in C_{p,n-2}(\bar{D})$, then D is pseudoconvex.*

In the general case we have

THEOREM 26.11 (Aĭzenberg). *If D is a bounded domain in $\mathbf{C}^n$ with boundary of class $C^{(2)}$, the compact set $\bar{D}$ has a one-sheeted envelope of holomorphy*(²) *$\mathcal{H}(\bar{D})$, and $\alpha \in C^{(\tau)}_{p,n-1}(\partial D)$, $0 < \tau \leq 1$, then α satisfies (26.1) if and only if there exist $\bar{\partial}$-closed forms $\alpha_1 \in \tilde{C}^{(0)}_{p,n-1}(\bar{D})$ and $\alpha_2 \in \tilde{C}^{(0)}_{(p,n-1)}(\mathbf{C}^n \setminus D)$ such that $\alpha = \alpha_1|_{\partial D} + \alpha_2|_{\partial D}$, where $\alpha_2 = O(|z|^{1-2n})$ as $z \to \infty$ and α_2 is $\bar{\partial}$-exact in $\mathbf{C}^n \setminus \mathcal{H}(\bar{D})$.*

6°. Forms orthogonal ot holomorphic functions are closely connected with logarithmic residue formulas and, in particular, with integral representations.

Let D be a bounded domain in $\mathbf{C}^n$ with smooth boundary, and suppose that the mapping $f: \bar{D} \to \mathbf{C}^n$ belongs to $A^n(\bar{D})$ and does not have zeros on ∂D. Then (see Theorem 2.4) f has finitely many zeros in D. Recall that E_f is the set of zeros of f, and $\mu_a(f)$ is the multiplicity of the zero $a \in E_f$. A *logarithmic residue formula* is understood to be a formula of the form

$$\int_{\partial D_\zeta} \varphi(\zeta)\kappa(\zeta) = \sum_{a \in E_f} \mu_a(f)\varphi(a), \tag{26.16}$$

valid for all $\varphi \in A_C(D)$. Here $\kappa \in C_{2n-1}(\partial D)$. An example of such a formula is (3.5). If f has a single zero in D of multiplicity 1, then the logarithmic residue formula becomes an *integral representation for holomophic functions*. Note that a form κ satisfying (26.16) is a solution of a nonhomogeneous system of equations of the form (26.16), while forms orthogonal to holomorphic functions are solutions of the corresponding homogeneous system of equations. Therefore, any form κ satisfying (26.16) can be obtained from a fixed form by adding a form orthogonal to the holomorphic functions, and, conversely, any form orthogonal to the holomorphic functions can be obtained as the difference between a fixed form κ_0 and some suitable form κ if κ_0 and κ satisfy (26.16).

(²) See §24 for the definition of the envelope of holomorphy of a compact set.

As an example of a form κ such that (26.16) holds in the case $n = 1$ is

$$\frac{1}{2\pi i} d \ln f = \frac{1}{2\pi i} \frac{df}{f},$$

where $f \in A(\bar{D})$. From the classical result mentioned in 1° it is possible to prove that any form κ such that (26.16) holds differs from $(2\pi i)^{-1} d(\ln f)$ by a term of the kind $g\,dz$, with $g \in A_C(D)$.

It is assumed that $n > 1$ in what follows. We determine conditions under which any form satisfying (26.16) looks like $\Omega(w^{(0)}, w^{(1)}, \dots, w^{(n-1)}, f)$ (see (3.3)) for suitable $w^{(0)}, w^{(1)}, \dots, w^{(n-1)}$.

The product $df_1 \wedge \cdots \wedge df_n$ is a factor in the form $\Omega(w^{(0)}, \dots, w^{(n-1)}, f)$. Therefore, if the Jacobian $\partial f/\partial z$ of f is degenerate at a point $z^0 \in \partial D$, then all forms of the kind (3.3) are degenerate at this point. By adding a $\bar{\partial}$-exact term of type $(n, n-1)$ to the form $\Omega(w^{(0)}, w^{(1)}, \dots, w^{(n-1)}, f)$, we can always find a κ satisfying (1) and nondegenerate at z^0. Therefore, it is natural to require that $\partial f/\partial z \neq 0$ on ∂D.

Any two forms of the kind (3.3) for one and the same mapping f differ by a $\bar{\partial}$-exact term (see Lemma 3.4). This implies that if any form satisfying (26.16) is of the kind (3.3), then any form orthogonal to the holomorphic functions when integrated over the boundary of the domain extends to D as a $\bar{\partial}$-exact form. For $n = 2$, Theorem 26.8 gives us that D is a pseudoconvex domain. We require somewhat more, namely, that D is strictly pseudoconvex.

THEOREM 26.12 (Dautov). *Suppose that D is a strictly pseudoconvex domain in $\mathbf{C}^n$ $(n \geq 2)$, and $f \in A^n(\bar{D})$ is such that $\partial f/\partial z \neq 0$ on ∂D. Then any form κ of degree $2n-1$ on ∂D that satisfies a Hölder condition with exponent $\tau > \frac{1}{2}$ and the condition* (26.16) *is of the kind $\Omega(w^{(0)}, w^{(1)}, \dots, w^{(n-1)}, f)$ for suitable vector-valued functions $w^{(j)} \in C^{(1)}(\partial D)$, $j = 0, 1, \dots, n-1$, in the following cases*:

1) $n = 2$;

2) $n \geq 3$, *and the complex tangents to the surface*(³) $f(\partial D)$ *do not go through the origin.*

Condition 2) can be relaxed, namely, it can be required that there be a vector-valued function $u^{(0)}$ that is smooth on ∂D and such that $\langle u^0, f \rangle = 1$ and any smooth form of type $(n, n-2)$ on ∂D looks like

$$\sum_{k<l} a_{kl}\, du^{(0)}_{[k,l]} \wedge df$$

for some smooth functions a_{kl}. Condition 2) or this relaxed condition most likely come from the method of proof.

We actually prove that κ has the form

$$\Omega\Big(\bar{f}, w, \underbrace{\bar{f}, \dots, \bar{f}}_{n-2}, f\Big)$$

for suitable w.

PROOF. Let $u^{(0)} = \bar{f}/|f|^2$. It is clear that $\langle u^{(0)}, f \rangle = 1$; therefore, the form $\Omega(u^{(0)}, \dots, u^{(0)}, f)$ satisfies (1). Consequently, if $\alpha = \kappa - \Omega(u^{(0)}, \dots, u^{(0)}, f)$, then (26.1) holds for α. By Theorem 26.7, $\alpha = \bar{\partial}\gamma|_{\partial D}$, where $\gamma \in C^{(1)}_{n,n-2}(\bar{D})$. We show that

$$\gamma|_{\partial D} = \frac{(n-2)!}{(2\pi i)^n} \sum_{k<l} a_{kl}\, du^{(0)}_{[k,l]} \wedge df, \tag{26.17}$$

where $a_{kl} \in C^{(1)}(\partial D)$.

(³) $f(\partial D)$ is a smooth surface (possibly with selfintersections), because ∂D is smooth and $\partial f/\partial z \neq 0$ on ∂D.

For $n = 2$, (26.17) is trivial. If $n \geqslant 3$, it suffices to prove (26.17) locally; the global expansion of γ can be obtained from the local expansions by using a partition of unity. Suppose that ζ^0 is a fixed point of ∂D, and let $z^0 = f(\zeta^0)$. Consider the surface $S = f(U \cap \partial D)$, where U is a neighborhood of ζ^0. Then S is a smooth surface given by the equation $\rho_1(z) = 0$, where $\rho_1(z) = \rho(f^{-1}(z))$, and ρ is a function defining the domain D. It can be assumed that $(\partial\rho_1/\partial z_1)(z) \neq 0$ and $z_1 \neq 0$ for $z \in S$. This situation can always be attained by rotating the coordinate system and making U smaller, because $z^0 = f(\zeta^0) \neq 0$ and $\operatorname{grad} \rho_1|_S \neq 0$. A direct computation yields

$$\overline{\left(z_1\frac{\partial\rho_1}{\partial z_1} + \cdots + z_n\frac{\partial\rho_1}{\partial z_n}\right)}\frac{d\bar z[1]\wedge dz}{|z|^{2n}(\partial\rho_1/\partial\bar z_1)}\bigg|_S = \frac{(2\pi i)^n}{(n-1)!}\Omega(\bar z,\ldots,\bar z, z)\bigg|_S = \frac{1}{z_1}d\left(\frac{\bar z}{|z|^2}\right)_{[1]}\wedge dz\bigg|_S.$$

The first equality shows that $\Omega(\bar z,\ldots,\bar z, z)|_S$ is not degenerate, since the complex tangents to $S = f(U \cap \partial D)$ do not go through 0 by hypothesis. The second equality means that the differentials

$$d\left(\frac{\bar z_2}{|z|^2}\right),\ldots,d\left(\frac{\bar z_n}{|z|^2}\right), dz_1,\ldots,dz_n$$

are linearly independent on S; consequently, the differentials

$$du_2^{(0)} = d\frac{\bar f_2}{|f|^2},\ldots,du_n^{(0)} = d\frac{\bar f_n}{|f|^2}, df_1,\ldots,df_n$$

are linearly independent on $\partial D \cap U$. This gives us (26.17) on $\partial D \cap U$.

Let $\eta_{kl} = f_l e_k - f_k e_l$, where e_k and e_l are the vectors with kth and lth coordinates, respectively, equal to 1 and other coordinates equal to 0, and let

$$w = u^{(0)} + \sum_{k<l} a_{kl}\eta_{kl}.$$

We have

$$\langle w, f\rangle = \langle u^{(0)}, f\rangle + \sum_{k<l} a_{kl}\langle \eta_{kl}, f\rangle = 1$$

and

$$\Omega\Big(u^{(0)}, w, \underbrace{u^{(0)},\ldots,u^{(0)}}_{n-2}, f\Big) - \Omega\Big(\underbrace{u^{(0)},\ldots,u^{(0)}}_{n-2}, f\Big)$$
$$= \frac{(-1)^{n(n-1)/2}}{(2\pi i)^n}\langle u^{(0)}, df\rangle \wedge d\langle w - u^{(0)}, df\rangle \wedge \left(d\langle u^{(0)}, df\rangle\right)^{n-2}$$
$$= \frac{(-1)^{n(n-1)/2}}{(2\pi i)^n}\sum_{k<l}\langle u^{(0)}, df\rangle \wedge \bar\partial a_{kl}\langle \eta_{kl}, df\rangle \wedge \left(\langle \bar\partial u^{(0)}, df\rangle\right)^{n-2}. \tag{26.18}$$

Consider the expression

$$\langle \theta^1, df\rangle \wedge \langle \theta^2, df\rangle \wedge \cdots \wedge \langle \theta^n, df\rangle, \tag{26.19}$$

where θ^1 is a vector-valued function such that $\langle \theta^1, \varphi\rangle = 1$, and $\theta^2,\ldots,\theta^n$ are vector-valued functions or forms such that $\langle \theta^2, f\rangle = \cdots = \langle \theta^n, f\rangle = 0$. Just as in Lemma 3.3 we can see that (26.19) does not change if θ^1 is replaced by any vector-valued function θ with $\langle \theta, f\rangle = 1$. Clearly, $\langle \eta_{kl}, f\rangle = 0$ and $\langle \bar\partial u^{(0)}, f\rangle = \bar\partial\langle u^{(0)}, f\rangle = \bar\partial 1 = 0$. Therefore, $\langle u^{(0)}, df\rangle$ can be replaced by

$\langle f_k^{-1}e_k, df\rangle = f_k^{-1}df_k$ in each term of (26.18). If $f_1\cdots f_n \neq 0$, then

$$\Omega\Big(u^{(0)}, w, \underbrace{u^{(0)},\ldots,u^{(0)}}_{n-2}, f\Big) - \Omega\Big(\underbrace{u^{(0)},\ldots,u^{(0)}}_{n}, f\Big)$$

$$= \frac{(-1)^{n(n-1)/2}}{(2\pi i)^n}\sum_{k<l}\frac{1}{f_k}df_k\wedge\bar\partial a_{kl}\wedge(f_l df_k - f_k df_l)\wedge(\langle du^{(0)}, df\rangle)^{n-2}$$

$$= \frac{(-1)^{n(n-1)/2}}{(2\pi i)^n}\sum_{k<l}\bar\partial a_{kl}\wedge df_k\wedge df_l\wedge(n-2)!\bigwedge_{\substack{m=1\\ m\neq k,l}}^{n} du_m^{(0)}\wedge df_m$$

$$= \frac{(n-2)!}{(2\pi i)^n}\sum_{k<l}\bar\partial a_{kl}\wedge du^{(0)}_{[k,l]}\wedge df = \bar\partial\gamma|_{\partial D} = \alpha.$$

Consequently, if $f_1\cdots f_n \neq 0$, then

$$\kappa = \alpha + \Omega\Big(\underbrace{u^{(0)},\ldots,u^{(0)}}_{n}, f\Big) = \Omega\Big(u^{(0)}, w, \underbrace{u^{(0)},\ldots,u^{(0)}}_{n-2}, f\Big). \tag{26.20}$$

The set $\partial D\cap\{f_1,\ldots,f_n\neq 0\}$ is dense in ∂D; therefore, (26.20) holds on the whole boundary. □

COROLLARY 26.13. *Suppose that D is a strictly pseudoconvex domain in $\mathbf{C}^2$ or a strictly pseudoconvex and linearly convex*([4]) *domain in $\mathbf{C}^n$, $n\geqslant 3$, and $\kappa_a(z)$ is a form in $z\in\partial D$ of degree $2n-1$ that satisfies a Hölder condition with exponent $\tau>\frac12$ and is such that for all $a\in D$ and all $\varphi\in A_C(D)$*

$$\varphi(a) = \int_{\partial D}\varphi(z)\kappa_a(z).$$

Then there exists a vector-valued function $w_a(z)$, smooth with respect to $z\in\partial D$, such that $\langle w_a(z), z-a\rangle\neq 0$ on ∂D and

$$\kappa_a(z) = \Omega(\bar z-\bar a, w_a(z), \bar z-\bar a,\ldots,\bar z-\bar a, z-a).$$

The corollary asserts that for the indicated class of domains with $n>1$ any integral representation for holomorphic functions with kernel satisfying a Hölder condition with exponent $\tau>\frac12$ is a generalized Cauchy-Fantappiè formula (formula (3.12)).

7°. We need some notation in order to give a local variant of Theorem 26.7. If D is a bounded domain in $\mathbf{C}^n$ with smooth boundary and $\alpha\in C^{(1)}_{n,q}(\partial D)$, $0\leqslant q\leqslant n-1$, then α can be extended to a form $\tilde\alpha\in C^{(1)}_{n,q}(\bar D)$. The form in $C_{n,q+1}(\partial D)$ obtained by restricting $\bar\partial\alpha$ to ∂D is denoted by $\bar\partial_b\alpha$. It is not hard to show that $\bar\partial_b\alpha$ does not depend on the extension $\tilde\alpha$ of α.

For $p=n$, Theorem 26.7 asserts that if α is a form of type $(n, n-1)$ on ∂D satisfying a Hölder condition with exponent $\tau>\frac12$ and condition (26.1), then there is a corresponding form $\gamma\in C^{(1)}_{n,n-2}(\partial D)$, that solves the equation

$$\bar\partial_b\gamma = \alpha. \tag{26.21}$$

([4]) See §24 for the definition of a linearly convex domain.

For $n = 2$, (26.21) is equivalent to the equation

$$Lg = \left(\frac{\partial \rho}{\partial \bar{z}_1} \frac{\partial}{\partial \bar{z}_2} - \frac{\partial \rho}{\partial \bar{z}_2} \frac{\partial}{\partial \bar{z}_1} \right) g = f, \qquad z \in \partial D, \tag{26.22}$$

where g is the coefficient of the form γ, while f is constructed from the form α, and ρ is a function defining D. The operator([5]) L was introduced by Lewy. He found that if D is a strictly convex domain in $\mathbf{C}^2$, then, generally speaking, the equation $Lg = f$ is not even locally solvable. More precisely, the set of functions such that $Lg = f$ is solvable is not a closed subspace of any natural function space.

We give a condition on the form α (and, consequently, also on f) that is necessary and sufficient for (26.21) (respectively, (26.22)) to have local solutions. Let D be a bounded strictly convex domain in $\mathbf{C}^n$ with $0 \in D$, and $\tilde{\tilde{D}} = \{w : \langle w, z \rangle \neq 1, z \in D\}$ the domain conjugate to the compact set $\bar{D}$ (see §24, 3°). Since D has a smooth boundary, to each point $z \in \partial D$ there corresponds a unique point $w(z) \in \partial \tilde{\tilde{D}}$ such that $\langle w(z), z \rangle = 1$. The strict convexity of D gives us that the mapping $z \mapsto w(z)$ is one-to-one on ∂D. The point $w(z)$ is determined by the equality

$$w(z) = \frac{P(z)}{\langle P(z), z \rangle},$$

where

$$P(z) = (P_1(z), \ldots, P_n(z)) = \left(\frac{\partial \rho}{\partial z_1}(z), \ldots, \frac{\partial \rho}{\partial z_n}(z) \right)$$

(see §25).

To a form $\alpha \in C_{n,n-1}(\partial D)$, as a functional on the space $A(\bar{D})$, there corresponds a holomorphic function in $\tilde{\tilde{D}}$

$$\alpha^*(w) = \int_{\partial D_\zeta} \frac{\alpha(\zeta)}{(1 - \langle \zeta, w \rangle)^n},$$

called the *Fantappiè indicatrix* of the form α (compare with §24, 4°).

THEOREM 26.14 (Khenkin). *Suppose that D is a strictly convex domain in $\mathbf{C}^n$ and $\alpha \in C_{n,n-1}(D)$ satisfies a Hölder condition with exponent $\tau > \frac{1}{2}$. Then* (26.21) *has a solution $\gamma \in C^{(1)}_{n,n-2}(\partial D)$ in some neighborhood of a point $z^0 \in \partial D$ if and only if the Fantappiè indicatrix α^* of the form α extends holomorphically to a neighborhood of the point $w(z^0) \in \partial \tilde{\tilde{D}}$.*

([5]) Note that the condition $Lg = 0$ means precisely that g satisfies the tangential Cauchy-Riemann conditions (see §24).

To prove this we need an assertion that is also of independent significance.

PROPOSITION 26.15. *Suppose that D is a bounded domain in $\mathbf{C}^n$ with smooth boundary, and $\alpha \in C_{n,n-1}(\partial D)$. If there exists a form $\tilde{\alpha} \in C_{n,n-1}(\overline{D})$ such that $\tilde{\alpha}|_{\partial D} = \alpha$ and $\bar{\partial}\tilde{\alpha} \in L^1_{n,n}(D)$, then the forms α^+ and α^- defined in D and $\mathbf{C}^n \setminus \overline{D}$, respectively, by*

$$\alpha^{\pm}(z) = \int_{\partial D_\zeta} \alpha(\zeta) \wedge U_{n,n-1}(\zeta, z) - \bar{\partial} \int_{D_\zeta} \tilde{\alpha}(\zeta) \wedge U_{n,n-2}(\zeta, z) \quad (26.23)$$

are $\bar{\partial}$-closed, extend continuously to $\overline{D}$, and satisfy the equation

$$\alpha = \alpha^+|_{\partial D} - \alpha^-|_{\partial D}.$$

PROOF. By the Martinelli-Bochner-Koppelman formula (7.1),

$$\alpha^+(z) = \tilde{\alpha}(z) + \int_{D_\zeta} \bar{\partial}\tilde{\alpha}(\zeta) \wedge U_{n,n-1}(\zeta, z), \qquad z \in D,$$

and

$$\alpha^-(z) = \int_{D_\zeta} \bar{\partial}\tilde{\alpha}(\zeta) \wedge U_{n,n-1}(\zeta, z), \qquad z \in \mathbf{C}^n \setminus D. \quad (26.24)$$

It is not hard to see that the integral

$$\int_{D_\zeta} \bar{\partial}\tilde{\alpha}(\zeta) \wedge U_{n,n-1}(\zeta, z)$$

is uniformly convergent for $z \in \mathbf{C}^n$ and thus defines a continuous form on $\mathbf{C}^n$. Consequently, α^+ and α^- extend continuously to ∂D, and

$$\alpha^+|_{\partial D} - \alpha^-|_{\partial D} = \tilde{\alpha}|_{\partial D} = \alpha.$$

PROOF OF THEOREM 26.14. *Necessity*. Suppose that (26.21) is solvable in some neighborhood U_{z^0} of z^0. By Stokes' formula,

$$\begin{aligned}\alpha^*(w) &= \int_{(\partial D \setminus U_{z^0})_\zeta} \frac{\alpha(\zeta)}{(1 - \langle w, \zeta \rangle)^n} + \int_{(w_{z^0} \cap \partial D)_\zeta} \frac{\bar{\partial}_b \gamma(\zeta)}{(1 - \langle w, \zeta \rangle)^n} \\ &= \int_{(\partial D \setminus U_{z^0})_\zeta} \frac{\alpha(\zeta)}{(1 + \langle w, \zeta \rangle)^n} + \int_{\partial(U_{z^0} \cap \partial D)_\zeta} \frac{\gamma(\zeta)}{(1 - \langle w, \zeta \rangle)^n}.\end{aligned}$$

Since D is strictly convex, the integrals on the right-hand side are defined not only in $\tilde{D}$ but also in some neighborhood of the point $w(z^0)$; therefore, α^* extends holomorphically through this point.

Sufficiency. By Lemma 26.3, there is a form $\tilde{\alpha} \in \tilde{C}^{(0)}_{n,n-1}(D)$ such that $\tilde{\alpha}|_{\partial D} = \alpha$, and $\bar{\partial}\tilde{\alpha}(\zeta) = \rho^{-1}(\zeta)\kappa(\zeta)$, where $\kappa \in \tilde{C}^{(0)}_{n,n}(D)$, $\kappa(\zeta) = 0$ for $\zeta \in \partial D$, and $\theta(t) = t^\tau$ is a modulus of continuity of the forms $\tilde{\alpha}$ and κ. As in the proof of Lemma 26.6, we get that $|\bar\partial\tilde\alpha(\zeta)| = O(1)[\rho(\zeta)]^{\tau-1}$. By Proposition 26.15, $\alpha = \alpha^+|_{\partial D} - \alpha^-|_{\partial D}$, where α^+ and α^- are given by (26.23).

Proposition 0.10 tells us that the first integral in (26.23) satisfies a Hölder condition in D with exponent τ', $\frac{1}{2} < \tau' < \tau$. Moreover, $U_{n,n-1}(\zeta, z)$ is a $\bar\partial$-closed form with respect to z; consequently, the first integral in (26.23) is a $\bar\partial$-closed form in D. Just as in Theorem 26.7 we can verify that this integral can be represented as $\bar\partial\gamma_1$ for some $\gamma_1 \in C^{(1)}_{n,n-1}(\bar D)$. The second term in (26.23) can be expressed as $\bar\partial\gamma_2$, where

$$\gamma_2 = \int_{D_\zeta} \tilde\alpha(\zeta) \wedge U_{n,n-2}(\zeta, z).$$

From Proposition 0.10 it follows that $\gamma_2 \in C^{(1)}_{n,n-2}(\bar D)$. Thus, $\alpha^+ = \bar\partial\gamma^+$ in D. Here $\gamma^+ = \gamma_1 + \gamma_2 \in C^{(1)}_{(n,n-2)}(\bar D)$.

It remains to see that the equation $\bar\partial_b\gamma^- = \alpha^-|_{\partial D}$ is solvable. Let Ω be a domain containing $\bar D$ for which (25.2) holds. Then

$$u^*(\zeta, z) = \frac{P(z)}{\Phi^*(\zeta, z)} = \frac{P(z)}{\langle P(z), \zeta - z\rangle}$$

is holomorphic in $\zeta \in \bar D$ for $z \in \Omega\setminus\bar D$, so (26.24) and Stokes' formula imply (compare with the proof of Lemma 26.5) that for $z \in \Omega\setminus\bar D$

$$\begin{aligned}\alpha^-(z) &= \int_{D_\zeta} \bar\partial\tilde\alpha(\zeta) \wedge W_{n,n-1}(u^*, \zeta, z)\\ &\quad - \bar\partial_z\int_{D_\zeta\times[0,1]_\lambda} \bar\partial\tilde\alpha(\zeta) \wedge \tilde W_{n,n-2}(u^*, \zeta, z, \lambda)\\ &= \int_{\partial D_\zeta} \alpha(\zeta) \wedge W_{n,n-1}(u^*, \zeta, z)\\ &\quad - \bar\partial_z\int_{D_\zeta\times[0,1]_\lambda} \bar\partial\tilde\alpha(\zeta) \wedge \tilde W_{n,n-2}(u^*, \zeta, z, \lambda). \qquad (26.25)\end{aligned}$$

We have

$$\begin{aligned}W_{n,n-1}(u^*, \zeta, z) &= \frac{\sum_{j=1}^n (-1)^j P_j(z)\, dP(z)_{[j]} \wedge dz}{\langle P(z), \zeta - z\rangle^n}\\ &= \frac{\langle P(z), z\rangle^n}{\langle P(z), \zeta\rangle - \langle P(z), z\rangle^n} \frac{\sum_{j=1}^n (-1)^{j-1} P_j(z)\, dP(z)_{[j]} \wedge dz}{\langle P(z), \zeta - z\rangle^n}\\ &= \frac{1}{(\langle w(z), \zeta\rangle - 1)^n}\,\omega(z, w(z)).\end{aligned}$$

From this and (26.25) it follows that for $z \in \Omega \setminus \bar{D}$

$$\alpha^-(z) = (-1)^n \alpha^*(w(z))\omega(z, w(z)) - \bar{\partial}_z \int_{D_\zeta \times [0,1]_\lambda} \bar{\partial}\tilde{\alpha}(\zeta) \wedge \tilde{W}_{n,n-2}(u^*, \zeta, z, \lambda). \tag{26.26}$$

By assumption, α^* extends holomorphically to some neighborhood of $w(z^0)$. Consequently, the form $\alpha^*(w(z))\omega(z, w(z))$ is defined in a (convex) neighborhood of z^0. Next,

$$\alpha^*(w) \sum_{j=1}^{n} (-1)^{j-1} w_j dw_{[j]} \wedge dz$$

is a holomorphic form of maximal dimension on the manifold $\{\langle w, z\rangle = 1\} \subset \mathbf{C}^{2n}$; therefore, it is closed, and hence the form $\alpha^*(w(z))\omega(z, w(z))$ is $\bar{\partial}$-closed. Moreover, it is smooth in a neighborhood of z^0; also, as in Theorem 26.7, it can be proved that $\alpha^*(w(z))\omega(z, w(z)) = \bar{\partial}\gamma_3$ in a neighborhood of z^0, where $\gamma_3 \in C^{(1)}_{(n,n-2)}(\bar{D})$.

Let γ_4 be the integral in the second term of (26.26). Just as in the proof of Theorem 26.7 we get that γ_4 is a linear combination of forms with coefficients of the kind

$$F_1(z) = \int_{D_\zeta} \frac{(\bar{\zeta}_j - \bar{z}_j)\varphi(z)}{|\zeta - z|^{2(n-1)}\Phi^*(\zeta, z)} \frac{\kappa(\zeta)}{\rho(\zeta)}, \qquad z \in \Omega \setminus \bar{D}.$$

Differentiating under the integral sign, we next prove with the help of (25.2) and the change of variables in Lemmas 10.6 and 26.7 that the second derivatives of F_1 do not grow more rapidly than $|\rho(z)|^{-1+\tau-1/2}$. From Lemma 25.8 it follows that the first derivatives of F_1 are continuous in $\Omega \setminus D$, i.e., the form γ_4 extennds smoothly to ∂D. From (26.26) we get that $\alpha^-|_{\partial D} = \bar{\partial}\gamma -$ in some neighborhood of z^0. Here $\gamma^- = (\gamma_3 - \gamma_4)|_{\partial D}$. $\square$

THEOREM 26.16. (Khenkin). *Suppose that D is a strictly convex domain and $f \in A_C(\bar{\bar{D}})$, and that f does not extend holomorphically through the point $w^0 \in \partial\bar{\bar{D}}$. Then the equation $\bar{\partial}_b\gamma = \alpha$, where*

$$\alpha = \frac{(n-1)!}{(2\pi i)^n} f(w(z)) \wedge \omega(z, w(z)),$$

does not have solutions in any neighborhood of the point $z^0 \in \partial D$ for which $w^0 = w(z^0)$.

PROOF. Theorem 26.14 shows that it suffices to prove that f is the Fantappiè indicatrix of the form α, i.e.,

$$f(w)=\int_{\partial D_\zeta}\frac{\alpha(\zeta)}{(1-\langle w,\zeta\rangle)^n}=\frac{(n-1)!}{(2\pi i)^n}\int_{\partial D_\zeta}\frac{f(w(\zeta))\omega(\zeta,w(\zeta))}{(1-\langle w,\zeta\rangle)^n}.$$

This is precisely formula (8.5) for the linearly convex domain $\tilde{\tilde{D}}$. □

§27. A basis in the space of holomorphic functions with fixed algebraic singularities

1°. The set $A(D)$ of functions holomorphic in a domain $D\subset \mathbf{C}^n$ forms a linear vector space. The notation $A(D)$ will also be used for this space with the topology of uniform convergence on compact subsets of D. A sequence $\{f_k\}_1^\infty\subset A(D)$ converges in this topology to a function $f\in A(D)$ if $f_n(z)\to f(z)$ uniformly in $z\in K$ for any compact set $K\subset D$. A sequence $\{\varphi_k\}_1^\infty\subset A(D)$ forms a basis for the space $A(D)$ if any $f\in A(D)$ can be represented in a unique way as $\sum_1^\infty c_k\varphi_k$, where $\{c_k\}$ is a sequence of complex numbers, and the series converges in the topology of $A(D)$.

EXAMPLE 1. A basis for the space $A(\mathbf{C}^1\setminus\{a_1,\dots,a_m\})$ is formed by the monomials $(z-a_j)^{-k}$, $j=1,\dots,m$, $k=1,2,\dots,$ and z^n, $n=0,1,\dots$. The expansion coefficients of an arbitrary function $f\in A(\mathbf{C}^1\setminus\{a_1,\dots,a_m\})$

$$f(z)=\sum_{j=1}^{m}\sum_{k=1}^{\infty}\frac{c_{jk}}{(z-a_j)^k}+\sum_{n=0}^{\infty}c_nz^n, \tag{27.1}$$

in the basis are uniquely determined by the formulas

$$c_{jk}=\frac{1}{2\pi i}\int_{|\zeta-a_j|=\varepsilon}f(\zeta)\zeta^{k-1}d\zeta,\qquad c_n=\frac{1}{2\pi i}\int_{|\zeta|=R}\frac{f(\zeta)\,d\zeta}{\zeta^{n+1}},$$

where $\varepsilon>0$ is sufficiently small, and $R>\max\{|a_1|,\dots,|a_m|\}$.

For holomorphic functions of several complex variables the zero sets of holomorphic functions play the role of isolated singular points. We consider a space $A(\mathbf{C}^n\setminus T)$ of holomorphic functions with algebraic singularities, where $T=\{z:\mathcal{P}(z)=0\}$, $\mathcal{P}$ being a polynomial.

2°. Let us first return to the Weil expansion (24.5). If $\chi_j(z)$, $j=1,\dots,N$, are polynomials or rational functions, then in (24.5) the function $\Omega_J(z)$, $J=(j_1,\dots,j_n)$, $1\leqslant j_1<\dots<j_n\leqslant N$, can be represented in the form

$$\Omega_J(\zeta,z)=\sum_{k=1}^{m(J)}g_{Jk}(\zeta)h_{Jk}(z),$$

where g_{Jk} and h_{Jk} are polynomials or rational functions, respectively. In this case the Weil expansion (24.5) can be represented in the form

$$f(z) = \sum_J \sum_{k=1}^{m(J)} \sum_{\alpha \geqslant 0} a_{Jk\alpha} h_{Jk}(z) \chi_J^\alpha(z), \tag{27.2}$$

where $\alpha = (\alpha_1, \ldots, \alpha_n)$, $\chi_J^\alpha = \chi_{j_1}^{\alpha_1} \cdots \chi_{j_n}^{\alpha_n}$;

$$a_{Jk\alpha} = \frac{1}{(2\pi i)^n} \int_{\gamma_J} \frac{f(\zeta) g_{Jk}(\zeta)\, d\zeta}{\chi_J^{\alpha+I}(\zeta)}. \tag{27.3}$$

The series (27.2), like the Weil series (24.5), converges uniformly on compact subsets of Δ. In the general case the functions $h_{Jk}(z)\chi_J^\alpha(z)$ do not form a basis for $A(D)$, since functions $f \in A(\Delta)$ are not always uniquely representable as a series (27.2).

EXAMPLE 2. For the Weil polyhedron $\Delta = \{z \in \mathbf{C}^2 : |z_1| < 1, |z_2| < 1, |z_1 + z_2| < 1\}$ every $f \in A(\Delta)$ can be expanded according to (27.2) and (27.3) in a series

$$f(z) = \sum_{m,n=0}^{\infty} a_{mn} z_1^m z_2^n + \sum_{m,n=0}^{\infty} b_{mn} z_1^m (z_1 + z_2)^n + \sum_{m,n=0}^{\infty} c_{mn} z_2^m (z_1 + z_2)^n.$$

However, this expansion is not unique, because the monomials

$$z_1^m z_2^n,\ z_1^m(z_1 + z_2) \quad \text{and} \quad z_2^m (z_1 + z_2)^n,$$

$m + n =$ const, are dependent.

We mention an important particular case where (27.2) can be unique. Consider the special Weil polyhedron $\Delta = \{z \in D : |\chi_j(z)| < R_j, j = 1, \ldots, n\}$, $\chi_j \in A(D)$, $\Delta \Subset D \subset \mathbf{C}^n$. In this case the skeleton of the polyhedron $\Delta_\varepsilon = \{z \in D : |\chi_j(z)| < R_j - \varepsilon_j, j = 1, \ldots, n\}$, consists of the single component $\gamma = \{z \in D : |\chi_j(z)| = R_j - \varepsilon_j,\ j = 1, \ldots, n\}$, and for rational functions χ_j the expansion (27.2) takes the form

$$f(z) = \sum_{|\alpha| \geqslant 0} \sum_{k=1}^{m} a_{k\alpha} h_k(z) \chi^\alpha(z), \tag{27.4}$$

where

$$a_{k\alpha} = \frac{1}{(2\pi i)^n} \int_\gamma \frac{f(\zeta) g_k(\zeta)\, d\zeta}{\chi^{\alpha+I} I(\zeta)}. \tag{27.5}$$

Recall that here

$$\chi_j(\zeta) - \chi_j(z) = \sum_{k=1}^{n} (\zeta_k - z_k) \mathcal{P}_{jk}(\zeta, z), \qquad j = 1, \ldots, n, \tag{27.6}$$

is a Hefer expansion in which the $\mathcal{P}_{jk}$ are polynomials or rational functions, and

$$\Omega(\zeta, z) = \begin{vmatrix} \mathcal{P}_{11}(\zeta, z) & \cdots & \mathcal{P}_{1n}(\zeta, z) \\ \cdot & \cdots & \cdot \\ \mathcal{P}_{n1}(\zeta, z) & \cdots & \mathcal{P}_{nn}(\zeta, z) \end{vmatrix} = \sum_{k=1}^{m} g_k(\zeta)h_k(z). \tag{27.7}$$

PROPOSITION 27.1 (Mikhaĭlova). *Any function f holomorphic in a special Weil polyhedron Δ has a unique expansion in a series* (27.4) *if and only if* $\det\|\lambda_{kj}\| \neq 0$, *where*

$$\lambda_{kj} = \frac{1}{(2\pi i)^n}\int_\gamma \frac{g_k(\zeta)h_j(\zeta)\,d\zeta}{\chi^{\mathrm{I}}(\zeta)}, \qquad k,j = 1,\ldots,m. \tag{27.8}$$

PROOF. *Necessity.* Take $f = h_j$, $j = 1,\ldots,m$. Then, since the expansion (27.4) is unique, the formulas (27.5) with $\alpha = 0 = (0,\ldots,0)$ give us that $\lambda_{kj} = a_{k0} = \delta_{kj}$, $k = 1,\ldots,m$. Consequently, $\det\|\lambda_{kj}\| \neq 0$.

Sufficiency. Suppose that $\det\|\lambda_{kj}\| \neq 0$. We show that if

$$0 \equiv \sum_{k=1}^{m}\sum_{\alpha \geqslant 0} a_{k\alpha}h_k(z)\chi^\alpha(z) \tag{27.9}$$

and the series converges uniformly on compact subsets of Δ, then $a_{k\alpha} = 0$ for all $k = 1,\ldots,m$ and $\alpha \geqslant 0$. Let us use induction on $p = |\alpha|$. Multiply (27.9) by the form $g_j(z)dz/(2\pi i)^n\chi^{\mathrm{I}}(z)$, $j = 1,\ldots,m$, and integrate over the cycle γ. This gives

$$0 = \sum_{k=1}^{m}\sum_{|\alpha| \geqslant 0} a_{k\alpha}\frac{1}{(2\pi i)^n}\int_\gamma h_k(z)g_j(z)\chi^{\alpha-\mathrm{I}}(z)\,dz.$$

By the Cauchy-Poincaré Theorem 0.6, $\int_\gamma h_k g_j \chi^{\alpha-\mathrm{I}}\,dz = 0$ for $|\alpha| > 0$, since in this case $\gamma \sim 0$ in the domain of regularity of the integrand form. Indeed, if $\alpha_q > 1$, then $\gamma = \pm\partial B$ for

$$B = \{z \in D : |\chi_j(z)| = R_j - \varepsilon, j = 1,\ldots [q]\ldots, n, |\chi_q(z)| \leqslant R_q - \varepsilon\},$$

and the form $h_k q_j \chi^{\alpha-\mathrm{I}}\,dz$ is holomorphic on $|B|$. Thus, the notation in (27.8),

$$\sum_{k=1}^{m} a_{k0}\lambda_{jk} = 0, \qquad j = 1,\ldots,m.$$

From the condition $\det\|\lambda_{jk}\| \neq 0$ it follows that $a_{k0} = 0$, $k = 1,\ldots,m$. Assume that $a_{k\alpha} = 0$ for $|\alpha| \leqslant p - 1$ in the expansion (27.9). Let us then show that $a_{k\beta} = 0$, $k = 1,\ldots,m$, for $|\beta| = p$. Multiplying (27.9) by the form

$g_j(z)dz/(2\pi i)^n\chi^{\beta+I}(z), j = 1,\dots,m$, we have

$$0 = \sum_{k=1}^{m} \sum_{|\alpha|\geqslant p} a_{k\alpha} \int_\gamma h_k(z)\chi^{\alpha-\beta-I}(z)\, dz.$$

The equality $\int_\gamma h_k g_j \chi^{\alpha-\beta-I}\, dz = 0$ holds for $\alpha \neq \beta$, $|\alpha| \geqslant p$, because $\alpha_q - \beta_q - 1 \geqslant 0$ for at least one index $q \in \{1,\dots,n\}$. Thus,

$$\sum_{k=1}^{m} a_{k\beta}\lambda_{jk} = 0, \qquad j = 1,\dots,n,$$

and so $a_{k\beta} = 0$, $k = 1,\dots,m$. □

REMARK. The Hefer expansion (27.6) and the representation (27.7) are not unique, so for a given special Weil polyhedron defined by rational functions χ_j ($j = 1,\dots,n$) it is possible to choose in different ways the system $\{h_k(z)\chi^\alpha(z), k = 1,\dots,m, \alpha \geqslant 0\}$ of functions for expanding functions $f \in A(\Delta)$. It may form a basis in $A(\Delta)$ for some choice of the functions $h_k(z)$ and $g_k(z)$, and it may not form a basis for others.

EXAMPLE 3. It is not hard to show that the special analytic polyhedron $\Delta = \{z \in \mathbf{C}^2 : |\chi_j(z)| < 1, j = 1,2\}$, where $\chi_1(z) = z_1^3 + z_2^3$ and $\chi_2(z) = z_1^2 + z_2^2$, is a Weil polyhedron. For the Hefer expansions

$$\chi_1(\zeta) - \chi_1(z) = \sum_{j=1}^{2} (\zeta_j - z_j)(z_j^2 + z_j\zeta_j + \zeta_j^2),$$

$$\chi_2(\zeta) - \chi_1(z) = \sum_{j=1}^{2} (\zeta_j - z_j)(\zeta_j + z_j)$$

the determinant

$$\Omega(\zeta, z) = \begin{vmatrix} z_1^2 + z_1\zeta_1 + \zeta_1^2 & z_2^2 + z_2\zeta_2 + \zeta_2^2 \\ z_1 + \zeta_1 & z_2 + \zeta_2 \end{vmatrix}$$

can be represented in the form

$$\Omega = (z_1^2 z_2 - z_2^2 z_1) + (z_1 z_2 - z_2^2)\zeta_1 + z_2(\zeta_1^2 + \zeta_1\zeta_2)$$
$$+ z_1(\zeta_1\zeta_2 - \zeta_2^2) + (z_1^2 - z_1 z_2)\zeta_2 + (\zeta_1^2\zeta_2 - \zeta_2^2\zeta_1)$$

and in the form

$$\Omega = (z_1^2 z_2 - z_2^2 z_1) + (z_1 z_2 - z_2^2)\zeta_1 + z_2\zeta_1^2 + (z_1^2 - z_1 z_2)\zeta_2$$
$$+ (z_1 - z_2)\zeta_1\zeta_2 + z_1\zeta_2^2 + (\zeta_1^2\zeta_2 - \zeta_1\zeta_2^2).$$

In the first case the expansion satisfies the conditions of Proposition 27.1, and in the second case it does not. Thus, the system

$$\{h_k(z)\chi_1^{\alpha_1}(z)\chi_2^{\alpha_2}(z), k = 1,\dots,6, \alpha_1, \alpha_2 \geqslant 0\},$$

where $h_1 = z_1^2 z_2 - z_1 z_2^2$, $h_2 = z_1 z_2 - z_2^2$, $h_3 = z_2$, $h_4 = z_1$, $h_5 = z_1^2 - z_1 z_2$, and $h_6 = 1$, forms a basis for $A(\Delta)$.

3°. We construct a basis for the space $A(\mathbf{C}^n/T)$, where $T = \{z \in \mathbf{C}^n, \mathcal{P}(z) = 0\}$, $\mathcal{P}$ being a polynomial. Without loss of generality it can be assumed that $\mathcal{P}$ has the form

$$\mathcal{P}(z) = z_n^m + A_1('z)z_n^{m-1} + \cdots + A_m('z), \tag{27.8}$$

where $'z = (z_1, \ldots, z_{n-1})$. Otherwise, we make a linear change of variables.

For any point $z \in \mathbf{C}^n \setminus T$ there is a number $R > 0$ such that $z \in \Delta_R \subset \mathbf{C}^n \setminus T$, where

$$\begin{aligned} \Delta_R &= \{z \in \mathbf{C}^n : |z_j| < R, j = 1, \ldots, n-1, 1/R < |\mathcal{P}(z)| < R\} \\ &= \{z : |z_j| < R, j = 1, \ldots, n-1, |\mathcal{P}(z)| < R, |1/\mathcal{P}(z)| < R\} \end{aligned} \tag{27.9}$$

It is not hard to show that Δ_R is a Weil polyhedron. The distinguished boundary of Δ_R consists of two components: $\gamma_R = \{z : |z_1| = \cdots = |z_{n-1}| = |\mathcal{P}(z)| = R\}$ and $\gamma_R^- = \{z : |z_1| = \cdots = |z_{n-1}| = |1/\mathcal{P}(z)| = R\}$. We observe that $\gamma_R^- \sim -\gamma_R$ in $\mathbf{C}^n \setminus T$. Let

$$\mathcal{P}(\zeta) - \mathcal{P}(z) = \sum_{j=1}^{n} (\zeta_j - z_j) Q_j(\zeta, z), \tag{27.10}$$

where the Q_j $(j = 1, \ldots, n)$ are polynomials. Then

$$\mathcal{P}^{-1}(\zeta) - \mathcal{P}^{-1}(z) = \sum_{j=1}^{n} (\zeta_j - z_j)\left[-Q_j(\zeta, z)\mathcal{P}^{-1}(\zeta)\mathcal{P}^{-1}(z)\right];$$

$$\Omega_1(\zeta, z) = \begin{vmatrix} 1 & 0 & \cdot & \cdot & \cdot & 0 \\ 0 & 1 & \cdot & \cdot & \cdot & 0 \\ \cdot & \cdot & \cdot & \cdot & \cdot & \cdot \\ 0 & 0 & \cdot & \cdot & 1 & 0 \\ Q_1 & Q_2 & \cdot & \cdot & \cdot & Q_n \end{vmatrix} = Q_n(\zeta, z),$$

$$\Omega_2(\zeta, z) = -Q_n(\zeta, z)\mathcal{P}^{-1}(\zeta)\mathcal{P}^{-1}(z).$$

Let

$$Q_n(\zeta, z) = \sum_{k=1}^{N} g_k(\zeta) h_k(z), \tag{27.11}$$

where the g_k and h_k are polynomials. According to (27.2) and (27.3), any function $f \in A(\mathbf{C}^n \setminus T)$ can be expanded in the Weil polyhedron Δ_R in a series

$$f(z) = \sum_{k=1}^{N} \sum_{\alpha \geqslant 0} a_{k\alpha} h_k(z) 'z'^{\alpha} \mathcal{P}^{\alpha_n}(z) + \sum_{k=1}^{N} \sum_{\alpha \geqslant 0} b_{k\alpha} h_k(z) 'z'^{\alpha} \mathcal{P}^{-\alpha_n - 1}(z),$$

where

$$a_{k\alpha} = \frac{1}{(2\pi i)^n} \int_{\gamma_R} \frac{f(\zeta) g_k(\zeta)\, d\zeta}{'\zeta'^{\alpha+'1} \mathcal{P}^{\alpha_n+1}(\zeta)}, \qquad b_{k\alpha} = \frac{1}{(2\pi i)^n} \int_{\gamma_R^-} \frac{f(\zeta) g_k(\zeta) \mathcal{P}^{\alpha_n}\, d\zeta}{'\zeta'^{\alpha+'1}}.$$

Passing to the uniform notation $a_{k\alpha_1 \cdots \alpha_{n-1}} - \alpha_n = b_{k\alpha}$ and taking into account the fact that $\gamma_R^- \sim -\gamma_R$, we write this expansion in the form

$$f(z) = \sum_{k=1}^{N} \sum_{'\alpha \geqslant '0} \sum_{\alpha_n = -\infty}^{\infty} a_{k\alpha} h_k(z) z_1^{\alpha_1} \cdots z_{n-1}^{\alpha_{n-1}} \mathcal{P}^{\alpha_n}(z), \tag{27.12}$$

where

$$a_{k\alpha} = \frac{1}{(2\pi i)^n} \int_{\gamma_R} \frac{f(\zeta) g_k(\zeta)\, d\zeta}{\zeta_1^{\alpha_1+1} \cdots \zeta_{n-1}^{\alpha_{n-1}+1} \mathcal{P}^{\alpha_n+1}}. \tag{27.13}$$

Since the cycle γ_R varies homotopically in $\mathbf{C}^n \setminus T$ as R changes, $R \to \infty$, it follows that the expansion (27.12) with the coefficients (27.13) is valid for an $f \in A(\mathbf{C}^n \setminus T)$ in the whole domain $\mathbf{C}^n \setminus T$. Moreover, (27.12) converges uniformly on any compact set $K \subset \mathbf{C}^n \setminus T$.

PROPOSITION 27.2. *Every function $f \in A(\mathbf{C}^n \setminus T)$ can be expanded uniquely in a series of the form* (27.12) *if and only if* $\det\|\lambda_{kj}\| \neq 0$, *where*

$$\lambda_{kj} = \frac{1}{(2\pi i)^n} \int_{\gamma_R} \frac{g_k(\zeta) h_j(\zeta)\, d\zeta}{\zeta_1 \cdots \zeta_{n-1} \mathcal{P}(\zeta)}. \tag{27.14}$$

PROOF. The necessity can be proved just as in Proposition 27.1.

We remark that, by Proposition 27.1, the condition $\det\|\lambda_{kj}\| \neq 0$ for the λ_{kj} defined by (27.14) is necessary and sufficient for the expansion of an arbitrary function $f \in A(\Delta_F^*)$ in a series

$$f(z) = \sum_{k=1}^{N} \sum_{\alpha \geqslant 0} a_{k\alpha} h_k(z) 'z'^{\alpha} \mathcal{P}^{\alpha_n}(z) \tag{27.15}$$

to be unique, where $\Delta_R^* = \{z : |z_j| < R,\ j = 1, \ldots, n-1,\ |\mathcal{P}(z)| < R\}$. This remark again implies the necessity in Proposition 27.2.

We prove the sufficiency. Suppose that $\det\|\lambda_{kj}\| \neq 0$ and that

$$0 \equiv \sum_{k=1}^{N} \sum_{'\alpha \geqslant '0} \sum_{\alpha_n = -\infty}^{\infty} a_{k\alpha} h_k(z) 'z'^{\alpha} \mathcal{P}^{\alpha_n}(z),$$

where the series converges uniformly on compact subsets of $\mathbf{C}^n \setminus T$. Let us show that all the coefficients $a_{k\alpha}$ are zero. Multiply the last equality by the form

$$g_j(z)\, dz / (2\pi i)^n 'z'^{\beta+'1} \mathcal{P}^{\beta_n+1}, \qquad j = 1, \ldots, n,$$

and integrate over the cycle γ_R. For $\alpha \neq \beta$

$$\frac{1}{(2\pi i)^n}\int_{\gamma_R}\frac{g_j(z)h_k(z)\,dz}{'z'^{\beta+'\alpha+'1}\mathcal{P}^{\beta_n-\alpha_n+1}}=0. \tag{27.16}$$

Indeed, if $\alpha_q \geqslant \beta_q$ for at least one index $q \in \{1,\dots,n\}$, then (27.16) follows from the Cauchy-Poincaré theorem. But if $\alpha_j \leqslant \beta_j$, $j=1,\dots,n$, then the integral on the left-hand side of (27.16) can be regarded as the coefficient of $h_k{'z}^{'\beta-'\alpha+'1}\mathcal{P}^{\beta_n-\alpha_n+1}$ in the expansion of the function $h_k(z) \in A(\overline{\Delta}^*_R)$ in a series of the form (27.15). This coefficient is equal to zero for $\beta-\alpha \neq 0$, since the expansion is unique. Thus, taking (27.14) into account, we have

$$\sum_{k=1}^{N} a_{k\beta}\lambda_{kj}=0, \qquad j=1,\dots,N.$$

Consequently, $a_{k\beta}=0$, $k=1,\dots,N$, because $\det\|\lambda_{kj}\| \neq 0$. □

THEOREM 27.3. *The functions*

$$z_n^{k-1}z_1^{\alpha_1}\cdots z_{n-1}^{\alpha_{n-1}}\mathcal{P}^{\alpha_n}(z), \tag{27.17}$$
$$k=1,\dots,m, \qquad \alpha_1,\dots,\alpha_{n-1}=0,1,\dots, \qquad \alpha_n=0,\pm1,\pm2,\dots,$$

form a basis in $A(\mathbf{C}^n\setminus T)$, *where* $T=\{z: P(z)=0\}$, $\mathcal{P}$ *being the polynomial* (27.8). *The expansion coefficients of a function* $f \in A(\mathbf{C}^n\setminus T)$ *in the basis* (27.17) *can be expressed by the formulas*

$$a_{k\alpha}=\frac{1}{(2\pi i)^n}\int_{\gamma_R}\frac{f(\zeta)\left[\zeta_n^{m-k}+\sum_{p=k}^{m-1}A_p('\zeta)\zeta_n^{p-k}\right]d\zeta}{\zeta_1^{\alpha_1+1}\cdots\zeta_{n-1}^{\alpha_{n-1}+1}\mathcal{P}^{\alpha_n+1}}, \tag{27.18}$$

where $\gamma_R=\{z:|z_1|=\cdots=|z_{n-1}|=|\mathcal{P}(z)|=R\}$, $R \geqslant 0$.

PROOF. The polynomial $Q_n(\zeta, z)$ in the Hefer expansion (27.10) can be taken equal to

$$Q_n(\zeta,z)=[\mathcal{P}(\zeta)-\mathcal{P}('\zeta,z_n)]/(\zeta_n-z_n)=\sum_{k=1}^{m} z_n^{k-1}\left[\zeta_n^{m-k}+\sum_{p=k}^{m-1}A_p('\zeta)\zeta_n^{p-k}\right].$$

From this it is clear that in (27.11) we can set

$$h_k(z)=z_n^{k-1}, \quad g_k(\zeta)=\zeta_n^{m-k}+\sum_{p=k}^{m-1}A_p('\zeta)\zeta_n^{p-k}, \qquad k=1,\dots,m. \tag{27.19}$$

Let us show that the conditions of Proposition 27.2 hold for this choice of the functions h_k and g_k. Indeed, for any fixed $'\zeta \in \mathbf{C}^{n-1}$

$$\frac{1}{2\pi i}\int_{|\mathcal{P}(\zeta)|=R}\frac{\zeta_n^k d\zeta_n}{\mathcal{P}(\zeta)}=\operatorname*{res}_{z_n=\infty}\left[\frac{z_n^k}{\mathcal{P}(\zeta)}\right]=\begin{cases}0, & k<m-1,\\ 1, & k=m-1.\end{cases}$$

Consequently,

$$\lambda_{kj} = \frac{1}{(2\pi i)^n} \int_{|\zeta_1|=R} \frac{d\zeta_1}{\zeta_1} \cdots \int_{|\zeta_{n-1}|=R} \frac{d\zeta_{n-1}}{\zeta_{n-1}} \int_{|P(\zeta)|=R} \frac{\zeta_n^{k-1} g_j(\zeta)\, d\zeta_n}{P(\zeta)} = \begin{cases} 0, & k<j, \\ 1, & k=j, \end{cases}$$

because $\deg_{z_n}[z_n^{k-1} g_j(z)] = m - j + k - 1$ and $\det\|\lambda_{jk}\| = 1 \neq 0$. □

In the case (27.19), i.e., under the conditions of Theorem 27.3, (27.12) is a generalization of the expansion (27.1) for holomorphic functions of n variables with algebraic singularities.

4°. In the case where the polynomial $\mathcal{P}(z)$ splits into linear factors it is possible to construct a more precise analogue of the expansion (27.1). Suppose that $\mathcal{P}(z) = \prod_{j=1}^N L_j(z)$, where $L_j(z) = \sum_{k=1}^n a_{jk} z_k + a_{j0}$, $j = 1, \dots, N$. Then $T = \bigcup_1^n E_j$, where $E_j = \{z : L_j(z) = 0\}$. Assume that $L_j \neq L_k$ for $k \neq j$, i.e., $\mathcal{P}$ does not contain multiple factors. Let Ω_r, $0 \leqslant r \leqslant p$, where p is the rank of the matrix $\|a_{jk}\|$, $j = 1, \dots, N$, $k = 1, \dots, n$, denote the family of all possible maximal subsets $\omega_r \subset \{1, \dots, N\}$ such that the ranks of the matrices

$$\|a_{jk}\|_{j \in \omega_r, k=1,\dots,n} \quad \text{and} \quad \|a_{jk}\|_{j \in \omega_r, k=0,1,\dots,n}$$

are equal to r. For each ω_r we choose $n - r$ linear functions $L_{k\omega_r}(z)$, $k = r+1, \dots, n$, such that the rank of the matrix of coefficients of the functions L_j, $j \in \omega_r$, and $L_{k\omega_r}$, $k = r+1, \dots, n$, is equal to n. Let $J(\omega_r)$ denote the set of multi-indices $\alpha = (\alpha_1, \dots, \alpha_n) \subset \omega_r$, for which the ordered systems of funcitons $L_{\alpha_1}, \dots, L_{\alpha_r}$ form a canonical family of r-systems of the system $\{L_j, j \in \omega_r\}$ (see §19, 5°). For $r = 0$ the sets $\Omega_0 = \{\omega_0 = \varnothing\}$ and $J(\omega_0) = \varnothing$ are assumed to be singletons. For each multi-index $\alpha \in J(\omega_r)$, $\omega_r \in \Omega_r$, $r = 0, \dots, p$, we construct the cycle

$$\Gamma_\alpha = \{z : |L_{\alpha_1}(z)| = \varepsilon_1, \dots, |L_{\alpha_r}(z)| = \varepsilon_r, |L_{k\omega_r}(z)| = R, k = 1, \dots, n - r\},$$

where R is a sufficiently large number, and ε_r, $\varepsilon_{r-1}/\varepsilon_r, \dots, \varepsilon_1/\varepsilon_2$ are sufficiently small positive numbers. The next result holds under these assumptions.

THEOREM 27.4 (Yuzhakov). *Every function $f \in A(\mathbf{C}^n \setminus \bigcup_1^N E_j)$ can be represented as a series*

$$f(z) = \sum_{r=0}^{p} \sum_{\omega_r \in \Omega_r} \sum_{\alpha \in J(\omega_r)} \sum_{\beta, r} c_{\alpha\beta} L^{\beta}_{\alpha\omega_r}(z),$$

where

$$\sum_{\beta, r} = \sum_{\beta_1, \dots, \beta_r = -\infty}^{\infty} \sum_{\beta_{r+1}, \dots, \beta_n = 0}^{\infty}, \qquad L^{\beta}_{\alpha\omega_r} = L^{\beta_1}_{\alpha_1} \cdots L^{\beta_r}_{\alpha_r} L^{\beta_{r+1}}_{r+1, \omega_r} \cdots L^{\beta_n}_{n, \omega_r}.$$

This series converges uniformly on compact subsets of $\mathbf{C}^n \setminus \bigcup_1^N E_j$, and its coefficients are uniquely determined by the formula

$$c_{\alpha\beta} = \frac{1}{(2\pi i)^n} \int_{\Gamma_\alpha} \frac{f(z)\, dL_{\alpha\omega_r}(z)}{L^{\beta+1}_{\alpha\omega_r}(z)},$$

where $dL_{\alpha\omega_r} = dL_{\alpha_1} \wedge \cdots \wedge dL_{\alpha_r} \wedge dL_{r+1, \omega_r} \wedge \cdots \wedge dL_{n\omega_r}$.

This theorem can be proved by using Theorem 19.2 along with the idea underlying the proof of Theorem 19.8.

Brief Historical Notes

Chapter I

§1. Formula (1.1) was obtained by Martinelli [308] and Bochner [235] independently and by different methods. Theorems 1.2 was stated for the case of a ball by Aronov [33], and in the general case by Aronov and Kytmanov [37]. Theorem 1.6 was given by Kytmanov and Aĭzenberg in [111] (see also [108]). The question of functions representable by the integral (1.1) was also examined in the erroneous articles [151] and [152].

§2. The concept of the index of a surface with respect to a point is due to Poincaré. The dynamic definition of the multiplicity of a zero is encountered in papers by Segre [343], Severi [344], and Martinelli [310]. Formula (2.3) for $n = 2$ was first obtained by Caccioppoli [242]. Here we mainly follow the monographs [123] (see Appendix B) and [201] (§48). However, we do not use the local theory of analytic functions in deriving the properties of the multiplicity of a zero nor in proving the discreteness of a compact set of zeros of a homolmorphic mapping. Theorem 2.5 (the Rouché principle) is a particular case of the Poincaré Bol theory (see [104], p. 14). See also [137] about the multiplicity of a zero of a holomorphic mapping. A generalization of the logarithmic residue to the case of incomplete intersections is considered in [347].

§3. The Cauchy-Fantappiè formula was obtained by Leray in [113] (see also [9], [15], [179], and [289]). The Leray-Koppelman formula (3.12) was mentioned by Roos [338]; however, it was actually contained in Koppelman's earlier article [289]. Formula (3.1) was derived by Yuzhakov [218], [221] and Roos [338], and the formula (3.18) by Roos [338]. Formula (3.5), which generalizes the Leray-Koppelman formula (3.12), is contained in [20]. For the Cauchy-Fantappiè formula (3.13) we can ask a question analogous to the one answered in §1 for formula (1.1): is a function holomorphic if it is representable by formula (3.13) with respect to its values on ∂D? The answer to this question is no (Kytmanov).

§4. Integrals of the form (4.3) were first studied by Didon in [247] for $n = 2$ and $\varphi \equiv 1$, but he did not discover their connection with the multiplicity of a zero. Theorem 4.5 and Proposition 4.11 for $n = 2$ were proved by Caccioppoli [242] ($\varphi \equiv 1$) and Martinelli [310], while for $n > 2$ they were proved by Sorani [349]. The multiplicity of an intersection of analytic surfaces and generalizations of the logarithmic residue for $n = 2$ were considered also by Wirtinger [366] and Bergman [234]. Theorem 4.13, which includes Theorem 4.6 and formula (3.1) as special cases, was proved in [221]. This theorem is a generalization of the integral representation of Sorani [351]. Theorem 4.8 is contained in [216] and [221], and Theorem 4.12 was proved in [194]. The formulas (3.1) and (4.3) were duplicated in [230].

§5. The Andreotti-Norguet formula (5.1) was published in [228] and [321] (see also [312]); it has important applications in the theory of q-pseudoconvex domains. The proof given is due to Aĭzenberg [21]; formulas (5.3), (5.5), the problem in 2°, and cases 1)–4) were taken from [20] and [21]. Proposition 5.4 is due to Bolotov.

§6. The concepts of the Bergman boundary, the Bergman kernel function (for $h \equiv 1$), the Bergman metric are encountered in a number of papers by Bergman (see, for example, [233] about the Bergman boundary). the concept of a ring boundary was introduced by Shilov (see, for example, [58]), and he proved the existence and uniqueness of this boundary for the space of maximal ideals of a normed ring. Theorem 6.1 and Corollary 6.2 are taken from [174] (§15) and [11]. The example of a domain of holomorphy for which the Shilov boundary and the Bergman boundary have different dimensions is due to Aĭzenberg (see [174], §16). Theorem 6.8 was stated by Gleason [260], Theorems 6.6 and 6.7 by Bungart [240], [241], and Theorem 6.9 by Aĭzenberg.

§7. Theorem 7.1 for forms of type $(0, q)$ was obtained by Koppelman [290], while in the case of forms of type (p, q) it was obtained by Aronov and Dautov [35] and by Roos [337]; however, there is an error in the latter. Theorems 7.4 and 7.5 are due to Dautov. The idea of introducing a function $\varphi(\zeta, z)$ with the property that $\varphi(z, z) \equiv 1$ into the integral formulas was used by Aĭzenberg [9], [15]. Formula (7.1) was extended to forms in the real space $\mathbf{R}^n$ by Norguet in the lecture notes [321]. If we represent a $\bar\partial$-closed form by formula (7.1), then the second term in (7.1) disappears. Conversely, suppose that a form α can be represented by the Martinelli-Bochner-Koppelman formula (7.1) without the second term. Is α $\bar\partial$-closed in this case? In other words, is the representability of a form α by formula (7.1) without the second term a characteristic property of $\bar\partial$-closed forms? The following is a particular case of this problem: is a function holomorphic if it is representable by a

Martinelli-Bochner integral (see §1)? Kytmanov gives a positive answer to this question in [110]. Tarkhanov obtained an analogous result for closed forms in $\mathbf{R}^n$ [162].

Chapter II

§8. Propositions 8.1–8.3 are taken from [9] and [15]. Formula (8.5) was obtained for convex domains by Leray [113]. It has been extended to regular linearly convex domains [13], [14]. The Bochner formula (8.7) was mentioned in [235], while the Bochner-Ono formula (8.8) was given in [235] (for a neighborhood of zero) and in [324]. The contents of 5° are due to Aĭzenberg.

§9. A formula for functions holomorphic in an analytic polyhedron of a very general form was published in a number of papers by Bergman (see, for example, [232]). The result in this section is due to Weil [363], but the idea for the proof presented is due to Norguet [318], [319].

§10. Theorems 10.4 and 10.5 were stated by Khenkin [178], [179]. A result close to Theorem 10.4 was obtained independently, but somewhat later, by Ramírez de Arellano [333] (see also [293] and [322]); see [276] in connection with Lemma 10.2.

§11. The Temlyakov integrals of the first and second kinds were obtained by Temlyakov in 1954 and definitely formulated in [164] and [165]. Formula (11.6) is due to Opial and Siciak [325]; see [4], [39], [40], and [115] for other extensions of the Temlyakov integral representation to the case of $n > 2$ complex variables. See [203] concerning extension to circular domains. A number of sources deal with applications of the Temlyakov integrals to diverse problems in function theory (see [2]–[4], [6] and [202]). Temlyakov integrals are the subject of [237], but there is an error there, and in [226] it is explained that Temlyakov integral representations cannot be extended to n-circular domains that are not convex. Theorem 11.2 and Examples 1) and 2) are taken from [10]; see also [7] and [17]. Example 3) is due to Zinov'ev [91]; see also [47]. The results in 3° are due to Aĭzenberg [10], [17]. A construction can be carried out similarly for the Bergman, Poisson, and Schwarz kernels [90]. Other Szegö and Bergman kernels are computed in [81], [83], [84], [91], and [92]. See the article [25] of Aĭzenberg and Egorychev about extension to unbounded domains. Formula (11.5) is implicitly contained in de Leeuw's article [296]. More precisely, he mentioned the existence of such a formula, but did not write out its kernel. This was done by Aĭzenberg in [10] for $n = 2$ under certain conditions (ensuring a unique choice of $\Delta(k)$) on the boundary ∂D, and then by Gindikin in [63] for any n and without any conditions on ∂D. He found (see [63]) that there is a problem in integral geometry which leads to formula

(11.15). It is required to recover a function on an n-dimensional torus if its integrals over the closed totally geodesic hypersurfaces on this torus are known. This problem is connected with Fourier series in the same way the familiar Radon problem is connected with Fourier integrals.

§12. See [55], Chapter III, on the role of holomorphic functions with nonnegative real part in mathematical physics. Example 1 is due to Korányi and Pukánszky [292]; a related result can be found in the article [56] of Vladimirov and Drozhzhinov, where holomorphic functions in a polydisk with nonnegative imaginary parts are studied. Example 3 is taken from [53] and [54]. The remaining results in this section were obtained by Aĭzenberg and Dautov [24]. In [52] and [54] Vladimirov gave characteristic integral formulas for functions with nonnegative imaginary part that are holomorphic in certain tubular (unbounded) domains. We do not know whether these results fit in the scheme of §12 when it is generalized to unbounded domains.

Much work relating to integral representations in the most divergent settings could not be included in this book. We mention the results of Bitsadze on analogues of the Cauchy integral for three-dimensional space [44]. A lot of work has been done on recovering a holomorphic function from its values on sufficiently small (in various senses) sets (results in the spirit of the Goluzin-Krylov formula in [143] (Chapter II, formula (5.9:3)), see [76], [93]–[95], [122], [129], [130] and [327]). Znamenskiĭ [95] has considered integral representations in the spirit of the Vostretsov formula (see [143]) for a family of holomorphic functions that is uniformly bounded on compact subsets of a domain D.

Chapter III

§13. As far back as 1830 Jacobi [277] actually used the concept of a multiple residue as the coefficient of the negative first powers in a multiple Laurent series. Attempts at generalizing the concept of a residue to functions of two complex variables are encountered in work of Cauchy, Didon [247], Picard [329], [330], and Poincaré [331]. However, the memoir [332] of Poincaré should be regarded as the beginning of the theory of multidimensional residues; there he generalized the Cauchy integral theorem to functions of two complex variables, considered the periods of rational functions of two variables (after proving that they can be expressed in terms of the periods of abelian integrals), and used double residues to substantiate a generalization found by Stieltjes [354] of the Lagrange expansion for two variables. The theory of multiple residues was subsequently extended in work of Lefschetz [297], de Rham [335], Segre [343], Severi [344], Leray [113], Dolbeault [248], [249], and others (see the remarks on §§14–19).

§14. The idea of applying the Alexander-Pontryagin duality is due to Martinelli, who used it in extending the Cauchy integral formula to holomorphic functions of n complex variables and to $(n + l)$-fold integrals $(0 \leqslant l \leqslant n - 1)$ [309], as well as in generalizing the logarithmic residue [310]. In [206]–[210] Yuzhakov developed this idea as proposed by Ivanov in connection with the general theory of residues and with residues of rational functions.

§15. A proof of the de Rham theorem can be found in [144] and [189]. The formulation presented here is taken from [113]. Theorem 15.6 of Serre is contained in [153]. Theorem 15.5 on residues was stated by Yuzhakov in [218] and [382].

§16. The exposition in §16 mainly follows Leray's book [113]. The operator δ^{-1} inverse to the Leray coboundary operator was actually used by Poincaré [332]. The theory of residues on complex analytic manifolds was largely created by Leray (see [113]). Its subsequent development and generalization was fostered by Norguet [316], [317], [320], Sorani [350], Dolbeault [250]–[252], Griffiths [267]–[269], Herrera [187], [273], [274], King [283], [284], and others [68], [103], [243], [246], [262], [342], [343], [345], [347]. Theorem 16.16 on the Froissart decomposition is contained in [192a], Part II, §3. Proposition 16.11 was proved in [268], Proposition 16.14 in [211], and Theorem 16.17 in [138] and [139].

§17. A method for finding the multiple residue class of semimeromorphic foms (Theorem 17.2) is described in [59] and [113], and the case of a simple residue class is analyzed in [172] (Chapter III, §4.3). Theorem 17.7 is contained in [268], and Theorem 17.5 in [112].

§18. The main results in this section were published by Yuzhakov in [207], [208], and [214]. Theorem 18.5 we first mentioned by Poincaré in [332]. Theorem 18.7 was given in [213], and Theorems 18.6 and 18.8 in [193] and [196].

§19. This section was written on the basis of [209], [210], and [213]. Multidimensional residues also form the subject of [253], [275], [385], [389], [391], [393], [395], [397], and [402].

Chapter IV

§20. Hermite gave a simple derivation of the Bürmann-Lagrange series (20.2) with the help of residues in 1865. The first generalization of the Lagrange expansion to functions of two variables (Theorem 20.7) was indicated by Stieltjes [354] and proved with the help of the two-dimensional residues of Poincaré [332]. Good [261], Sack [339], and others (see [67] and [359]) have considered Theorem 20.7 for $n \geqslant 2$, as well as certain other modifications of it.

The formula for inverting a vector-valued function, generalizing (20.2), was obtained by Percus [328] by operator methods. Formula (20.19) was obtained in [216] by using logarithmic residues. Results close to Proposition 20.8 were obtained in [105]. Generalizations of the Lagrange expansion to arbitrary implicit functions (Theorems 20.2 and 20.12, Propositions 20.3 and 20.4) can be found in [215]–[217].

A formula for inverting a power series (the formula for the nth derivative of the inverse function) was found by Sylvester [360] in 1854, and rediscovered in [326]. Its generalization for systems of power series (Theorem 20.10) was obtained for $n = 2$ and 3 by Cayley [244] and Sylvester [361]. For $n \geqslant 2$ Theorem 20.10 was proved by Sack [340] by operator methods. Theorem 20.9 was proved by Bolotov and Yuzhakov in [46]. Yuzhakov obtained Theorem 20.14 and Theorems 20.15–20.20 on estimates of the domain of convergence and the remainder for a multiple Lagrange series.

Other estimates of the domain of convergence for inversion of a holomorphic mapping were determined by Shoĭkhet [204].

§21. Some results on systems of nonlinear algebraic equations were found in the last century by Jacobi [278], Hermite [272], and Kronecker. Theorems 21.1 and 21.3 was obtained by Aĭzenberg [19], [20], [21], Theorems 21.4 and 21.5 by Aĭzenberg and Tsikh [28], Theorem 21.6 by Bolotov, and Propositions 21.7 and 21.8 by Aĭzenberg. Example 7 was computed by at the request of the authors by M. Mkrtchyan. We did not discuss the various iteration methods for the approximate solution of systems of nonlinear equations; a detailed survey of these methods and a bibliography can be found in [135].

§22. Theorem 22.8 was published in [106], and a modification of it for the Milnor numbers in [107]. Theorem 22.10 was stated and proved on the basis of a completely analogous theorem of Bernstein [42] on the number of roots of a system of Laurent polynomials. Theorem 22.4 was proved for $n = 2$ in [195], and for $n > 2$ in [223]. Theorem 22.6 is also contained in [223]. Theorem 22.7 was obtained by Tsikh [197].

§23. Theorem 23.1 was stated in [85] and [86]. The method of computing combinatorial sums with the help of residues is thoroughly worked out in the books [82] and [84] of Egorychev. A great variety of examples are analyzed in these references, and a number of problems of combinatorial nature arising in group theory, graph theory, and function theory are solved.

Chapter V

§24. Much work has dealt with generalization of the classical theorems of Hartogs, Osgood and Braun (Theorem 24.3) and Bochner and Severi (Theorem 24.2): see [41], [50], [227], [235], [239], [256], [288], [311], [314], [341], [344],

[364], and [365]. Theorem 24.1 is due to Aronov and Kytmanov [37], [38], and Theorem 24.4 and Corollary 24.5 to Kytmanov and Aĭzenberg [11]. Theorems 7 and 10, which concern the holomorphic extension of functions in the class $L^p(\partial D)$ into D, are taken from Khenkin and Chirka [186] (Chapter I). Khenkin showed that the assertion of Theorem 24.4 is valid also for functions in $L^1(\partial D)$ (here it is necessary to speak of $H^1(D)$ in place of $A_C(D)$.) Theorem 24.7 for the case of a ball is contained in Agranovskiĭ and Val'skiĭ [1] (see also [313]), and for the general case in Stout [358]. In [1] it is actually proved that $f \in C(\partial D)$ extends holomorphically into the ball D if the integrals of f over the sections of ∂D by complex lines are equal to zero. Lemma 24.6 was proved by Kytamonov, following a suggestion of Khenkin. Theorem 24.8 is due to Aĭzenberg and Dautov [24].

Theorem 24.9 and Corollary 24.10 were formulated by Weil. Linear convexity for $n = 2$ was introduced in an equivalent form by Behnke and Peschl in [231]; it was considered and investigated in detail in the general case by Aĭzenberg [13], [14]. A closely related but not equivalent concept was investigated by Martineau in [304]. Theorem 24.14 was erroneously announced by Martineau in [305] for compact sets and domains that are linearly convex in the Martineau sense. Valid results for compact convex sets and convex domains were obtained by Martineau [304] and Aĭzenberg [12] and, in the form presented in §24, by Aĭzenberg [13]. Theorem 24.11 and Corollary 24.17 are taken from [14]. Theorem 24.16 is due to Trutnev [169], and Theorem 24.15 to Znamenskiĭ. See [22], [27], [29], [89], [116], [117], [159], [160], [167]–[169], [220] and [305]–[307] on other applications of the concept of linear convexity and on its properties. Much work has been done on diverse applications of integral representations to the study of spaces of holomorphic functions: see [13], [60], [87], [88], [124], [125], [167], [169], [177]–[179], [186], [281], [285], [304]–[307], [353] and [362].

§25. The Levi problem, whose solution is given by Oka's theorem, was posed in 1910 in [198]. Oka proved his theorem for domains in $\mathbf{C}^2$ in 1942. Later, proofs appeared for $\mathbf{C}^n$ by Oka [323], Bremermann [238], and Norguet [315]. The best-known proofs are due to Grauert [264] and Hörmander [189]. §25 was written by Dautov on the basis of a manuscript of Khenkin (1973), where Oka's main theorems are derived in an elementary fashion from integral formulas for convex domains. Theorem 25.10, which is the basis of §25, was proved by Romanov and Khenkin [147]. It refines previous results of Khenkin [180], Grauert and Lieb [265], Øvrelid [322], Kerzman [280], and Lieb [302]. Theorems 25.14 and 25.16, which take into account the different behavior of the right-hand side of the $\bar{\partial}$-equation along the complex tangent and the complex normal, were proved by Khenkin [185], Skoda [346], and Dautov and

Khenkin [77]. The formula presented in this section (formula (25.12)) for the solution of the $\bar{\partial}$-problem was obtained in [77] for the purpose of proving Theorem 25.16. Theorems 25.14 and 25.16 form the basis for Theorem 25.15 (a description of the sets of zeros of functions of Nevanlinna class; Khenkin [185] and Skoda [346]) and Theorem 25.17 (a description of the sets of zeros of functions of Nevanlinna-Dzhrbashyan class; Dautov and Khenkin [77]). Other formulas for the solution of the $\bar{\partial}$-equation, along with estimates of them and applications, can be found in [142], [179], [180], [182], [183], [322], [344] and [357]. A more detailed survey concerned the $\bar{\partial}$-problem is given in [184] and [186].

§26. The first description of smooth forms orthogonal to holomorphic functions is strictly pseudoconvex domains was obtained by Dautov [71] (see also [23]). The corresponding extension in the generalized sense is a consequence of the uniform estimates of Khenkin [180] for the $\bar{\partial}$-operator (see [184]). Theorem 26.1 was proved by Dautov according to a suggestion of Khenkin, and Theorem 26.2 was proved by Khenkin [185]. Theorem 26.7 is actually contained in [184]. Theorems 26.8, 26.10, and 26.11 are Theorems 6.2, 8.1, and 7.1, respectively, in [23] with refinement of certain formulations (see also [18], [72] and [74]). Theorem 26.9 was communicated by Khenkin. Theorem 26.12 and Corollary 26.13 were obtained by Dautov. For $n = 2$ Corollary 26.13 coincides with Theorem 13.2 in [23]; for $n > 2$ a general form of integral representation different from that in Corollary 26.13 is given in Theorem 13.4 of [23].

The "tangential" Cauchy-Riemann operator $\bar{\partial}_b$ was introduced by Kohn [287] and Kohn and Rossi [288]. It is also defined for forms in $C^{(1)}_{p,q}(\partial D)$ for any p, $0 \leqslant p \leqslant n$, and not just for $p = n$, as in §26.7°. The Lewy equation (26.22) was considered by him in [299] and [300]. Theorems 26.14 and 26.16 are due to Khenkin. In the case when $n = 2$ and D is a ball, Theorem 26.14 is just a reformulation of results of Sato and Greiner, Kohn, and Stein [266]. Theorem 26.16 is a refinement of the classical assertions of Lewy [300] and Hörmander [188] on the existence of unsolvable Lewy equations.

Proposition 26.15 is a theorem on jumps for forms. The first such assertions can be found in Aĭzenberg [18] and Andreotti and Hill [172] (see also [23]).

§27. The expansion (27.1) was noted for polynomials and rational functions by Aĭzenberg. Theorem 27.4 was proved by Yuzhakov [127]. Proposition 27.1, along with certain other conditions for uniqueness of the expansion (27.1), were obtained by Mikhaĭlova [126]. Yuzhakov suggested the idea for constructing a basis for the space $A(\mathbf{C}^n \setminus T)$, where $T = \{z\colon P(z) = 0\}$, P a polynomial, by approximating the domain $\mathbf{C}^n \setminus T$ by Weil polyhedra and applying Proposition 27.1 to them. This idea was implemented (Theorem 27.3) in Mikhaĭlova's article [126]. The simplified proof presented here for Theorem 27.3 was suggested by Zakharyuta. There is a generalization of Theorem 27.3 to the case of pseudopolynomials and of certain classes of entire functions in [127].

Bibliography

1. M. L. Agranovskiĭ and R. E. Val'skiĭ, *Maximality of invariant algebras of functions*, Sibirsk. Mat. Zh. **12** (1971), 3–12; English transl. in Siberian Math. J. **12** (1971).
2. L. A. Aĭzenberg, *Temlyakov integrals and the boundary properties of analytic functions of two complex variables*, Dokl. Akad. Nauk SSSR **120** (1958), 935–938. (Russian)
3. ______, *Pluriharmonic functions*, Dokl. Akad. Nauk SSSR **124** (1959), 967–969. (Russian)
4. ______, *On Temlyakov integrals and boundary properties of analytic functions of several complex variables*, Moskov. Oblast. Ped. Inst. Uchen. Zap. **77** (Trudy Kafedr. Mat. **5**) (1959), 13–35. (Russian)
5. ______, *The spaces of functions analytic in (p, q)-circular domains*, Dokl. Akad. Nauk SSSR **136** (1961), 521–524; English transl. in Soviet Math. Dokl. **2** (1961).
6. ______, *On a theorem of B. A. Vostretsov*, Moskov. Oblast. Ped. Inst. Uchen. Zap. **110** (Mat. **7**) (1962), 37–46. (Russian)
7. ______, *Integral representations for functions holomorphic in multiply circular domains*, Dokl. Akad. Nauk SSSR **138** (1961), 9–12; English transl. in Soviet Math. Dokl. **2** (1961).
8. ______, *Integral representations of functions which are holomorphic in convex domains of C^n*, Dokl. Akad. Nauk SSSR **151** (1963), 1247–1249; English transl. in Soviet Math. Dokl. **4** (1963).
9. ______, *Integral representations of holomorphic functions of several complex variables*, Dokl. Akad. Nauk SSSR **155** (1964), 9–12; English transl. in Soviet Math. Dokl. **5** (1964).
10. ______, *Integral representations of functions holomorphic in n-circular domains* ("*Continuation*" *of Szegö kernels*), Mat. Sb. **65(107)** (1964), 104–143. (Russian)
11. ______, *A theorem on the Shilov boundary for functions of several complex variables*, Moskov. Oblast. Ped. Inst. Uchen. Zap. **166** (Mat. Anal. **10**) (1966), 175–176. (Russian)
12. ______, *The general form of a linear continuous functional in spaces of functions that are holomorphic in convex domains of C^n*, Dokl. Akad. Nauk SSSR **166** (1966), 1015–1018; English transl. in Soviet Math. Dokl. **7** (1966).
13. ______, *Linear convexity in C^n and the distribution of the singularities, of holomorphic functions*, Bull. Acad. Polon. Sci. Sér. Sci. Math. Astronom. Phys. **15** (1967), 487–495. (Russian; French summary)
14. ______, *The expansion of holomorphic functions of several complex variables in partial fractions*, Sibirsk. Mat. Zh. **8** (1967), 1124–1142; English transl. in Siberian Math. J. **8** (1967).
15. ______, *Integral representations of holomorphic functions of several complex variables*, Trudy Moskov. Mat. Obshch. **21** (1970), 3–26; English transl. in Trans. Moscow Math. Soc. **21** (1970).
16. ______, *Polynomials orthogonal to holomorphic functions of several complex variables and an analogue to the Riesz Theorem*, Dokl. Akad. Nauk SSSR **199** (1971), 255–257; English transl. in Soviet Math. Dokl. **12** (1971).
17. ______, *Expansion of integral representations of Szegö kernels and quasikernels for n-circular domains*, Properties of Holomorphic Functions of Several Complex Variables, Inst. Fiz. Sibirsk. Otdel. Akad. Nauk SSSR, Krasnoyarsk, 1973, pp. 3–34. (Russian)

18. ______, *On $\bar\partial$-closed exterior differential forms of type $(p, n-1)$ in the space C^n*, Dokl. Akad. Nauk SSSR **209** (1973), 1009–1012; English transl. in Soviet Math. Dokl. **14** (1973).

19. ______, *On a formula for the generalized multidimensional logarithmic residue, and its application to the solution of systems of nonlinear equations*, Preprint IFSO-7M, Inst. Fiz. Sibirsk. Otdel. Akad. Nauk SSSR, Krasnoyarsk, 1976. (Russian) R. Zh. Mat. **1977** #4Б157.

20. ______, *On a formula for the generalized multidimensional logarithmic residue and the solution of systems of nonlinear equations*, Dokl. Akad. Nauk SSSR **234** (1977), 505–508; English transl. in Soviet Math. Dokl. **18** (1977).

21. ______, *A generalized multidimensional logarithmic residue formula, integral representations of holomorphic functions of several complex variables, and some of their applications*, Complex Analysis and Manifolds, Inst. Mat. Akad. Nauk Ukrain. SSR, Kiev, 1978, pp. 7–46. (Russian)

22. L. A. Aĭzenberg and A. S. Gubanova, *The domains of holomorphy of functions with real or nonnegative Taylor coefficients*, Teor. Funktsiĭ, Funktsional. Anal. i Prilozhen. No. 15 (1972), 50–55. (Russian)

23. L. A. Aĭzenberg and Sh. A. Dautov, *Differential forms orthogonal to holomorphic functions or forms, and their properties*, "Nauka", Novosibirsk, 1975; English transl., Amer. Math. Soc., Providence, R. I., 1983.

24. ______, *Holomorphic functions of several complex variables with nonnegative real part. Traces of holomorphic and pluriharmonic functions on the Shilov boundary*, Mat. Sb. **99(141)** (1976), 342–355; English transl. in Math. USSR Sb. **28** (1976).

25. L. A. Aĭzenberg and G. P. Egorychev, *Integral representations with Szegő kernels for functions holomorphic in unbounded n-circular domains*, Dokl. Akad. Nauk SSSR **231** (1976), 265–268; English transl. in Soviet Math. Dokl. **17** (1976).

26. L. A. Aĭzenberg and B. S. Zinov'ev, *Elementary properties and integral representations of holomorphic functions of several complex variables*, Krasnoyarsk. Gos. Univ., Krasnoyarsk, 1975. (Russian)

27. L. A. Aĭzenberg and V. M. Trutnev, *A certain Borel summability method for formal power series*, Sibirsk. Mat. Zh. **12** (1971), 1398–1404; English transl. in Siberian Math. J. **12** (1971).

28. L. A. Aĭzenberg and A. K. Tsikh, *On the application of a multidimensional logarithmic residue to a system of nonlinear algebraic equations*, Preprint No. IFSO-9M, Inst. Fiz. Sibirsk. Otdel. Akad. Nauk SSSR, Krasnoyarsk, 1977. (Russian) R. Zh. Mat. **1978** #5Б134.

29. L. A. Aĭzenberg, A. P. Yuzhakov and L. Ya. Makarova, *Linear convexity in C^n*, Sibirsk. Mat. Zh. **9** (1968), 731–746; English transl. in Siberian Math. J. **9** (1968).

30. P. S. Aleksandrov, *Combinatorial topology*, OGIZ, Moscow, 1947; English transl., Vols. 1, 2, 3, Graylock Press, Albany, N. Y., 1956, 1957, 1960.

31. ______, *Topological duality theorems*. I: *Closed sets*, Trudy Mat. Inst. Steklov. **48** (1955); English transl. in Amer. Math. Soc. Transl. (2) **30** (1963).

32. A. M. Aronov, *The Martinelli-Bochner integral formula for exterior differential forms*, Holomorphic Functions of Several Complex Variables, Inst. Fiz. Sibirsk. Otdel. Akad. Nauk SSSR, Krasnoyarsk, 1972, pp. 167–170. (Russian)

33. ______, *Functions that can be represented by a Martinelli-Bochner integral*, Properties of Holomorphic Functions of Several Complex Variables, Inst. Fiz. Sibirsk. Otdel. Akad. Nauk SSSR, Krasnoyarsk, 1973, pp. 35–39. (Russian)

34. ______, *On the boundary values of the derivatives of a Martinelli-Bochner integral*, On Holomorphic Functions of Several Complex Variables, Inst. Fiz. Sibirsk. Otdel. Akad. Nauk SSSR, Krasnoyarsk, 1976, pp. 20–28. (Russian)

35. A. M. Aronov and Sh. A. Dautov, *Integral representations for complex differential forms*, Sibirsk. Mat. Zh. **13** (1972), 739–747; English transl. in Siberian Math. J. **13** (1972).

36. ______, *On a result of B. Weinstock*, Properties of Holomorphic Functions of Several Complex Variables, Inst. Fiz. Sibirsk. Otdel. Akad. Nauk SSSR, Krasnoyarsk, 1973, pp. 193–195. (Russian)

37. A. M. Aronov and A. M. Kytmanov, *The holomorphy of functions that are representable by the Martinelli-Bochner integral*, Funktsional. Anal. i Prilozhen. **9** (1975), no. 3, 83–84; English transl. in Functional Anal. Appl. **9** (1975).
38. ______, *A criterion for the existence of a holomorphic extension of a smooth function given on the boundary of a domain in* $\mathbf{C}^n$, On Holomorphic Functions of Several Complex Variables, Inst. Fiz. Sibirsk. Otdel. Akad. Nauk SSSR, Krasnoyarsk, 1976, pp. 13–19. (Russian)
39. I. I. Bavrin, *Integral representations of holomorphic functions of several complex variables*, Dokl. Akad. Nauk SSSR **169** (1966), 495–498; English transl. in Soviet Math. Dokl. **7** (1966).
40. ______, *Integral representations of holomorphic functions*, Moskov. Oblast. Ped. Inst. Uchen. Zap. **166** (Mat. Anal. **10**) (1966), 3–47. (Russian)
41. V. A. Baĭkov, *On a boundary theorem for holomorphic functions of two complex variables*, Uspekhi Mat. Nauk **24** (1969), no. 5(149), 221–222. (Russian)
42. D. N. Bernsteĭn, *The number of roots of a system of equations*, Funktsional. Anal. i Prilozhen. **9** (1975), no. 3, 1–4; English transl. in Functional Anal. Appl. **9** (1975).
43. O. V. Besov, V. P. Il'in and S. M. Nikol'skiĭ, *Integral representations of functions and imbedding theorems*, "Nauka", Moscow, 1975; English transl., Vol. 1, 2, Wiley, 1979.
44. A. V. Bitradze, *Foundations of the theory of analytic functions of a complex variable*, "Nauka", Moscow, 1969; German Transl., Akademie-Verlag, Berlin, 1973.
45. V. A. Bolotov, *On Taylor's formula and an analog for systems of implicit functions*, On Holomorphic Functions of Several Complex Variables, Inst. Fiz. Sibirsk. Otdel. Akad. Nauk SSSR, Krasnoyarsk, 1976, pp. 29–38. (Russian)
46. V. A. Volotov and A. P. Yuzhakov, *A generalization of the inversion formulas for systems of power series to systems of implicit functions*, Mat. Zametki **23** (1978), 47–54; English transl. in Math. Notes **23** (1978).
47. M. A. Borodin, *Certain integral representations of functions which are holomorphic in doubly circular domains*, Sibirsk. Mat. Zh. **10** (1969), 287–296; English transl. in Siberian Math. J. **10** (1969).
48. B. L. van der Waerden, *Moderne Algebra*. Vol. II, 2nd ed., Springer-Verlag, 1940; English transl., Ungar, New York, 1950.
49. V. S. Vinogradov, *An analog of the Cauchy-Riemann system in four-dimensional space*, Dokl. Akad. Nauk SSSR **154** (1964), 16–19; English transl. in Soviet Math. Dokl. **5** (1964).
50. ______, *An analog of an integral of Cauchy type for analytic functions of several complex variables*, Dokl. Akad. Nauk SSSR **178** (1968), 282–285; English transl. in Soviet Math. Dokl. **9** (1968).
51. V. S. Vladimirov, *Methods of the theory of functions of many complex variables*, "Nauka", Moscow, 1964; English transl., M.I.T. Press, Cambridge, Mass., 1966.
52. ______, *Holomorphic functions with nonnegative imaginary part in a tubular domain over a cone*, Mat. Sb. **79(121)** (1969), 128–152; English transl. in Math. USSR Sb. **8** (1969).
53. ______, *Holomorphic functions with positive imaginary part in the future tube*. I, Mat. Sb. **93(135)** (1974), 3–17; English transl. in Math. USSR Sb. **22** (1974).
54. ______, *Holomorphic functions with positive imaginary part in the future tube*. IV, Mat. Sb. **104(146)** (1977), 341–370; English transl. in Math. USSR Sb. **33** (1977).
55. ______, *Generalized functions in mathematical physics*, "Nauka", Moscow, 1976; rev. English transl., "Mir", Moscow, 1979.
56. V. S. Vladimirov and Yu. N. Drozhzhinov, *Holomorphic functions in a polydisc with nonnegative imaginary part*, Mat. Zametki **15** (1974), 55–61; English transl. in Math. Notes **15** (1974).
57. Robert C. Gunning and Hugo Rossi, *Analytic functions of several complex variables*, Prentice-Hall, Englewood Cliffs, N. J., 1965.
58. I. M. Gel'fand, D. A. Raĭkov and G. E. Shilov, *Commutative normed rings*, Fizmatgiz, Moscow, 1960; English transl., Chelsea, New York, 1964.
59. I. M. Gel'fand and G. E. Shilov, *Generalized functions*. Vol. I: *Operations on them*, Fizmatgiz, Moscow, 1958; English transl., Academic Press, 1964.

60. S. G. Gindikin, *Analytic functions in tubular domains*, Dokl. Akad. Nauk SSSR **145** (1962), 1205–1208; English transl. in Soviet Math. Dokl. **3** (1962).

61. ______, *Analysis in homogeneous domains*, Uspekhi Mat. Nauk **19** (1964), no. 4 (118), 3–92; English transl. in Russian Math. Surveys **19** (1964).

62. ______, *The Radon transformation for holomorphic functions of several variables, and related integral formulas*, Uspekhi Mat. Nauk **20** (1965), no. 3 (123), 237–239. (Russian)

63. ______, *A universal formula for multicircular domains in n-dimensional complex space*, Sibirsk. Mat. Zh. **7** (1966), 708–712; English transl. in Siberian Math. J. **7** (1966).

64. V. A. Golubeva, *Some problems in the analytic theory of Feynman integrals*, Uspekhi Mat. Nauk **31** (1976), no. 2(188), 135–202; English transl. in Russian Math. Surveys **31** (1976).

65. G. M. Goluzin, *Geometric theory of functions of a complex variable*, 2nd ed., "Nauka", Moscow, 1966; English transl., Amer. Math. Soc., Providence, R. I., 1969.

66. Adolf Hurwitz and R. Courant, *Vorlesungen über allgemeinen Funktionentheorie und elliptische Funktionen*, 3rd ed., Springer-Verlag, 1929; rev. Russian transl., "Nauka", Moscow, 1968.

67. Édouard Goursat, *Cours d'analyse mathématique*. Tome II, 5th ed., Gauthier-Villars, Paris, 1929; English transl. of 2nd ed., Parts 1, 2, Ginn, Boston, Mass., 1916, 1917; reprint, Dover, New York, 1959.

68. V. L. Gutenmakher, *Integration of multivalued forms and weighted homogeneous polynomials*, Funktsional. Anal. i Prilozhen. **6** (1972), no. 4, 79–80; English transl. in Functional Anal. Appl. **6** (1972).

69. N. M. Gunther, *La theorie du potentiel et ses applications aux problèmes fondamentaux de la physique mathématique*, Gauthier-Villars, Paris, 1934; English transl. [N. M. Günter], Ungar, New York, 1967.

70. Nelson Dunford and Jacob T. Schwartz, *Linear operators*. Vol. I, Interscience, 1958.

71. Sh. A. Dautov, *On forms orthogonal to holomorphic functions on integration along the boundary of strictly pseudoconvex domains*, Dokl. Akad. Nauk SSSR **203** (1972), 16–18; English transl. in Soviet Math. Dokl. **13** (1972).

72. ______, *The forms that are orthogonal to the holomorphic functions, and related questions*, Funktsional. Anal. i Prilozhen. **6** (1972), no. 4, 81–82; English transl. in Functional Anal. Appl. **6** (1972).

73. ______, *$\bar\partial$-closed forms of type $(p, n-1)$ as an analog of holomorphic functions of a single complex variable*, Holomorphic Functions of Several Complex Variables, Inst. Fiz. Sibirsk. Otdel. Akad. Nauk SSSR, Krasnoyarsk, 1972, pp. 21–36. (Russian)

74. ______, *The general form of the integral representation of holomorphic functions in* $\mathbf{C}^n$, Holomorphic Functions of Several Complex Variables, Inst. Fiz. Sibirsk. Otdel. Akad. Nauk SSSR, Krasnoyarsk, 1972, pp. 37–45. (Russian)

75. Sh. A. Dautov and A. M. Kytmanov, *The boundary values of an integral of Martinelli-Bochner type*, Properties of Holomorphic Functions of Several Complex Variables, Inst. Fiz. Sibirsk. Otdel. Akad. Nauk SSSR, Krasnoyarsk, 1973, pp. 49–54. (Russian)

76. Sh. A. Dautov and E. S. Mkrtchyan, *Reconstruction of a function from H^1 in a polycylinder*, On Holomorphic Functions of Several Complex Variables, Inst. Fiz. Sibirsk. Otdel. Akad. Nauk SSSR, Krasnoyarsk, 1976, pp. 39–45. (Russian)

77. Sh. A. Dautov and G. M. Khenkin, *Zeros of holomorphic functions of finite order, and weighted estimates of solutions of the $\bar\partial$-equation*, Preprint No. IFSO-8M, Inst. Fiz. Sibirsk. Otdel. Akad. Nauk SSSR, Krasnoyarsk, 1977. (Russian) R. Zh. Mat. **1978** #7Б257.

78. M. M. Dzhrbashyan, *Integral transforms and representations of functions in the complex domain*, "Nauka", Moscow, 1966. (Russian)

79. A. S. Dynin, *An algebra of pseudodifferential operators on the Heisenberg group. Symbolic calculus*, Dokl. Akad. Nauk SSSR **227** (1976), 792–795; English transl. in Soviet Math. Dokl. **17** (1976).

80. J. Dieudonné, *Foundations of modern analysis*, Academic Press, 1960.

81. G. P. Egorychev, *The application of multiple residues to the determination of Szegő kernels for certain n-circular domains*, Properties of Holomorphic Functions of Several Complex Variables, Inst. Fiz. Sibirsk. Otdel. Akad. Nauk SSSR, Krasnoyarsk, 1973, pp. 55–76. (Russian)
82. ______, *Combinatorial sums and the method of generating functions*, Krasnoyarsk. Gos. Univ., Krasnoyarsk, 1974. (Russian)
83. ______, *Computation of Szegő and Bergman kernels for certain n-circular domains by means of multidimensional residues*, Izv. Vyssh. Uchebn. Zaved. Matematika **1975**, no. 1(152), 101–103; English transl. in Soviet Math. (Iz. VUZ) **19** (1975).
84. ______, *Integral representation and the computation of combinatorial sums*, "Nauka", Novosibirsk, 1977; English transl., Amer. Math. Soc., Providence, R. I. (to appear).
85. G. P. Egorychev and A. P. Yuzhakov, *An application of multidimensional residues to the determination of generating functions and the computation of combinatorial sums*, Holomorphic Functions of Several Complex Variables, Inst. Fiz. Sibirsk. Otdel. Akad. Nauk SSSR, Krasnoyarsk, 1972, pp. 47–61. (Russian)
86. ______, *The determination of generating functions and combinatorial sums by means of multidimensional residues*, Sibirsk. Mat. Zh. **15** (1974), 1049–1060; English transl. in Siberian Math. J. **15** (1974).
87. V. P. Zakharyuta, *Bases and isomorphism of spaces of functions analytic in convex domains of several variables*, Teor. Funktsiĭ, Funktsional. Anal. i Prilozhen. Vyp. **5** (1967), 5–12. (Russian)
88. ______, *On the Weil integral formula and some of its applications*, All-Union Conf. Theory of Functions of a Complex Variable (Kharkov, 1971), Abstracts of Reports, Fiz.-Tekhn. Inst. Nizkikh Temperatur Akad. Nauk Ukrain. SSR, Kharkov, 1971, pp. 78–80. (Russian)
89. B. S. Zinov'ev, *Analytic conditions and certain questions of the approximation of linearly convex domains with smooth boundaries in the space* $\mathbf{C}^n$, Izv. Vyssh. Uchebn. Zaved. Matematika **1971**, no. 6(109), 61–69. (Russian)
90. ______, *On the Bergman, Poisson and Schwarz kernels for multicircular domains of holomorphy*, All-Union Conf. Theory of Functions of a Complex Variable (Kharkov, 1971), Abstracts of Reports, Fiz.-Tekhn. Inst. Nizkikh Temperatur Akad. Nauk Ukrain. SSR, Kharkov, 1971, pp. 83–85. (Russian)
91. ______, *Reproducing kernels for multicircular domains of holomorphy*, Sibirsk. Mat. Zh. **15** (1974), 35–48; English transl. in Siberian Math. J. **15** (1974).
92. B. S. Zinov'ev and G. P. Svistkova, *Finding the reproducing Bergman and Szegő kernels for a certain class of multicircular domains*, Properties of Holomorphic Functions of Several Complex Variables, Inst. Fiz. Sibirsk. Otdel. Akad. Nauk SSSR, Krasnoyarsk, 1973, pp. 209–214. (Russian)
93. S. V. Znamenskiĭ, *A connection between analytic functions of several variables and Dirichlet series of one variable. Applications to the representation of nonlinear analytic operators*, Dokl. Akad. Nauk SSSR **223** (1975), 544–547; English transl. in Soviet Math. Dokl. **16** (1975).
94. ______, *A connection between holomorphic functions of several variables, analytic functionals, and operators with Dirichlet series of one variable; and some of its applications*, On Holomorphic Functions of Several Complex Variables, Inst. Fiz. Sibirsk. Otdel. Akad. Nauk SSSR, Krasnoyarsk, 1976, pp. 46–59. (Russian)
95. ______, *Integral formulas for a family of holomorphic functions that are uniformly bounded in the interior of a domain* $D \subset \mathbf{C}^n$, Uspekhi Mat. Nauk **33** (1978), no. 3(201), 168; English transl. in Russian Math. Surveys **33** (1978).
96. V. A. Kakichev, *Boundary properties of a Cauchy-type integral of several variables*, Shakhty Gos. Ped. Inst. Uchen. Zap. **2** (1959), no. 6, 25–90. (Russian)
97. ______, *An integral of Cauchy type for a topological product of two-dimensional analytic surfaces*, Dokl. Akad. Nauk SSSR **129** (1959), 1218–1221. (Russian)
98. ______, *The Schwarz integral and Hilbert's formulas for analytic functions of several complex variables*, Izv. Vyssh. Uchebn. Zaved. Matematika **1959**, no. 2 (9), 80–93. (Russian)

99. ______, *Character of the continuity of the boundary values of a Martinelli-Bochner integral*, Moskov. Oblast. Ped. Inst. Uchen. Zap. **96** (Trudy Mat. Kafedr **6**) (1960), 145–150. (Russian)
100. ______, *Properties of a meromorphic Hilbert integral, and some of their applications*, Moskov. Oblast. Ped. Inst. Uchen. Zap. **110** (Mat. **7**) (1962), 163–181. (Russian)
101. ______, *Limit values for a special case of a Weil integral*, Moskov. Oblast. Ped. Inst. Uchen. Zap. **137** (Mat. **8**) (1964), 121–132. (Russian)
102. Henri Cartan, a) *Calcul différentiel*, Hermann, Paris, 1967; English transl., Hermann, Paris, and Houghton-Mifflin, Boston, Mass., 1971.
b) *Formes différentielles*, Hermann, Paris, 1967; English transl., Hermann, Paris, and Houghton-Mifflin, Boston, Mass., 1971.
103. V. A. Krasnov, *Cohomology of complexes of meromorphic forms and residues*, Izv. Akad. Nauk SSSR Ser. Mat. **36** (1972), 1237–1268; English transl. in Math. USSR Izv. **6** (1972).
104. M. A. Krasnosel'skiĭ and P. P. Zabreĭko, *Geometric methods of nonlinear analysis*, "Nauka", Moscow, 1975. (Russian)
105. A. V. Kuprikov, *The Lagrange series for a multivariable inverse mapping in* $\mathbf{C}^n$, Combinatorial and Asymptotic Analysis, Krasnoyarsk. Gos. Univ., Krasnoyarsk, 1975, pp. 163–167. (Russian)
106. A. G. Kushnirenko, *The Newton polygon and the number of solutions of a system of k equations with k unknowns*, Uspekhi Mat. Nauk **30** (1975), no. 2(182), 266–267. (Russian)
107. ______, *The Newton polygon, the Milnor numbers*, Funktsional. Anal. i Prilozhen. **9** (1975), no. 1, 74–75; English transl. in Functional Anal. Appl. **9** (1975).
108. A. M. Kytmanov, *A criterion for the holomorphy of an integral of Martinelli-Bochner type*, Combinatorial and Asymptotic Analysis, Krasnoyarsk. Gos. Univ., Krasnoyarsk, 1975, pp. 169–177. (Russian)
109. ______, *On some properties of exterior differential forms of class* **B**, On Holomorphic Functions of Several Complex Variables, Inst. Fiz. Sibirsk. Otdel. Akad. Nauk SSSR, Krasnoyarsk, 1976, pp. 184–189. (Russian)
110. ______, *On a characteristic property of* $\bar{\partial}$*-closed exterior differential forms*, Uspekhi Mat. Nauk **31** (1976), no. 2(188), 217–218. (Russian)
111. A. M. Kytmanov and L. A. Aĭzenberg, *On the holomorphy of continuous functions that are representable by the Martinelli-Bochner integral*, Preprint No. IFSO-5M, Inst. Fiz. Sibirsk. Otdel. Akad. Nauk SSSR, Krasnoyarsk, 1975. (Russian) R. Zh. Mat. **1975** #8Б166.
112. E. K. Leĭnartas, *Factorization of rational functions of several variables into partial fractions*, Izv. Vyssh. Uchebn. Zaved. Matematika **1978**, no. 10(197), 47–51; English transl. in Soviet Math. (Iz. VUZ) **22** (1978).
113. Jean Leray, *Le calcul différentiel et intégral sur une variété analytique complexe* (*Problème de Cauchy*, III), Bull. Soc. Math. France **87** (1959), 81–180.
114. Solomon Lefschetz, *Algebraic topology*, Amer. Math. Soc., Providence, R. I., 1942.
115. Li Chae Kon, *Integral representation of functions of complex variables*, Su-hak kwa Mul-li **3** (1959), no. 1, 27–30. (Korean) R. Zh. Mat. **1961** #2Б94.
116. L. Ya. Makarova, *Supplementary linearly convex envelopes*, Sibirsk. Mat. Zh. **11** (1970), 547–551; English transl. in Siberian Math. J. **11** (1970).
117. L. Ya. Makarova, G. Khudaĭberganov and V. P. Cherkashin, *A remark on polynomial and linear convexity*, Izv. Akad. Nauk Uz SSR Ser. Fiz.-Mat. Nauk **17** (1973), 36–41. (Russian)
118. V. A. Malyshev, *An analytic method in the theory of two-dimensional positive random walks*, Sibirsk. Mat. Zh. **13** (1972), 1314–1329; English transl. in Siberian Math. J. **13** (1972).
119. B. Malgrange, *Ideals of differentiable functions*, Tata Inst. Fund. Res., Bombay, and Oxford Univ. Press, London, 1967.
120. A. I. Markushevich, *Theory of analytic functions*, 2nd rev. aug. ed., Vol. I, "Nauka", Moscow, 1967; English transl. of 1st ed., Vols. 1, 2, Prentice-Hall, Englewood Cliffs, N. J., 1965.

121. V. P. Maslov, *Operational methods*, "Nauka", Moscow, 1973; English transl., "Mir", Moscow, 1976.

122. A. V. Mesis and B. I. Odvirko-Budko, *The reconstruction of a function of two complex variables from its boundary values*, Ural. Gos. Univ. Mat. Zap. **6** (1967), tetrad' 2, 102–113 (1968). (Russian)

123. John Milnor, *Singular points of complex hypersurfaces*, Princeton Univ. Press, Princeton, N. J., and Univ. of Tokyo Press, Tokyo, 1968.

124. B. S. Mityagin, *Approximate dimension and bases in nuclear spaces*, Uspekhi Mat. Nauk **16** (1961), no. 4(100), 63–132; English transl. in Russian Math. Surveys **16** (1961).

125. B. S. Mityagin and G. M. Khenkin, *Linear problems of complex analysis*, Uspekhi Mat. Nauk **26** (1971), no. 4(160), 93–152; English transl. in Russian Math. Surveys **26** (1971).

126. T. S. Mikhaĭlova, *Construction of a basis in spaces of functions that are holomorphic in certain domains of* $\mathbf{C}^n$, Sibirsk. Mat. Zh. **13** (1972), 1387–1394; English transl. in Siberian Math. J. **13** (1972).

127. ______, *Spaces of functions that are holomorphic in certain domains of* $\mathbf{C}^n$, Sibirsk. Mat. Zh. **14** (1973), 806–815; English transl. in Siberian Math. J. **14** (1973).

128. A. P. Mishina and I. V. Proskuryakov, *Higher algebra*, Fizmatgiz, Moscow, 1962; English transl., Pergamon Press, 1965.

129. E. S. Mkrtchyan, *The reconstruction of a holomorphic function from its values on certain uniqueness sets*, Akad. Nauk Armyan. SSR Dokl. **62** (1976), 199–202. (Russian)

130. ______, *On an integral representation for functions that are holomorphic in star-shaped domains of* $\mathbf{C}^n$, On Holomorphic Functions of Several Complex Variables, Inst. Fiz. Sibirsk. Otdel. Akad Nauk SSSR, Krasnoyarsk, 1976, pp. 97–106. (Russian)

131. Paul Montel, *Leçons sur les familles normales de fonctions analytiques et leurs applications*, Gauthier-Villars, 1927.

132. N. I. Muskhelishvili, *Singular integral equations*, 2nd ed., Fizmatgiz, Moscow, 1962; English transl. of 1st ed., Noordhoff, 1953.

133. Raghavan Narasimhan, *Analysis on real and complex manifolds*, Masson, Paris, and North-Holland, Amsterdam, 1968.

134. Rolf Nevanlinna, *Eindeutige analytische Funktionen*, Springer-Verlag, 1936; rev. English transl., 1970.

135. J. M. Ortega and W. C. Rheinboldt, *Iterative solution of nonlinear equations in several variables*, Academic Press, 1970.

136. V. P. Palamodov, *Linear differential operators with constant coefficients*, "Nauka", Moscow, 1967; English transl., Springer-Verlag, 1970.

137. ______, *The multiplicity of a holomorphic transformation*, Funktsional. Anal. i Prilozhen. **1** (1967), no. 3, 54–65; English transl. in Functional Anal. Appl. **1** (1967).

138. Yu. V. Pedan, *On an analogue of the Froissart decomposition theorems for two complex variables*, Dokl. Akad. Nauk SSSR **208** (1973), 46–47; English transl. in Soviet Math. Dokl. **14** (1973).

139. ______, *On an improvement for two complex variables of the Froissart decomposition theorem*, Teor. Funktsiĭ, Funktsional. Anal. i Prilozhen. Vyp. **19** (1974), 33–44. (Russian)

140. ______, *A study of the Riemann surfaces of certain double integrals that depend on one complex parameter*. II, Izv. Vyssh. Uchebn. Zaved. Matematika **1976**, no. 12(175), 66–76; English transl. in Soviet Math. (Iz. VUZ) **20** (1976).

141. A. I. Petrosyan, *Uniform approximation of functions by polynomials on Weil polyhedra*, Izv. Akad. Nauk SSSR Ser. Mat. **34** (1970), 1241–1261; English transl. in Math. USSR Izv. **4** (1970).

142. P. L. Polyakov, *Banach cohomology of piecewise strictly pseudoconvex domains*, Mat. Sb. **88(130)** (1972), 238–255; English transl. in Math. USSR Sb. **17** (1972).

143. I. I. Privalov, *Boundary properties of analytic functions*, 2nd ed., GITTL, Moscow, 1950; German transl., VEB Deutscher Verlag Wiss., Berlin, 1956.

144. Georges de Rham, *Variétés différentiables. Formes, courants, formes harmoniques*, Actualités Sci. Indust., No. 1222, Hermann, Paris, 1955.

145. A. V. Romanov, *A formula and estimates for the solutions of the tangential Cauchy-Riemann equation*, Dokl. Akad. Nauk SSSR **220** (1975), 532–535; English transl. in Soviet Math. Dokl. **16** (1975).

146. ______, *A formula and estimates for the solutions of the tangential Cauchy-Riemann equation*, Mat. Sb. **99(141)** (1976), 58–83; English transl. in Math. USSR Sb. **28** (1976).

147. A. V. Romanov and G. M. Khenkin, *Exact Hölder esti]mates for the solutions of the $\bar\partial$-equation*, Izv. Akad. Nauk SSSR Ser. Mat. **35** (1971), 1171–1183; English transl. in Math. USSR Izv. **5** (1971).

148. L. I. Ronkin, *Elements of the theory of analytic functions of several variables*, "Naukova Dumka", Kiev, 1977. (Russian)

149. Walter Rudin, *Function theory in polydiscs*, Benjamin, 1969.

150. A. A. Savelov, *Plane curves*, Fizmatgiz, Moscow, 1960. (Russian)

151. A. I. Serbin, *On functions representable by a Martinelli-Bochner integral*, Dokl. Akad. Nauk SSSR **196** (1971), 1276–1279; English transl. in Soviet Math. Dokl. **12** (1971).

152. ______, *Conditions for the representation of functions by the Martinelli-Bochner integral*, Izv. Vyssh. Uchebn. Zaved. Matematika **1975**, no. 6(157), 136–142; English transl. in Soviet Math. (Iz. VUZ) **19** (1975).

153. Jean-Pierre Serre, *Quelques problèmes globaux relatifs aux variétés de Stein*, Colloq. Fonctions de Plusieurs Variables (Bruxelles, 1953), Georges Thone, Liège, 1953, pp. 57–68.

154. A. S. Smogorzhevskiĭ and E. S. Stolova, *Handbook of the theory of plane curves of third order*, Fizmatgiz, Moscow, 1961. (Russian)

155. S. L. Sobolev, *Equations of mathematical physics*, 3rd ed., GITTL, Moscow, 1954; English transl., Pergamon Press, Oxford, and Addison-Wesley, Reading, Mass., 1964.

156. ______, *Introduction to the theory of cubature formulas*, "Nauka", Moscow, 1974. (Russian)

157. Edwin H. Spanier, *Algebraic topology*, McGraw-Hill, 1966.

158. Michael Spivak, *Calculus on manifolds*, Benjamin, 1965.

159. V. A. Stepanenko, *Generalization of the concept of linear convexity and its applications*, Properties of Holomorphic Functions of Several Complex Variables, Inst. Fiz. Sibirsk. Otdel. Akad. Nauk SSSR, Krasnoyarsk, 1973, pp. 123–137. (Russian)

160. ______, *On an example of a bounded linearly convex but not convex domain in* $\mathbf{C}^n$, On Holomorphic Functions of Several Complex Variables, Inst. Fiz. Sibirsk. Otdel. Akad. Nauk SSSR, Krasnoyarsk, 1976, pp. 200–202. (Russian)

161. Shlomo Sternberg, *Lectures on differential geometry*, Prentice-Hall, Englewood Cliffs, N. J., 1964.

162. N. N. Tarhanov, *A characteristic property of closed exterior differential forms*, On Holomorphic Functions of Several Complex Variables, Inst. Fiz. Sibirsk. Otdel. Akad. Nauk SSSR, Krasnoyarsk, 1976, pp. 203–209. (Russian)

163. C. Teleman, *Elements of topology and differentiable manifolds*, Ed. Didact. şi Pedagog., Bucharest, 1964 (Romanian); German transl., VEB Deutscher Verlag Wiss., Berlin, 1968.

164. A. A. Temlyakov, *Integral representation of functions of two complex variables*, Izv. Akad. Nauk SSSR Ser. Mat. **21** (1957), 89–92. (Russian)

165. ______, *Representations of functions of two complex variables by integrals*, Dokl. Akad. Nauk SSSR **120** (1958), 976–979. (Russian)

166. ______, *Integral representations*, Dokl. Akad. Nauk SSSR **131** (1960), 263–264; English transl. in Soviet Math. Dokl. **1** (1960).

167. V. M. Trutnev, *Some properties of functions that are holomorphic on strongly linear convex sets*, Uspekhi Mat. Nauk **27** (1972), no. 5(167), 273–274. (Russian)

168. ______, *On an analog of the Laurent series for functions of several complex variables that are holomorphic on strongly linearly convex sets*, Holomorphic Functions of Several Complex Variables, Inst. Fiz. Sibirsk. Otdel. Akad. Nauk SSSR, Krasnoyarsk, 1972, pp. 139–152. (Russian)

169. ______, *Properties of functions that are holomorphic on strongly linearly convex sets*, Properties of Holomorphic Functions of Several Complex Variables, Inst. Fiz. Sibirsk. Otdel. Akad. Nauk SSSR, Krasnoyarsk, 1973, pp. 139–155. (Russian)

170. E. T. Whittaker and G. N. Watson, *A course of modern analysis*, 4th ed., Cambridge Univ. Press, 1927.

171. J. L. Walsh, *Interpolation and approximation by rational functions in the complex domain*, 2nd ed., Amer. Math. Soc., Providence, R. I., 1956.

172. F. Pham, *Introduction à l'étude topologique des singularités de Landau*, Mém. Sci. Math., fasc. 164, Gauthier-Villars, Paris, 1967.

173. B. A. Fuks, *Introduction to the theory of analytic functions of several complex variables*, Fizmatgiz, Moscow, 1962; English transl., Amer. Math. Soc., Providence, R. I., 1963.

174. ______, *Special chapters in the theory of analytic functions of several complex variables*, Fizmatgiz, Moscow, 1963; English transl., Amer. Math. Soc., Providence, R. I., 1965.

175. D. B. Fuks, A. T. Fomenko and V. L. Gutenmakher, *Homotopic topology*, Izdat. Moskov. Gos. Univ., Moscow, 1969. (Russian)

176. V. P. Khavin, *On a problem of V. V. Golubev*, Dokl. Akad. Nauk SSSR **126** (1959), 511–513. (Russian)

177. ______, *Spaces of analytic functions*, Itogi Nauki: Mat. Anal. 1964, VINITI, Moscow, 1966, pp. 76–164; English transl. in Progress in Math., Vol. 1, Plenum Press, 1968.

178. G. M. Khenkin, *The Banach spaces of analytic functions in a ball and in a bicylinder are nonisomorphic*, Funktsional. Anal. i Prilozhen. **2** (1968), no. 4, 82–91; English transl. in Functional Anal. Appl. **2** (1968).

179. ______, *Integral representations of functions holomorphic in strictly pseudoconvex domains and some applications*, Mat. Sb. **78(120)** (1969), 611–632; English transl. in Math. USSR Sb. **7** (1969).

180. ______, *Integral representation of functions in strictly pseudoconvex domains and applications to the $\bar{\partial}$-problem*, Mat. Sb. **82(124)** (1970), 300–308; English transl. in Math. USSR Sb. **11** (1970).

181. ______, *The approximation of functions in pseudoconvex domains and a theorem of Z. L. Leĭbenzon*, Bull. Acad. Polon. Sci. Sér. Sci. Math. Astr. Phys. **19** (1971), 37–42. (Russian)

182. ______, *A uniform estimate for the solution of the $\bar{\partial}$-problem in a Weil domain*, Uspekhi Mat. Nauk **26** (1971), no. 3(159), 211–212. (Russian)

183. ______, *Continuation of bounded holomorphic functions from submanifolds in general position to strictly pseudoconvex domains*, Izv. Akad. Nauk SSSR Ser. Mat. **36** (1972), 540–567; English transl. in Math. USSR Izv. **6** (1972).

184. ______, *The Lewy equation and analysis on pseudoconvex manifolds*. Uspekhi Mat. Nauk **32** (1977), no. 3(195), 57–118; English transl. in Russian Math. Surveys **32** (1977).

185. ______, *H. Lewy's equation and analysis on a pseudoconvex manifold*. II, Mat. Sb. **102(144)** (1977), 71–108; English transl. in Math. USSR Sb. **31** (1977).

186. G. M. Khenkin and E. M. Chirka, *Boundary properties of holomorphic functions of several complex variables*, Itogi Nauki: Sovremennye Problemy Mat., Vol. 4, VINITI, Moscow, 1975, pp. 12–142; English transl. in J. Soviet Math. **5** (1976), no. 5.

187. M. Herrera, *Residues on complex spaces*, Several Complex Variables. I (Proc. Conf., College Park, Md., 1970), Lecture Notes in Math., vol. 155, Springer-Verlag, 1970, pp. 110–114.

188. Lars Hörmander, *Linear partial differential operators*, Springer-Verlag, Berlin, and Academic Press, New York, 1963.

189. ______, *An introduction to complex analysis on several variables*, Van Nostrand, Princeton, N. J., 1966.

190. A. G. Khovanskiĭ, *Newton polyhedra, and toroidal varieties*, Funktsional. Anal. i Prilozhen. **11** (1977), no. 4, 56–64; English transl. in Functional Anal. Appl. **11** (1977).

191. Hua Luo-geng, *Harmonic analysis of functions of several complex variables in the classical domains*, Science Press, Peking, 1958; English transl., Amer. Math. Soc., Providence, R. I., 1963.

192. Rudolph C. Hwa and Vigdor L. Teplitz, *Homology and Feynman integrals*, Benjamin, 1966.

192a. Dimitri Fotiadi et al., *Applications of an isotopy theorem*, Topology **4** (1965), 159–191.

193. A. K. Tsikh, *An extension of Froissart's decomposition and its application to residues of several rational functions*, Properties of Holomorphic Functions of Several Complex Variables, Inst. Fiz. Sibirsk. Otdel. Akad. Nauk SSSR, Krasnoyarsk, 1973, pp. 167–180. (Russian)

194. ______, *Cycles that separate the zeros of analytic functions in* $\mathbf{C}^n$, Sibirsk. Mat. Zh. **16** (1975), 1118–1121; English transl. in Siberian Math. J. **16** (1975).

195. ______, *Multiplicity of the zero of a holomorphic mapping in* $\mathbf{C}^2$, On Holomorphic Functions of Several Complex Variables, Inst. Fiz. Sibirsk. Otdel. Akad. Nauk SSSR, Krasnoyarsk, 1976, pp. 131–138. (Russian)

196. ______, *The two-dimensional homology group of the complement of an algebraic curve in* $\mathbf{C}P^2$, Izv. Vyssh. Uchebn. Zaved. Matematika **1977**, no. 5(180), 122–124; English transl. in Soviet Math. (Iz. VUZ) **21** (1977).

197. ______, *Milnor's number and the multiplicity of the zero of a holomorphic mapping*, Combinatorial and Asymptotic Analysis, No. 2, Krasnoyarsk. Gos. Univ., Krasnoyarsk, 1977, pp. 170–173. (Russian)

198. N. G. Chebotarev, *Theory of algebraic functions*, OGIZ, Moscow, 1948. (Russian)

199. E. M. Chirka, *Analytic representation of CR-functions*, Mat. Sb. **98(140)** (1975), 591–623; English transl. in Math. USSR Sb. **27** (1975).

200. B. V. Shabat, *Introduction to complex analysis*, 2nd ed., Part I, "Nauka", Moscow, 1976. (Russian)

201. ______, *Introduction to complex analysis*, 2nd ed., Part II, "Nauka", Moscow, 1976. (Russian)

202. V. P. Sheĭnov, *Extension of the Krylov-Goluzin formula to the case of two complex variables*, Moskov. Oblast. Ped. Inst. Uchen. Zap. **110** (Mat. **7**) (1962), 133–139. (Russian)

203. ______, *An integral representation for circular domains in* $\mathbf{C}^2$, *and some applications of it*, Sibirsk. Mat. Zh. **8** (1967), 1423–1431; English transl. in Siberian Math. J. **8** (1967).

204. D. M. Shoĭkhet, *On some estimates of the radius of univalence of holomorphic mappings in* $\mathbf{C}^n$, *and an analog of Carathéodory's theorem*, On Holomorphic Functions of Several Complex Variables, Inst. Fiz. Sibirsk. Otdel. Akad. Nauk SSSR, Krasnoyarsk, 1976, pp. 139–148. (Russian)

205. M. Hervé, *Several complex variables. Local theory*, Tata Inst. Fund. Res., Bombay, and Oxford Univ. Press, London, 1963.

206. A. P. Yuzhakov, *Computation of integrals of functions of two complex variables along a closed two-dimensional contour*, Ural. Gos. Univ. Uchen. Zap. **23** (1960), no. 2, 73–85. (Russian)

207. ______, *On the theory of residues of functions of two complex variables*, Moskov. Oblast. Ped. Inst. Uchen. Zap. **110** (Mat. **7**) (1962), 153–162. (Russian)

208. ______, *Some instances of calculating residues of functions of two complex variables by expanding into Laurent series*, Izv. Akad. Nauk Armyan. SSR Ser. Fiz.-Mat. Nauk **15** (1962), no. 4, 13–24. (Russian)

209. ______, *On the residues of functions of several complex variables*, Izv. Vyssh. Uchebn. Zaved. Matematika **1964**, no. 5(42), 149–161. (Russian)

210. ______, *Calculation of residues of a meromorphic function of several complex variables whose denominator can be decomposed into linear factors*, Ural. Gos. Univ. Mat. Zap. **5** (1966), tetrad' 4, 116–124 (1967). (Russian)

211. ———, *A certain coboundary condition, in the sense of Leray, and its application to the logarithmic residue*, Sibirsk. Mat. Zh. **11** (1970), 708–711; English transl. in Siberian Math. J. **11** (1970).

212. ———, *Homology groups of a reducible algebraic curve in* $\mathbf{C}P^2$, All-Union Conf. Theory of Functions of a Complex Variable (Kharkov, 1971), Abstracts of Reports, Fiz.-Tekhn. Inst. Nizkikl Temperatur Akad. Nauk Ukrain. SSR, Kharkov, 1971, pp. 242–243. (Russian)

213. ———, *A sufficient condition for the separation of analytic singularities in* $\mathbf{C}^n$ *and a basis of a certain space of holomorphic functions*, Mat. Zametki **11** (1972), 585–596; English transl. in Math. Notes **11** (1972).

214. ———, *The residues of rational functions of two variables*, Holomorphic Functions of Several Complex Variables, Inst. Fiz. Sibirsk. Otdel. Akad. Nauk SSSR, Krasnoyarsk, 1972, pp. 153–163. (Russian)

215. ———, *A generalization of the Lagrange expansion to arbitrary implicit functions*, Dokl. Akad. Nauk SSSR **219** (1974), 822–824; English transl. in Soviet Math. Dokl. **15** (1974).

216. ———, *On an application of the multiple logarithmic residue to the expansion of implicit functions in power series*, Mat. Sb. **97(139)** (1975), 177–192; English transl. in Math. USSR Sb. **26** (1975).

217. ———, *The representation of implicit functions in the forms of function series*, Combinatorial and Asymptotic Analysis, Krasnoyarsk. Gos. Univ., Krasnoyarsk, 1975, pp. 179–183. (Russian)

218. ———, *Principles of the theory of multidimensional residues*, Krasnoyarsk. Gos. Univ., Krasnoyarsk, 1975. (Russian)

219. ———, *Facts from the theory of multidimensional residues*, Supplement to G. P. Egorychev [84], Russian pp. 224–261 (English, to appear).

220. A. P. Yuzhakov and V. P. Krivokolesko, *Certain properties of linearly convex domains with smooth boundaries in* $\mathbf{C}^n$, Sibirsk. Mat. Zh. **12** (1971), 452–458; English transl. in Siberian Math. J. **12** (1971).

221. A. P. Yuzhakov and A. V. Kuprikov, *The logarithmic residue in* $\mathbf{C}^n$, Properties of Holomorphic Functions of Several Complex Variables, Inst. Fiz. Sibirsk. Otdel. Akad. Nauk SSSR, Krasnoyarsk, 1973, pp. 181–191. (Russian)

222. A. P. Yuzhakov and Z. B. Orekhovskiĭ, *The analytic function that is defined by the integral of a certain rational form in* $\mathbf{C}^n$, Sibirsk. Mat. Zh. **16** (1975), 191–194; English transl. in Siberian Math. J. **16** (1975).

223. A. P. Yuzhakov and A. K. Tsikh, *The multiplicity of the zero of a system of holomorphic functions*, Sibirsk. Mat. Zh. **19** (1978), 693–697; English transl. in Siberian Math. J. **19** (1978).

224. Sh. Yarmukhamedov, *A generalization of the Martinelli-Bochner integral representation*, Mat. Zametki **15** (1974), 739–747; English transl. in Math. Notes **15** (1974).

225. Patrick Ahern and Robert Schneider, *The boundary behavior of Henkin's kernel*, Pacific J. Math. **66** (1976), 9–14.

226. L. A. Aĭzenberg and R. B. Braun, *Über das Integral von Temlyakov*, Math. Ann. **224** (1976), 281–285.

227. Aldo Andreotti and C. Denson Hill, *E. E. Levi convexity and the Hans Lewy problem*. I: *Reduction to vanishing theorems*, Ann. Scuola Norm. Sup. Pisa (3) **26** (1972), 325–363.

228. Aldo Andreotti and François Norguet, *Problème de Levi pour les classes de cohomologie*, C. R. Acad. Sci. Paris **258** (1964), 778–781.

229. Richard F. Basener, *Peak points, barriers and pseudoconvex boundary points*, Proc. Amer. Math. Soc. **65** (1977), 89–92.

230. Dominique Bescheron, *Multiplicité d'intersection et formules intégrales*, Fonctions de Plusieurs Variables Complexes. II (Sém. F. Norguet, 1974/75), Lecture Notes in Math., Vol. 482, Springer-Verlag, 1975, pp. 168–179.

231. H. Behnke and E. Peschl, *Zur Theorie der Funktionen mehrerer komplexer Veränderlichen. Konvexität in Bezug auf analytische Ebenen im Kleinen und Grossen*, Math. Ann. **111** (1935), 158–177.
232. Stefan Bergmann [Bergman], *Über eine Integraldarstellung von Funktionen Zweier komplexer Veränderlichen*, Mat. Sb. **1(43)** (1936), 851–861.
233. ______, *Über die ausgezeichnete Randflächen in der Theorie der Funktionen von zwei komplexen Veränderlichen*, Math. Ann. **104** (1931), 611–636.
234. ______, *The number of intersection points of two analytic surfaces in the space of two complex variables*, Math. Z. **72** (1959/60), 294–306.
235. S. Bochner, *Analytic and meromorphic continuation by means of Green's formula*, Ann. of Math. (2) **44** (1943), 652–673.
236. ______, *Group invariance of Cauchy's formula in several variables*, Ann. of Math. (2) **45** (1944), 686–707.
237. Robert Burkhard Braun, *Temlyakovsche Integraldarstellungen*, Math. Ann. **207** (1974), 337–347; erratum, ibid. **221** (1976), 291.
238. Hans. J. Bremermann, *Über die Äquivalenz der pseudokonvexen Gebiete und der Holomorphiegebiete in Raum von n komplexen Veränderlichen*, Math. Ann. **128** (1954), 63–91.
239. ______, *On a generalized Dirichlet problem for plurisubharmonic functions and pseudo-convex domains. Characterization of Šilov boundaries*, Trans. Amer. Math. Soc. **91** (1959), 246–276.
240. Lutz Bungart, *Holomorphic functions with values in locally convex spaces and applications to integral formulas*, Trans. Amer. Math. Soc. **111** (1964), 317–344.
241. ______, *Boundary kernel functions for domains on complex manifolds*, Pacific J. Math. **14** (1964), 1151–1164.
242. Renato Caccioppoli, *Residui di integrali doppi e intersezioni di curve analitiche*, Ann. Mat. Pura Appl. (4) **29** (1949), 1–14.
243. J.-M. Kantor, *Introduction à la théorie des résidus*, Sèminaire Pierre Lelong (Analyse), Année 1970, Lecture Notes in Math., Vol. 205, Springer-Verlag, 1971, pp. 1–10.
244. A. Cayley, *Note sur une formule pour la reversion des séries*, J. Reine Angew. Math. **52** (1856), 276–284.
245. Brian Cole and R. Michael Range, *A-measures on complex manifolds and some applications*, J. Functional Anal. **11** (1972), 393–400.
246. Johannes G. van der Corput, *Sul calcolo neutralizzato dei residui*, Atti Accad. Naz. Lincei Rend. Cl. Sci. Fis. Mat. Natur. (8) **38** (1965), 166–170.
247. F. Didon, *Note sur une formule de calcul intégral*, Ann. Sci. École Norm. Sup. (2) **2** (1873), 31–48.
248. Pierre Dolbeault, *Formes différentielles et cohomologie sur une variété analytique complexe*. I, Ann. of Math. (2) **64** (1956), 83–130.
249. ______, *Formes différentielles et cohomologie sur une variété analytique complexe*. II, Ann. of Math. (2) **65** (1957), 282–330.
250. ______, *Theory of residues and homology*, Sèminaire Pierre Lelong (Analyse), Année 1969, Lecture Notes in Math., Vol. 116, Springer-Verlag, 1970, pp. 152–163.
251. ______, *Courants résidus des formes semi-méromorphes*, Séminaire Pierre Lelong (Analyse), Année 1970, Lecture Notes in Math., Vol. 205, Springer-Verlag, 1971, pp. 56–70.
252. ______, *Valeurs principales et résidus sur les espaces analytiques complexes (d'après M. Herrera et D. Lieberman)*, Séminaire Pierre Lelong (Analyse), Année 1970/71, Lecture Notes in Math., Vol. 275, Springer-Verlag, 1972, pp. 14–26.
253. Fouad El Zein, *Comparison des résidus de Grothendieck et de Herrera*, C. R. Acad. Sci. Paris Sér. A-B **278** (1974), 863–866.
254. Rainer Epe, *Charakterisierung des Schilowrandes von Holomorphiegebieten*, Schr. Math. Inst. Univ. Münster No. 25 (1963).

255. Charles Fefferman, *On the Bergman kernel and biholomorphic mappings of pseudoconvex domains*, Bull. Amer. Math. Soc. **80** (1974), 667–669.

256. Gaetano Fichera, *Caratterizzazione della traccia, sulla frontiera di un campo, di una funzione analitica di più variabili complesse*, Atti Accad. Naz. Lincei Rend. Cl. Sci. Fis. Mat. Natur. (8) **22** (1957), 706–715.

257. Wolfgang Fischer and Ingo Lieb, *Lokale Kerne und beschränkte Lösungen für den* $\bar\partial$*-Operator auf q-konvexen Gebieten*, Math. Ann. **208** (1974), 249–265.

258. G. B. Folland and J. J. Kohn, *The Neumann problem for the Cauchy-Riemann complex*, Princeton Univ. Press, Princeton, N. J.; and Univ. of Tokyo Press, Tokyo, 1972.

259. Frank Forelli and Walter Rudin, *Projections on spaces of holomorphic functions in balls*, Indiana Univ. Math. J. **24** (1974/75), 593–602.

260. Andrew M. Gleason, *The abstract theorem of Cauchy-Weil*, Pacific J. Math. **12** (1962), 511–525.

261. I. J. Good, *Generalizations to several variables of Lagrange's expansion, with applications to stochastic processes*, Proc. Cambridge Philos. Soc. **56** (1960), 367–380.

262. Gerald Gordon, *The residue calculus in several complex variables*, Fonctions de Plusieurs Variables Complexes (Sém. Norguet 1970–73, à la Mémoire d'Andre Martineau), Lecture Notes in Math., Vol. 409, Springer-Verlag, 1974, pp. 430–438.

263. Bernhard Gramsch, *Über das Cauchy-Weil-Integral für Gebiete mit beliebigem Rand*, Arch. Math. (Basel) **28** (1977), 409–421.

264. Hans Grauert, *On Levi's problem and the imbedding of real-analytic manifolds*, Ann. of Math. (2) **68** (1958), 460–472.

265. Hans Grauert and Ingo Lieb, *Das Ramirezsche Integral und die Lösung der Gleichung* $\bar\partial f = \alpha$ *im Bereich der beschränkten Formen*, Complex Analysis (Proc. Conf., Houston, Texas, 1969), Rice Univ. Studies **56** (1970), no. 2, 29–50 (1971).

266. P. C. Greiner, J. J. Kohn and E. M. Stein, *Necessary and sufficient conditions for solvability of the Lewy equation*, Proc. Nat. Acad. Sci. U.S.A. **72** (1975), 3287–3289.

267. Phillip A. Griffiths, *The residue calculus and some transcendental results in algebraic geometry*. I, Proc. Nat. Acad. Sci. U.S.A. **55** (1966), 1303–1309.

268. ______, *On the periods of certain rational integrals*. I, II, Ann. of Math. (2) **90** (1969), 460–495, 496–541.

269. ______, *Periods of integrals on algebraic manifolds: Summary of main results and discussion of open problems*, Bull. Amer. Math. Soc. **76** (1970), 228–296.

270. F. Reese Harvey and H. Blaine Lawson, Jr., *Boundaries of complex analytic varieties*, Bull. Amer. Math. Soc. **80** (1974), 180–183.

271. ______, *On boundaries of complex analytic varieties*. I, Ann. of Math. (2) **102** (1975), 223–290.

272. Hermite, *Sur l'extension du théorème de M. Sturm à un système d'équations simultanées*, C. R. Acad. Sci. Paris **35** (1852), 52–54.

273. Miguel Herrera, *Residues on complex spaces*, Several Complex Variables. I (Proc. Conf., College Park, Md., 1970), Lecture Notes in Math., Vol. 155, Springer-Verlag, 1970, pp. 110–114.

274. ______, *Residues of forms with logarithmic singularities*, Rev. Un. Mat. Argentina **25** (1970/71), 379–387.

275. M. Herrera and D. Lieberman, *Residues and principal values on complex spaces*, Math. Ann. **194** (1971), 259–294.

276. Lars Hörmander, *L^p estimates for (pluri-) subharmonic functions*, Math. Scand. **20** (1967), 65–78.

277. C. G. J. Jacobi, *De resolutione aequationum per series infinitas*, J. Reine Angew. Math. **6** (1830), 257–286.

278. ______, *Theoremata nova algebraica circa systema duarum aequationum inter duas variabiles propositarum*, J. Reine Angew. Math. **14** (1835), 281–288.

279. Friedrich Kambartel, *Orthonormale Systems und Randintegralformeln in der Funktionentheorie mehrerer Veränderlichen*, Schr. Math. Inst. Univ. Münster No. 18 (1960).

280. Norberto Kerzman, *Hölder and L^p estimates for solutions of $\bar{\partial} u = f$ in strongly pseudoconvex domains*, Comm. Pure Appl. Math. **24** (1971), 301–379.

281. ______, *The Bergman kernel funciton. Differentiability at the boundary*, Math. Ann. **195** (1972), 149–158.

282. N. Kerzman and E. M. Stein, *The Szegő kernel in terms of Cauchy-Fantappiè kernels*, Duke Math. J. **45** (1978), 197–224.

283. James R. King, *Global residues and intersections on a complex manifold*, Trans. Amer. Math. Soc. **192** (1974), 163–199.

284. ______, *Residues and Chern classes*, Differential Geometry, Proc. Sympos. Pure Math., Vol. 27, Part 2, Amer. Math. Soc., Providence, R. I., 1975, pp. 91–97.

285. J. J. Kohn, *Harmonic integrals on strongly pseudo-convex manifolds*. I, Ann. of Math. (2) **78** (1963), 112–148.

286. ______, *Harmonic integrals on strongly pseudo-convex manifolds*. II, Ann. of Math. (2) **79** (1964), 450–472.

287. ______, *Boundaries of complex manifolds*, Proc. Conf. Complex Anal. (Minneapolis, Minn., 1964), Springer-Verlag, 1965, pp. 81–94.

288. J. J. Kohn and Hugo Rossi, *On the extension of holomorphic functions from the boundary of a complex manifold*, Ann. of Math. (2) **81** (1965), 451–472.

289. Walter Koppelman, *The Cauchy integral for functions of several complex variables*, Bull. Amer. Math. Soc. **73** (1967), 373–377.

290. ______, *The Cauchy integral for differential forms*, Bull. Amer. Math. Soc. **73** (1967), 554–556.

291. Adam Korányi, *The Poisson integral for generalized half-planes and bounded symmetric domains*, Ann. of Math. (2) **82** (1965), 332–350.

292. A. Korányi and L. Pukánszky, *Holomorphic functions with positive real part on polycylinders*, Trans. Amer. Math. Soc. **108** (1963), 449–456.

293. A. Korányi and S. Vági, *Singular integrals on homogeneous spaces and some problems of classical analysis*, Ann. Scuola Norm. Sup. Pisa (3) **25** (1971), 575–648 (1972).

294. Guy Laville, *Résolution du $\partial\bar{\partial}$ avec croissance dans les ouverts pseudoconvexes ėtiolės de $\mathbf{C}^n$*, C. R. Acad. Sci. Paris Sér. A-B **274** (1972), A554–A556.

295. ______, *Décomposition des fonctions dans un domaine strictement pseudoconvexe*, C. R. Acad. Sci. Paris Sėr A-B **279** (1974), A549–A550.

296. K. de Leeuw, *Functions on circular subsets of the space of n complex variables*, Duke Math. J. **24** (1957), 415–431.

297. S. Lefschetz, *On the residues of double integrals belonging to an algebraic surface*, Quart. J. Pure Appl. Math. **47** (1916), 333–343.

298. E. E. Levi, *Studii sui punti singolari essenziali delle funzioni analitiche di due o più variabili complesse*, Ann. Mat. Pura Appl. (3) **17** (1910), 61–87.

299. Hans Lewy, *On the local character of the solutions of an atypical linear differential equation in three variables and a related theorem for regular functions of two complex variables*, Ann. of Math. (2) **64** (1956), 514–522.

300. ______, *An example of a smooth linear partial differential equation without solution*, Ann. of Math. (2) **66** (1957), 155–158.

301. Ingo Lieb, *Die Cauchy-Riemannschen Differentialgleichungen auf streng pseudokonvexen Gebieten. Beschränkte Lösungen*, Math. Ann. **190** (1970/71), 6–44.

302. ______, *Die Cauchy-Riemannschen Differentialgleichungen auf streng pseudokonvexen Gebieten. Stetige Randwerte*, Math. Ann. **199** (1972), 241–256.

303. F. S. Macaulay, *The abstract theory of modular systems*, Cambridge Univ. Press, 1916.

304. André Martineau, *Sur la topologie des espaces de fonctions holomorphes*, Math. Ann. **163** (1966), 62–88.

305. ______, *Indicatrices des fonctionnelles analytiques et inversion de la transformation de Fourier-Borel par la transformation de Laplace*, C. R. Acad. Sci. Paris **255** (1962), 1845–1847, 2888–2890.

306. ______, *Sur la notion d'ensemble fortement linéellement convexe*, Preprint, Univ. Montpellier, Montpellier, 1966.

307. ______, *Sur la notion d'ensemble fortement linéellement convexe*, An. Acad. Brasil. Ci. **40** (1968), 427–435.

308. Enzo Martinelli, *Alcuni teoremi integrali per le funzioni analitiche di più variabili complesse*, Reale Accad. d'Italie Mem. Cl. Sci. Fis. Mat. e Nat. **9** (1938), no. 7.

309. ______, *Sulle estensioni della formula integrale di Cauchy alle funzioni analitiche di più variabili complesse*, Ann. Mat. Pura Appl. (4) **34** (1953), 277–347.

310. ______, *Contributi alla teoria dei residui per le funzioni di due variabili complesse*, Ann. Mat. Pura Appl. (4) **39** (1955), 335–343.

311. ______, *Sulla determinazione di una funzione analitica di più variabili complesse in un campo, assegnatane la traccia sulla frontiera*, Ann. Mat. Pura Appl. (4) **55** (1961), 191–202.

312. ______, *Sopra una formula di Andreotti-Norguet*, Boll. Un. Mat. Ital. (4) **11** (1975), 455–457.

313. Alexander Nagel and Walter Rudin, *Moebius-invariant function spaces on balls and spheres*, Duke Math. J. **43** (1976), 841–865.

314. Ricardo Nirenberg, *On the H. Lewy extension phenomenon*, Trans. Amer. Math. Soc. **168** (1972), 337–356.

315. François Norguet, *Sur les domaines d'holomorphie des fonctions uniformes de plusieurs variables complexes*, Bull. Soc. Math. France **82** (1954), 137–159.

316. ______, *Dérivées partielles et résidus de formes différentielles sur une variété analytique complexe*, Sém. P. Lelong 1958/59, Exposé 10, Fac. Sci. Paris, Paris, 1959.

317. ______, *Sur la theorie des résidus*, C. R. Acad. Sci. Paris **248** (1959), 2057–2059.

318. ______, *Représentations intégrales des fonctions de plusieurs variables complexes*, C. R. Acad. Sci. Paris **250** (1960), 1780–1782.

319. ______, *Problèmes sur les formes différentielles et les courants*, Ann. Inst. Fourier (Grenoble) **11** (1961), 1–82.

320. ______, *Introduction à la theorie cohomologique des résidus*, Séminaire Pierre Lelong (Analyse), Année 1970, Lecture Notes in Math., Vol. 205, Springer-Verlag, 1971, pp. 34–55.

321. ______, *Introduction aux fonctions de plusieurs variables complexes: représentations intégrales*, Fonctions de Plusieurs Variables Complexes (Sém. Norguet 1970/73; à la mémoire d'Andre Martineau), Lecture Notes in Math., Vol. 409, Springer-Verlag, 1974, pp. 1–97.

322. Nils Øvrelid, *Integral representation formulas and L^p-estimates for the $\bar\partial$-equation*, Math. Scand. **29** (1971), 137–160.

323. Kiyoshi Oka, *Sur les fonctions analytiques de plusieurs variables*. IX, *Domaines finis sans point critique intérieur*, Japanese J. Math. **23** (1953), 97–155 (1954).

324. Isao Ono, *An analytic kernel in several complex variables*, Sci. Rep. Tokyo Kyoiku Daigaku Sect. A. **5** (1956), 260–266.

325. Zdzisław Opial and Józef Siciak, *Integral formulas for functions holomorphic in cconvex n circular domains*, Zeszyty Nauk. Univ. Jagiello. Prace Mat. No. 9 (1963), 67–75.

326. Alexandre Ostrowski, *Le développement de Taylor de la fonction inverse*, C. R. Acad. Sci. Paris **244** (1957), 429–430.

327. D. J. Patil, *Recapturing H^2-functions on a polydisc*, Trans. Amer. Math. Soc. **188** (1974), 97–103.

328. J. K. Percus, *A note on extension of the Lagrange inversion formula*, Comm. Pure Appl. Math. **17** (1964), 137–146.

329. E. Picard, *Sur les périodes des intégrales doubles*, C. R. Acad. Sci. Paris **102** (1886), 349–350.

330. ______, *Sur le calcul des périodes des intégrales doubles*, C. R. Acad. Sci. Paris **102** (1886), 410–412.
331. H. Poincaré, *Sur les résidus des intégrales doubles*, C. R. Acad. Sci. Paris **102** (1886), 202–204.
332. ______, *Sur les résidus des intégrales doubles*, Acta Math. **9** (1887), 321–380.
333. Enrique Ramírez de Arellano, *Ein Divisionsproblem und Randintegraldarstellungen in der komplexen Analysis*, Math. Ann. **184** (1969/70), 172–187.
334. R. Michael Range and Yum Tong Siu, *Uniform estimates for the* $\bar{\partial}$*-equation on domains with piecewise smooth strictly pseudoconvex boundaries*, Math. Ann. **206** (1973), 325–354.
335. Georges de Rham, *Relations entre le topologie et la théorie des intégrales multiples*, L'Enseignement Math. **35** (1936), 213–228.
336. G. B. Rizza, *Su diverse estensioni dell'invariante di E. E. Levi nella teoria delle funzioni di più variabili complesse*, Ann. Mat. Pura Appl. (4) **44** (1957), 73–89.
337. Guy Roos, *Formules intégrales pour les formes différentielles sur* $\mathbf{C}^n$. I, Ann. Scuola Norm. Sup. Pisa (3) **26** (1972), 171–179.
338. ______, *L'intégrale de Cauchy dans* $\mathbf{C}^n$, Fonctions de Plusieurs Variables Complexes (Sém. Norguet, 1970/73; à la Memoire d'André Martineau), Lecture Notes in Math., Vol. 409, Springer-Verlag, 1974, pp. 176–195.
339. R. A. Sack, *Interpretation of Lagrange's expansion and its generalization to several variables as integration formulas*, J. Soc. Indust. Appl. Math. **13** (1965), 47–59.
340. ______, *Generalization of Lagrange's expansion for functions of several implicitly defined variables*, J. Soc. Indust. Appl. Math. **13** (1965), 913–926.
341. David G. Schaeffer, *An extension of Hartogs' theorem for domains whose boundary is not smooth*, Proc. Amer. Math. Soc. **25** (1970), 714–715.
342. Marcos Sebastiani, *Un exposé de la formule de Leray-Norguet*, Bull. Soc. Brasil. Mat. **1** (1970), no. 2, 47–57.
343. Beniamino Segre, *Sull'estensione della formula integrale di Cauchy e sui residui degli integrali n-plia, nella teoria delle funzioni di n variabili complesse*, Atti Primo Congr. Un. Mat. Ital. (Firenze, 1937), Zanichelli, Bologna, 1938, pp. 174–180.
344. Francesco Severi, *Funzioni analitiche e forme differenziali*, Atti Quarto Congr. Un. Mat. Ital. (Taormina, 1951), Vol. I, Perrella, Rome, 1953, pp. 125–140.
345. Weishu Shih, *Une remarque sur la formule de résidus*, Bull. Amer. Math. Soc. **76** (1970), 717–718.
346. Henri Skoda, *Valeurs au bord pour les solutions de l'opérateur d'', et caractérisation des zéros des fonctions de la classe de Nevanlinna*, Bull. Soc. Math. France **104** (1976), 225–299.
347. Jorge E. Solomin, *Le résidu logarithmique dans les intersections non complètes*, C. R. Acad. Sci. Paris Sér. A-B **284** (1977), A1061–A1064.
348. Friedrich Sommer, *Über die Integralformeln in der Funktionentheorie mehrerer komplexer Veränderlichen*, Math. Ann. **125** (1952), 172–182.
349. Giuliano Sorani, *Sull'indicatore logaritmico per le funzioni di più variabili complesse*, Rend. Mat. e Appl. (5) **19** (1960), 130–142.
350. ______, *Sui residui delle forme differenziali di una varietà analitica complessa*, Rend. Mat. e Appl. (5) **22** (1963), 1–23.
351. ______, *Integral representations of holomorphic functions*, Amer. J. Math. **88** (1966), 737–746.
352. E. M. Stein, *Boundary behavior of holomorphic functions of several complex variables*, Princeton Univ. Press, Princeton, N. J., and Univ. of Tokyo Press, Tokyo, 1972.
353. ______, *Singular integrals and estimates for the Cauchy-Riemann equations*, Bull. Amer. Math. Soc. **79** (1973), 440–445.
354. T.-J. Stieltjes, *Sur une généralisation de la serie de Lagrange*, Ann. Sci. École Norm. Sup. (3) **2** (1885), 93–98.
355. M. Stoll, *Integral formulae for pluriharmonic functions on bounded symmetric domains*, Duke Math. J. **41** (1974), 393–404.

356. Edgar Lee Stout, *An integral formula for holomorphic functions on strictly pseudoconvex hypersurfaces*, Duke Math. J. **42** (1975), 347–356.
357. ______, *H^p-functions on strictly pseudoconvex domains*, Amer. J. Math. **98** (1976), 821–852.
358. ______, *The boundary values of holomorphic functions of several complex variables*, Duke Math. J. **44** (1977), 105–108.
359. P. A. Sturrock, *Generalization of the Lagrange expansion with application to physical problems*, J. Mathematical Phys. **1** (1960), 405–408.
360. J. J. Sylvester, *Note on Burman's law for the inversion of the independent variable*, Philosophical Magazine (4) **8** (1854), 535–540.
361. ______, *On the change of systems of independent variables*, Quart. J. Pure Appl. Math. **1** (1857), 43–56.
362. U. Venugopalkrishna, *Fredholm operators associated with strongly pseudoconvex domains in* $\mathbf{C}^n$, J. Functional Anal. **9** (1972), 349–373.
363. Andre Weil, *L'intégrale de Cauchy et les fonctions de plusieurs variables*, Math. Ann. **111** (1935), 178–182.
364. Barnet M. Weinstock, *Continuous boundary values of analytic functions of several complex variables*, Proc. Amer. Math. Soc. **21** (1969), 463–466.
365. ______, *An approximation theorem for ∂-closed forms of type $(n, n-1)$*, Proc. Amer. Math. Soc. **26** (1970), 625–628.
366. Wilhelm Wirtinger, *Ein Integralsatz über analytische Gebilde im Gebiete von mehreren komplexen Veränderlichen*, Monatshefte Math. Phys. **45** (1937), 418–431.

Supplementary Bibliography

367. L. A. Aĭzenberg and A. K. Tsikh, *Application of a multidimensional logarithmic residue to systems of nonlinear algebraic equations*, Sibirsk. Mat. Zh. **20** (1979), 699–707; English transl. in Siberian Math. J. **20** (1979).
368. V. S. Vladimirov, *On Cauchy-Bochner representations*, Izv. Akad. Nauk SSSR Ser. Mat. **36** (1972), 534–539; English transl. in Math. USSR Izv. **6** (1972).
369. A. Gaziev, *Limit values of a Martinelli-Bochner integral*, Izv. Vyssh. Uchebn. Zaved. Matematika **1978**, no. 9(196), 25–30; English transl. in Soviet Math. (Iz. VUZ) **21** (1978).
370. S. G. Gindikin and G. M. Khenkin, *Integral geometry for $\bar\partial$-cohomology in q-linearly concave domains in* CP^n, Funktsional. Anal. i Prilozhen. **12** (1978), no. 4, 6–23; English transl. in Functional Anal. Appl. **12** (1978).
371. Sh. A. Dautov and G. M. Khenkin, *Zeros of holomorphic functions of finite order and weighted estimates for solutions of the $\bar\partial$-equation*, Mat. Sb. **107(149)** (1978), 163–174; English transl. in Math. USSR Sb. **35** (1979), no. 4.
372. V. A. Kakichev, *Representation of functions holomorphic in circular bicylindrical domains by integrals of Cauchy type with a special form of density*. I, II, Teor. Funktsiĭ, Funktsional. Anal. i Prilozhen. Vyp. **19** (1974), 3–12, 12–23. (Russian)
373. ______, *A two-dimensional boundary value problem of the type of the Riemann problem that contains integrodifferential operators*, Izv. Akad. Nauk Armyan. SSR Ser. Mat. **12** (1977), 189–203. (Russian)
374. ______, *Methods for solving certain boundary value problems for analytic functions of two complex variables*, Tyumen. Gos. Univ., Tyumen, 1978. (Russian)
375. V. A. Kakichev and Le Din'Zon, *The Schwarz integral and the Schwarz problem for the bi-halfplane*, Math. Anal. and Appl., Izdat. Rostov. Gos. Univ., Rostov-on-Don, 1978, pp. 29–38. (Russian)

376. V. A. Kakichev and V. M. Shelkovich, *Solution of boundary problems for analytic functions of several variables in the algebras of Vladimirov*, Mat. Zametki **22** (1977), 51–60; English transl. in Math. Notes **22** (1977).
377. A. M. Kytmanov, *An integral characteristic property of $\bar{\partial}$-closed complex differential forms*, Sibirsk. Mat. Zh. **19** (1978), 788–792; English transl. in Siberian Math. J. **19** (1978).
378. A. M. Kytmanov and L. A. Aĭzenberg, *The holomorphy of continuous functions that are representable by the Martinelli-Bochner integral*, Izv. Akad. Nauk Armyan. SSR Ser. Mat. **13** (1978), 158–169. (Russian)
379. A. V. Romanov, *Convergence of iterates of the Martinelli-Bochner operator, and the Cauchy-Riemann equation*, Dokl. Akad. Nauk SSSR **242** (1978), 780–783; English transl. in Soviet Math. Dokl. **19** (1978).
380. _____, *Spectral analysis of the Martinelli-Bochner operator for the ball in $\mathbf{C}^n$, and its applications*, Funktsional. Anal. i Prilozhen. **12** (1978), no. 3, 86–87; English transl. in Functional Anal. Appl. **12** (1978).
381. G. M. Khenkin, *Solutions with estimates of the H. Lewy and Poincaré-Lelong equations. The construction of functions of a Nevanlinna class with given zeros in a strictly pseudoconvex domain*, Dokl. Akad. Nauk SSSR **224** (1975), 771–774; English transl. in Soviet Math. Dokl. **16** (1975).
382. A. P. Yuzhakov, *Application of de Rham duality to local residues*, On Holomorphic Functions of Several Complex Variables, Inst. Fiz. Sibirsk. Otdel. Akad. Nauk SSSR, Krasnoyarsk, 1976, pp. 149–160. (Russian)
383. A. I. Yanushauskas [Algimantas Janušauskas], *The analytic theory of elliptic equations*, "Nauka", Novosibirsk, 1979. (Russian)
384. Sh. Yarmukhamedov, *The Martinelli-Bochner integral formula and the Phragmén-Lindelöf principle*, Dokl. Akad. Nauk SSSR **243** (1978), 1414–1417; English transl. in Soviet Math. Dokl. **19** (1978).
385. Nachman Aronszajn, *Calculus of residues and general Cauchy formulas in $\mathbf{C}^n$*, Bull. Sci. Math. (2) **101** (1977), 319–352.
386. Paolo de Bartolomeis and Mario Landucci, *Régularisation au bord et problème de Cauchy pour l'opérateur $\bar{\partial}$ à croissance de "type Hardy"*, C. R. Acad. Sci. Paris Sér. A-B **285** (1977), A937–A939.
387. A. Beauville, *Une notion de résidu en geometrie analytique*, Séminaire Pierre Lelong (Analyse), Année 1970, Lecture Notes in Math., Vol. 205, Springer-Verlag, 1971, pp. 183–203.
388. Aline Bonami and Noël Lohoué, *Noyaux de Szegö de certains domaines do $\mathbf{C}^n$ et inégalités L^p*, C. R. Acad. Sci. Paris Sér. A-B **285** (1977), A699–A702.
389. James B. Carrell, *A remark on the Grothendieck residue map*, Proc. Amer. Math. Soc. **70** (1978), 43–48.
390. R. R. Coifman, R. Rochberg and Guido Weiss, *Factorization theorems for Hardy spaces of complex spheres*, Several Complex Variables, Proc. Sympos. Pure Math., Vol. 30, Part 1, Amer. Math. Soc., Providence, R. I., 1970, pp. 119–121.
391. N. Coleff and M. Herrera, *Fibering of residual currents*, Several Complex Variables, Proc. Sympos. Pure Math., Vol. 30, Part 1, Amer. Math. Soc., Providence, R. I., 1977, pp. 11–19.
392. Klas Diederich, *Some recent developments in the theory of the Bergman kernel function: A survey*, Several Complex Variables, Proc. Sympos. Pure Math., Vol. 30, Part 1, Amer. Math. Soc., Providence, R. I., 1970, pp. 127–137.
393. Pierre Dolbeault and Jean Poly, *Differential forms with subanalytic singularities; integral cohomology; residues*, Several Complex Variables, Proc. Sympos. Pure Math., Vol. 30, Part 1, Amer. Math. Soc., Providence, R. I., 1977, pp. 255–261.
394. John P. D'Angelo, *A note on the Bergman kernel*, Duke Math. J. **45** (1978), 259–265.
395. Fouad Elzein [El Zein], *Résidus en géométrie algébrique*, Compositio Math. **23** (1971), 379–405.

396. Simon [S. G.] Gindikin and Guennadi Henkin [G. M. Khenkin], *Transformation de Radon pour la d''-cohomologie des domaines q-linéairement concaves*, C. R. Acad. Sci. Paris Sér. A-B **287** (1978), A209–A212.

397. Phillip Griffiths and Joseph Harris, *Residues and zero-cycles on algebraic varieties*, Ann. of Math. (2) **108** (1978), 461–505.

398. Gennadi M. Henkin [Khenkin] and Jürgen Leiterer, *Global integral formulas for solving the $\bar\partial$-equation on Stein manifolds*, Ann. Polon. Math. **39** (1981), 93–116.

399. Norberto Kerzman, *Remarks on estimates for the $\bar\partial$-equation*, L'analyse Harmonique dans le Domaine Complexe (Actes Table Ronde Internat., Centre Nat. Recherche Sci., Montpellier, 1972), Lecture Notes in Math., Vol. 336, Springer-Verlag, 1973, pp. 111–124.

400. Mario Landucci, *Uniform bounds on derivatives for the $\bar\partial$-problem in the polydisk*, Several Complex Variables, Proc. Sympos. Pure Math., Vol. 30, Part 1, Amer. Math. Soc., Providence, R. I., 1977, pp. 177–180.

401. Lu Qi-keng, *On the Cauchy-Fantappiè formula*, Acta Math. Sinica **16** (1966), 344–363; English transl. in Chinese Math. Acta **8** (1966).

402. Zoghman Mebkhout, *Valeur principale et résidu simple des formes à singularités essentielles*, Fonctions de Plusieurs Variables Complexes. II (Sém. Norguet, 1974/75), Lecture Notes in Math., Vol. 482, Springer-Verlag, 1975, pp. 190–215.

403. N. R. O'Brian, *Zeroes of holomorphic vector fields and the Grothendieck residue*, Bull. London Math. Soc. 7 (1975), 33–38.

404. A. M. Ostrowski, *On Kronecker's elimination theory*, J. Reine Angew. Math. **296** (1977), 37–57.

405. Gilles Raby, *Formes méromorphes et semi-méromorphes sur une surface analytique complexe*, C. R. Acad. Sci. Paris Sér. A-B **287** (1978), A125–A128.

406. R. Michael Range, *Hölder estimates for $\bar\partial$ on convex domains in $\mathbf{C}^2$ with real analytic boundary*, Several Complex Variables, Proc. Sympos. Pure Math., Vol. 30, Part 2, Amer. Math. Soc., Providence, R. I., 1977, pp. 31–33.

407. Guy Roos, *Formules intégrales pour les formes différentielles sur $\mathbf{C}^n$*. II, Fonctions de Plusieurs Variables Complexes. III (Sém. Norguet, 1975/77), Lecture Notes in Math., Vol. 670, Springer-Verlag, 1978, pp. 31–52.

408. Mikio Sato, *Regularity of hyperfunction solutions of partial differential equations*, Proc. Internat. Congr. Math. (Nice, 1970), Vol. 2, Gauthier-Villars, Paris, 1971, pp. 785–794.

409. L. M. Sibner and R. J. Sibner, *A note on the Atiyah-Bott fixed point formula*, Pacific J. Math. **53** (1974), 605–609.

410. Henri Skoda, *Zéros des fonctions de la classe de Nevanlinna dans les ouverts strictement pseudoconvexes*, C. R. Acad. Sci. Paris Sér. A-B **280** (1975), A1677–A1680.

411. Domingo Toledo and Yue Lin L. Tong, *The holomorphic Lefschetz formula*, Bull. Amer. Math. Soc. **81** (1975), 1133–1135.

412. Wolfgang Tutschke, *Die Cauchysche Integralformel für morphe Funktionen mehrerer komplexen Variabler*, Math. Nachr. **54** (1972), 385–391.

413. V. S. Vladimirov, *Holomorphic functions of several complex variables with nonnegative imaginary part and some applications*, Complex Analysis and Its Applications (Lectures, Internat. Sem., Trieste, 1975), Vol. III, Internat. Atomic Energy Agency, Vienna, 1976, pp. 259–287.

414. Barnet M. Weinstock, *Uniform approximation and the Cauchy-Fantappiè integral*, Several Complex Variables, Proc. Sympos. Pure Math., Vol. 30, Part 2, Amer. Math. Soc., Providence, R. I., 1977, pp. 187–191.

Subject Index

Index of Notation